Mutation-Driven Evolution

Masatoshi Nei
Pennsylvania State University

Mutation-Driven Evolution. First Edition. Masatoshi Nei.

OXFORD
UNIVERSITY PRESS

Great Clarendon Street, Oxford, OX2 6DP,
United Kingdom

Oxford University Press is a department of the University of Oxford.
It furthers the University's objective of excellence in research, scholarship,
and education by publishing worldwide. Oxford is a registered trade mark of
Oxford University Press in the UK and in certain other countries

First published 2013
First published in paperback 2014

Published in the United States of America by Oxford University Press
198 Madison Avenue, New York, NY 10016, United States of America

British Library Cataloguing in Publication Data
Data available

ISBN 978–0–19–872410–0

Contents

Preface

I started my career with theoretical population genetics in the 1960s after a short period of field work experience concerning quantitative genetics. At that time evolutionary studies were conducted primarily by comparing phenotypic characters among individuals within and between species. These studies did not give clear insights into the cause and the effect of evolution, because the genetic basis of phenotypic characters was not well understood.

In theoretical population genetics, we could consider a set of alleles at one or a few loci and study the theoretical changes of genotype frequencies due to mutation, natural selection, and genetic drift. These studies gave only possible evolutionary changes of populations, but they were still much better than intuitive arguments. For this reason, a large body of mathematical theories of evolution was developed. These theories depended on many simplifying assumptions about the breeding system, population structure, selection coefficients, gene interaction, etc., and different assumptions about these factors often generated very different predictions of evolutionary changes. This resulted in many controversies which could not be resolved easily because of the difficulty of doing experimental studies. At that time, population genetics was dominated by neo-Darwinism with the idea of pervasive natural selection, and I was working within the framework of neo-Darwinism. Furthermore, because it was difficult to identify the homologous genes between different species, population genetics studies were primarily concerned with the gene frequency changes within species.

In the early 1960s a number of molecular biologists were working on the evolutionary changes of genes and proteins at the molecular level, and this approach of studying evolution was integrated with population genetics theory in the latter half of the 1960s and in the 1970s. This integration transformed the study of evolution profoundly. First, we could now identify homologous genes in different species and study long-term evolution of genes by comparing the nucleotide or amino acid sequences from different species. Second, molecular data on the evolutionary change of genes soon indicated the importance of mutation in evolution. Third, comparison of the DNA contents of different species suggested that gene or genome duplication occurred frequently in the process of evolution. Because gene or genome duplication is a form of mutation in the broad sense, I realized that mutation is the driving force of evolution. Yet, this view was regarded as a heresy at the time when neo-Darwinism dominated the field. In the meantime the neutral theory of molecular evolution was proposed to explain the evolution of genes and proteins. This theory clearly showed that the evolution of nucleotide sequences has occurred mainly by random fixation of neutral mutations. However, most neo-Darwinians did not pay much attention to this discovery, because they believed that neutral evolution has nothing to do with phenotypic evolution, in which most evolutionists are interested. In fact, even the proponents of the neutral theory of molecular evolution stated that phenotypic evolution occurs mostly by natural selection, as will be mentioned later.

By the early 1970s, I came to believe that the principle of phenotypic evolution must be the same as that of molecular evolution because both types of evolution are controlled by mutation at the DNA level. I briefly presented this idea in my 1975 book *Molecular Population Genetics and Evolution*. However, few people paid attention to this view. I elaborated this idea in several publications in the 1980s including my 1987 book *Molecular Evolutionary Genetics*, but the response was not great. The problem was that the molecular biology of morphogenesis was not well developed at that time and it was not easy to show the roles of mutation in phenotypic evolution convincingly.

In the past two or three decades, this situation has changed dramatically, and it is now possible to evaluate the roles of mutation and selection in phenotypic evolution at the molecular level. In the 1980s I became interested in understanding the evolution of the adaptive immune system of vertebrates by studying evolutionary changes of immunoglobulins, major histocompatibility complex genes, T-cell receptor genes, etc. In later years I also studied the evolution of genes controlling body segmentation (HOX genes), flowering in plants (MADS-box genes), sensory receptors, microRNAs, etc. in collaboration with graduate students and postdoctorals in my laboratory. These studies have been very helpful in clarifying my view that the driving force of evolution is mutation and natural selection is of secondary importance. This view is different from Hugo de Vries's mutation theory, and I previously called it the new mutation theory of evolution or neomutationism. In this book, it will be called the theory of mutation-driven evolution to convey the message that the importance of natural selection is duly appreciated in conjunction with the role of mutation.

During this period, I came to realize that evolutionary biology must be rebuilt upon the knowledge of molecular biology. Every biological process involved in metabolism and reproduction of organisms is governed by the function of DNAs and RNAs at the most fundamental level. Environmental effects on the formation of organisms can also be studied by using the knowledge of epigenetics. Natural selection and genetic drift are ultimately determined by the differential rates of birth and death of individuals, which are again the consequences of metabolism and reproduction. Population genetics and ecology are useful for visualizing the long-term change in populations and for understanding the consequences of population size change and competition and cooperation of organisms. However, the prediction of population genetics is always abstract, and it does not give any explanations about how a particular character such as mammalian sex determination or vertebrate brain has evolved. To answer these types of question, we must use the molecular biology approach. For this reason, I have become interested in explaining the evolution of specific characters and attempted to give molecular answers to some of the questions Charles Darwin and other investigators have posed in the past.

Evolution is a broad subject encompassing many areas of biology such as molecular biology, genetics, ecology, sociobiology, and paleontology. In this book, however, I will be concerned primarily with the mechanism of evolution with an emphasis on genetic and molecular aspects. I have decided to do this, because this is the backbone of evolutionary biology and it has been controversial ever since Darwin's publication of *Origin of Species*. I will discuss this problem with a historical perspective to understand various theories of evolution presented in the past 150 years and their relationships with the new theory of mutation-driven evolution. The historical perspective presented here may not necessarily be the same as that of currently popular books such as Mayr's *Growth of Biological Thoughts*, because I found some misconceptions in these books when I examined the original sources. I have tried to present the views of the original authors as much as possible.

However, the main purpose of this book is to present a comprehensive theory of mutation-driven evolution considering the latest information on molecular and phenotypic evolution. We are all aware that phenotypic characters show an enormous amount of variation both within and between species. All this variation is ultimately caused by differences in the structure and function of DNAs and RNAs, whether the characters are affected by environmental factors or not. Therefore, phenotypic evolution must be ultimately explained in terms of molecular biology. For this reason, I will consider the evolutionary change of molecules, genes, and genomes and then the molecular basis of phenotypic evolution. Of course, our knowledge of the molecular basis of phenotypic evolution is quite limited. We have little idea about how the human brain, the elephant's trunk, the body structures of whales, etc. have evolved. Nevertheless, we are beginning to understand the molecular basis of many complex phenotypic characters, and the future of the study of phenotypic evolution is bright. However, these problems have to be studied by using new technologies and new evolutionary concepts.

In this book, I have tried to cover these new developments in evolutionary biology and critically examine both old and recent findings to establish the general principles of evolution. In my view evolution is not the enhancement of fitness of individuals or populations, but it represents the increase (or decrease) of phenotypic complexity, which may be measured by the number of cell types or some other quantity. For this reason, my conception of evolutionary biology is different from the currently popular views. In this book, I hope I have explained the theory of mutation-driven evolution in a logical fashion.

This book is written as a monograph rather than a textbook. Therefore, the topics covered are not necessarily comprehensive, and it is assumed that the readers are acquainted with the basic knowledge of genetics and molecular biology. Nevertheless, I have presented essential aspects of these disciplines that are required for understanding my arguments. I have used many examples to illustrate the importance of mutation in evolution, and in this case I have often presented the studies which have been conducted in my laboratory. I have done this because I am familiar with them and therefore I can avoid serious errors. However, this book is based on the knowledge accumulated by numerous investigators over the past several decades though the sources are not always clearly mentioned.

In Chapter 1 a brief history of scientific studies of evolution is presented, starting with Charles Darwin's work on evolution and covering subsequent development of studies that have led to the idea of mutation-driven evolution. Chapters 2 and 3 are devoted to the development of neo-Darwinism and its significance and limitations for the study of evolution. Chapters 4 and 5 present evolutionary changes in genes and genomic structures and their relationships with phenotypic evolution. The molecular basis of phenotypic evolution is presented in Chapter 6 with an emphasis on the mechanism of gene expression in the development of phenotypic characters and the interaction between genes and environmental factors. In Chapter 7 various mechanisms of generating hybrid sterility and inviability are discussed in relation to the formation of new species at both genic and genomic level. Here a new view of speciation is presented. Chapter 8 is devoted to the evolution of several important phenotypic characters such as sex determination and insect caste systems. Chapter 9 presents the general concept of mutation-driven evolution and its significance in the study of evolution. The final chapter presents a general summary of this book and conclusions.

I am deeply indebted to my colleagues and students who have collaborated with me during the last three decades. Some of them helped me in developing statistical methods that have been used in our data analysis, whereas others conducted time-consuming data analysis. I am particularly grateful to Takashi Gojobori, Austin Hughes, Tatsuya Ota, Koichiro Tamura, Sudhir Kumar, George Zhang, Alex Rooney, Yoshiyuki Suzuki, Helen Piontkivska, Jongmin Nam, Yoshihito Niimura, Nikolas Nikolaidis, Dimitra Chalkia, Zhenguo Lin, Masafumi Nozawa, and Sabyasachi Das. Hie Lim Kim, Zhenguo Zhang, and Sayaka Miura also helped me in preparing several figures used in this book. I also would like to express my gratitude to Jan Klein, Pekka Pamilo, Wojtek Makalowski, Tatsuya Ota, and Alex Rooney, who read the entire manuscript of the book and provided constructive comments. I am extremely grateful to Tina Kushner, who prepared the final manuscript with great care and helped me in organizing the References list.

The work in this book was partly supported by research grants from the National Institute of Health and the National Science Foundation. Part of the book was written when I was a visiting professor at the Tokyo Institute of Technology, Tokyo, Japan. I am grateful to Norihiro Okada for his hospitality during my visit.

Masatoshi Nei
Pennsylvania State University

CHAPTER 1

Selectionism and Mutationism

1.1. Darwin's Theory of Evolution

In the mid-nineteenth century, Charles Darwin (1859) published a book called *The Origin of Species*, which is regarded as one of the greatest books ever written in the history of science. Through this book, he could convince the world that all living organisms are not independent creations but they were derived from a single common ancestor by descent with modification. He did this by assembling massive data on evolution from various fields of biology and geology and considering the mechanism of evolution materialistically. It is often said that he could accomplish this achievement because he discovered natural selection.

The full title of the original edition (1859) of Darwin's book was "*On the Origin of Species by Means of Natural Selection or the Preservation of Favoured Races in the Struggle for Life.*" It is interesting to see that he defined the then new terminology natural selection within the title of the book, apparently because he wanted to avoid any misunderstanding of the terminology. This definition implies that natural selection is a mechanism to save favorable individuals for the next generation and has nothing to do with creation of innovative characters or variations. In his time the mechanism of generation of heritable variations was not known, and apparently for this reason Darwin considered natural selection as the major force of evolution. However, he never implied that natural selection has creative power. In the sixth edition of *The Origin of Species* (Darwin 1872, p. 63), he stated: "*Several writers have misapprehended or objected to the term Natural Selection. Some have even imagined that natural selection induces variability, whereas it implies only the preservation of such variations as arise and are beneficial to the being under the conditions of life.*"

If we use the current terminology in biology, the essence of his theory of evolution may be summarized as follows. (1) Most natural populations contain a large amount of phenotypic variation on which natural selection may operate. (2) Phenotypic variation is continuous rather than discontinuous and includes heritable components. (3) Darwin did not know the cause of phenotypic variation but suggested that it is generated by use and disuse of characters, climatic changes, correlation of growth (modification of a character generated as a consequence of natural selection for another character), and chance effects (random changes). (4) Evolution occurs gradually by means of natural selection, which inevitably causes extinction of less improved forms of life. (5) Accumulation of the results of natural selection gradually increases interpopulational differences in morphological and physiological characters and eventually generates new species, genera, families, etc. (6) All organisms on earth were derived from a single proto-organism through a slow process of descent with modification. (7) The similar organisms currently observed in different parts of the world have been generated by recent migration. (8) The discontinuity of paleontological data in different geological periods does not indicate that evolution occurred discontinuously but that the fossil record is incomplete and there are many gaps in the record.

The above statements are a brief summary of Darwin's theory of evolution. In practice, he was not as straightforward as mentioned above. He was a humble and cautious scientist and tried to avoid any dogmatic statements. This made his book acceptable for a wide range of biologists, but it

Mutation-Driven Evolution. First Edition. Masatoshi Nei.

also made his statements often ambiguous. In fact, he did not always distinguish between the causes of variations and the results of natural selection. If one accepts any form of inheritance of acquired characters (Lamarck 1809) as he did, the difference between the cause of variation and the result of natural selection necessarily becomes unclear. This is because the Lamarckian doctrine generates new heritable variations which are similar to the results of natural selection and the variations may be reverted to original characters when the environmental condition changes back to the original status. This is also true with the effects of climatic changes. This situation becomes worse if one accepts the blending inheritance to which Darwin subscribed (see Section 1.2). Darwin (1859) occasionally referred to sports as a source of new variation, but he did not think this was an important factor.

Conceptually, his theory of evolution consists of two processes: (1) generation of new variations and (2) natural selection of favored variations. In this sense his view is similar to the modern concept of evolution. Darwin provided strong arguments for the second process (not evidence), but he could not give satisfactory explanations for the first evolutionary process. Therefore, he simply assumed that sufficient amounts of variations always exist within populations and did not worry about the cause too much. This treatment of the first evolutionary process was unsatisfactory to many biologists (e.g. Thomas Huxley and Francis Galton), and various new hypotheses were proposed later. In fact, most of the later controversies on evolution were concerned with the generation of new variations, which are now known as mutations, and their relationships with the second process of natural selection. Around 1910, it was often said that "*natural selection may explain the survival of the fittest, but it cannot explain the arrival of the fittest*" (de Vries 1912, p. 827). The same criticism was raised repeatedly in the past (Morgan 1903, 1932), and it has become louder and louder recently (e.g. Ohno 1970; Nei 1987; Kirschner and Gerhart 2005; Stoltzfus 2006).

1.2. Criticisms of Darwin's Theory

Darwin's theory of natural selection was supported enthusiastically by some distinguished biologists such as August Weismann and Alfred Wallace, but there were many biologists or paleontologists who opposed gradual evolution by natural selection and proposed alternative theories. For example, Herbert Spencer and Ernst Heckel accepted the inheritance of acquired characters as an evolutionary force at least to the same extent as natural selection (Bowler 1983). In fact, Lamarckism was considered as an important alternative theory in the post-*Origin* era, and this view was maintained even after genetics was firmly established in the 1910s and 1920s. The final rejection of Lamarckism occurred only when Luria and Delbruck (1943) showed that bacterial resistance to bacteriophages is caused by spontaneous mutation rather than by induction through phage exposure. Later Lederberg and Lederberg (1952) conducted indirect (sib) selection experiments with the bacterium *Escherichia coli* and showed that drug resistance in bacteria evolves by pre-adaptive mutations without being exposed to any drugs. Here sib selection refers to selection in which the drug resistance of a colony is judged by the result of the assay of its sibling colony, so that the selected colony line is never exposed to the drug. Similar results were obtained with respect to DDT resistance in *Drosophila melanogaster* (Crow 1957). Other popular theories at that time were saltationism, in which new species or subspecies are suddenly produced in geological time, and orthogenesis, which claims that each organism is destined to follow a certain course of evolution by some internal force. However, since these views were based on old mystical concepts about unknown laws of inheritance or evolution, they gradually disappeared. Some authors emphasized the importance of geographical isolation for speciation, whereas others suggested that hybridization is an important factor for the origin of new species (see Mayr 1982; Bowler 1983). These views were clearly concerned with short-term evolution, and they were not real criticisms of Darwinism, which dealt with long-term evolution. With the rediscovery of Mendel's law of inheritance in 1900, it became clear that they are not fundamentally important for explaining evolution.

One of the most serious objections raised against Darwin's theory of evolution was concerned with his claim that natural selection operates on continuous or fluctuating variation and this is sufficient for generating new species. This claim was questioned by Thomas Huxley and Francis Galton, who were otherwise staunch supporters of Darwin's evolutionary theory. In the middle nineteenth century, many biologists including Charles Darwin believed in the theory of blending inheritance, in which the characters of offspring of a pair of male and female parents tend to be intermediate between those of the two parents. Fleeming Jenkin (1867), who was a professor of engineering at the University of Edinburgh, indicated that Darwin's theory based on this idea creates various problems about effectiveness of natural selection because blending inheritance is expected to reduce the extent of variation every generation. He indicated that in the theory of blending inheritance even discontinuous variations such as sports (large effect variants) cannot be established in the population and therefore no new species can be generated. The idea of blending inheritance was abandoned only after Mendel's laws of inheritance were rediscovered (Morgan 1925, pp. 139–140; Fisher 1930, pp. 1–7). Another comment made by the physicist Jenkin was that Darwin's theory of natural selection was too speculative and experimental verification was necessary. In his 1869 letter to his friend J. D. Hooker, Darwin stated that "*Fleeming Jenkin has given me much trouble but has been of more real use than any other essay or review*" (Mayr 1982, p. 512). For these reasons, Darwin weakened his claim of the efficacy of natural selection, particularly about discontinuous variations. (Ironically, the significance of Jenkin's paper in the development of evolutionary theory is still debated by historians and philosophers (e.g. Gayon 1998; Bulmer 2004), but this is out of the scope of this book.)

To avoid the reduction of variation, there was a need for finding mechanisms that generate new variation. One such mechanism Darwin proposed was the hypothesis of pangenesis, in which hereditary substances called gemmules or pangenes were assumed to be shed by all parts of the organism and carried in the bloodstream or some other agencies to the germline cells with uneven contributions from different organs. When the contribution from a particular organ is large, the organ is manifested disproportionately in the next generation. In this way new variations are expressed in the next generation to offset the dissipation of variations by blending inheritance (Darwin 1868). This hypothesis was previously used by Lamarck (1809) to explain the mechanism of acquired characters. However, when Galton conducted blood transfusion experiments using different varieties of rabbits, no evidence of pangenesis was obtained (see Provine 1971). For the above reason, Huxley, Galton, and Bateson indicated that a new evolutionary theory based on discontinuous variations, which would not disappear by mixing with the original variations, should be developed.

At the end of the nineteenth century, the number of supporters of Darwin's theory of natural selection dwindled substantially, and the only major supporters were August Weismann, Alfred Wallace, Walter Weldon, and Karl Pearson. Around this time, Weldon and Pearson initiated a biometrical study of Darwinian evolution, but this study was severely criticized by William Bateson, who supported evolution by discontinuous variation. When Mendel's law of inheritance was rediscovered in 1900, Bateson took the law as support of his theory of discontinuous evolution and attacked the biometricians. This event contributed to an eclipse of Darwinism (Bowler 1983; Gayon 1998). Bateson is also known to have coined the word genetics.

1.3. Evolution by Discontinuous Variation

As mentioned above, Huxley and Galton did not believe that natural selection acting on continuous variation is sufficient to generate the large morphological and physiological differences between species. A number of evolutionists such as Bateson and de Vries therefore proposed the theory of evolution by discontinuous variation. They emphasized the importance of new discrete variations (mutations in the present terminology) in evolution. To support this idea, Bateson (1894) compiled and described various morphological oddities or discrete variations in the animal kingdom in his 598-page long book, *Materials for the Study of Variation*. These oddities included a bumblebee with legs attached to the antennae, butterflies with extra eyespots on the wings, frogs with extra vertebrae, a man with extra nipples, and many others (886 examples). He called the abnormality with one body part transformed into another a homeotic transformation. The purpose of this compilation was to show that there are many discontinuous variations in animal species and they could be the materials for generating new species. He believed that to understand the origin of variations and evolution one must study newly arisen variations rather than the variations that are observable in natural populations, with which Darwin was concerned. He also rejected Lamarckian inheritance. Unfortunately, it was later shown that most of the morphological oddities he compiled were not heritable and therefore they did not contribute to evolution. For this reason, his theory of evolution by discontinuous variation was not widely accepted. Interestingly, however, his homeotic transformation was later observed in many different groups of animals, and the recent genetic and molecular studies of formation of these abnormal animals led to the currently burgeoning field of developmental biology.

In the nineteenth century various forms of intraspecific variation were discussed in relation to evolution, but the discussion was always vague because Mendelian inheritance was not known and the concept of heritable and nonheritable variation had not been well established. As mentioned earlier, Darwin did not know how new variations occur. Therefore, his argument of evolution was concentrated on natural selection. Bateson was opposite to Darwin and studied mutational variants that were observable in natural populations. Unlike popular accounts, he was fully aware of the importance of natural selection for the establishment of new mutations in populations. He stated "*In the view of the phenomenon of variation here outlined, there is nothing which is any way opposed to the theory of the origin of species by means of natural selection or the preservation of favoured races in the struggle for life*" (Bateson 1894, pp. 80). He simply emphasized the importance of the study of discrete variations in evolution and was critical of panselectionism, which was popular at that time. Unfortunately, he did not provide any clear answer to the mechanism of occurrence of discrete variations. He stated "*Inquiry into the cause of variation is as yet, in my judgment, premature.*" A few years later, de Vries (1901–1903, 1909, 1910) proposed that morphological and physiological characters are subject to random mutations and these mutations generate discontinuous variations. However, this was not a complete answer to Bateson's inquiry about the formation of mutant phenotypes. To answer his question, biologists had to wait for 100 years. Only in the last few decades, developmental and evolutionary biologists have begun to study this important problem. This issue will be discussed in Chapters 5–9.

In the beginning of the twentieth century, de Vries proposed a theory of formation of new species (elementary species by his terminology) or varieties by means of mutations in his big book composed of two volumes, *The Mutation Theory* (de Vries 1901–1903). (He also rediscovered Mendel's law of inheritance.) He classified new variations into two types: (1) individual variations and (2) mutations. Individual variations correspond to Darwin's continuous variation, whereas mutations are discrete changes of phenotypic characters and may generate new elementary species by single steps of changes. He asserted that individual variations may be useful for producing new varieties or breeds within species but they never lead to new species and that to generate new species mutations are necessary. This assertion is based on the results of his extensive studies of hybridization and artificial selection in various plants and an evaluation of similar studies conducted by other investigators. A good example of the first part of his assertion is the existence of many different breeds of dogs which have been generated by artificial selection in the last 10 000 years. The extent of morphological variation among these breeds of dogs has become so extensive that many of the breeds would be regarded as different species if they were found in the wild without human intervention. In fact, de Vries stated that the two extreme ends of a character showing individual variation can be sometimes greater than the difference between two different species. However, because hybridization of these breeds produces healthy offspring, they are still regarded as members of the same species. The reason for this has not been well studied, but we are now beginning to understand it thanks to the recent progress of developmental biology, as will be discussed later.

To prove the second part of his assertion, de Vries compiled a large number of examples of elementary species in his book, *The Mutation Theory*.

Examples well known to us are cauliflowers derived from the cabbage *Brassica oleracia*, awnless oats, and strawberries without runners. These are elementary species recently generated by mutations and breed true when outcrossing is prevented. According to de Vries, however, these elementary species are not always purebred and may segregate a certain proportion of non-pure individuals in each generation.

The most famous examples of elementary species are those derived from the American evening primrose *Oenothera lamarckiana*, which he found in an abandoned potato field near Amsterdam. He first observed that there were conspicuously different forms of individuals growing in the wild populations of this species in the Netherlands. Conducting a breeding experiment for 14 years, he showed that *O. lamarckiana* continually produced small proportions of variant forms and that these variant forms either bred true or segregated into *O. lamarckiana* and the new types. Some of the new forms were morphologically quite different from *O. lamarckiana*, so that new (elementary) species names were assigned. Especially, one of the new elementary species, called *O. gigas*, was bigger and more vigorous than its parental species, *O. lamarckiana*. This form appeared only once among about 50 000 plants he examined in a 14-year period of study. Because his experimental results were very convincing, his mutation theory was welcomed by many biologists at the time of publication of his book (Allen 1969). However, de Vries's contention that new species arise by single mutational changes was later questioned. Davis (1912), Renner (1917) and Cleland (1923) showed that the strain of *O. lamarckiana* de Vries studied was apparently a permanent heterozygote for two chromosomal complexes and that most of de Vries's mutants were segregants from this unusual genetic form. This finding was a serious blow to de Vries's mutation theory (see Cleland 1972 for details), and some historians stated that de Vries's theory did not pass the test of time (Allen 1969) and was refuted resoundingly (Provine 1980, p. 55). Stebbins (1950, p. 102), a leading plant evolutionist in the twentieth century, also stated that "*de Vries' elementary species is either a figment of the imagination or a phenomenon peculiar to plants with self-pollination and an anomalous cytological condition*," despite the fact that de Vries was the first person to develop an evolutionary theory based on experimental studies.

However, de Vries's interest was not in the instability of *O. lamarckiana* but the generation of new variants by mutation. Later studies have shown that his mutants were mostly due to chromosomal changes and included polyploids, aneuploids, translocations, inversions, and single gene mutations. Notably, *O. gigas* was a tetraploid (Lutz 1907; Gates 1908) and bred true. It is now well known that these chromosomal variants often form new species particularly in plants. Therefore, de Vries's claim was not really incorrect, as Wright (1977, pp. 411–412) emphasized. Dobzhansky (1951, pp. 287–294) also considered polyploidization as an important mechanism of generating new species. However, in the middle twentieth century speciation by polyploidization was thought to be rare, and this was certainly true in animals. For this reason, de Vries's mutation theory of evolution was discredited. In recent years, however, this view has changed dramatically as the genomic study of evolution progressed (Doyle et al. 2008; Velasco et al. 2010). It is now known that new species have arisen many times by polyploidization as a single step. This issue will be discussed later (Chapter 7).

de Vries was a visionary scientist and introduced experimental study of evolution for the first time. In the preface of his 1901 book (de Vries 1909), he stated: "*A knowledge of the laws of mutation must sooner or later lead to the possibility of inducing mutations at will and so of originating perfectly new characters in animals and plants. And just as the process of selection has enabled us to produce improved races, greater in value and in beauty, so a control of the mutative process will, it is hoped, place in our hands the power of originating permanently improved species of animals and plants.*" As every geneticist knows, this prediction was brought to its fruition in the latter half of the twentieth century.

At this stage, it should also be indicated that de Vries was fully aware of the necessity of natural selection for new mutant forms to be established in natural populations (de Vries 1909, p. 212). Because mutation is assumed to occur at random, disadvantageous mutations are obviously eliminated and only mutants that are competitively stronger than the original types can survive. He simply emphasized the importance of understanding the first process of evolution (generation of innovative characters) as envisioned by Darwin (1859). The fact that de Vries was not antagonistic to natural selection was also emphasized by Wright (1960) and Gayon (1998).

1.4. Mutationism

de Vries's mutation referred to any heritable change of phenotypic characters. With our current knowledge, this means any change of genetic materials including nucleotide substitutions, nucleotide insertions/deletions, gene duplications/deletions, gene transposition, changes in gene interactions, various types of chromosomal changes, and genome duplication. In de Vries's time, none of these detailed genetic changes was known, and only thing he could study was heritable changes of morphological or physiological characters.

As Mendelian genetics progressed, genes were identified as the unit of inheritance, and mutations generally implied heritable changes of genes. In the meantime chromosomal changes were discovered and shown to affect some morphological characters, but these changes were not always specific and therefore they were not studied on a regular basis. Polyploidization was also shown to generate new genetic variation, but its effect on phenotypic evolution was thought to be minor (Stebbins 1950).

For these reasons, mutations generally referred to changes in protein-coding genes rather than to chromosomal changes. This was particularly so in the work of Thomas Morgan (1916, 1925, 1932) and his school. During this gene-centric era, de Vries's idea that a new species can be generated by a single mutational event was ignored. Instead, a new school of thought advocating evolution by genic mutation became predominant, and a leader of this school was Thomas Morgan. Morgan's view is often called mutationism, somewhat inappropriately.

During the first 15 years after the rediscovery of Mendelism in 1900, there was leaping progress in the study of genetics. First, blending inheritance was disproved, and Mendelian inheritance of discrete characters was shown to apply to many characters in plants and animals. This indicated that genetic variability does not decay by bisexual reproduction but can be maintained in populations (Castle 1903; Hardy 1908; Weinberg 1908, 1963). Second, Johannsen's (1909) pureline theory showed that Darwin's continuous variation is composed of heritable and nonheritable variation and selection on the latter variation is ineffective. Third, Nilsson-Ehre (1909), East (1910), and Emerson and East (1913) showed that the inheritance of quantitative characters can be explained by the independent segregation of alleles at multiple loci. Weinberg (1910, 1984) and Fisher (1918) also showed that the correlation of quantitative characters between relatives as observed by the biometricians Galton, Weldon, and Pearson can be explained by Mendelian inheritance. These findings ended the heated controversy between the biometricians and the Mendelians. Fourth, Morgan, Sturtevant, Muller, and Bridges (1915) and Muller and Altenburg (1919) showed that new mutations arise spontaneously with a very low but measurable frequency and they are inherited as Mendelian characters. Most of the mutations were deleterious, but some of them appeared to be virtually neutral or slightly advantageous.

Morgan's theory of evolution (1916, 1932) was based on the above findings by geneticists. He separated the process of generating innovative characters and the preservation of these characters in evolution, as was done by Bateson and de Vries. In his view the first process is accomplished by random mutations that occur at each genetic locus whether the character is continuous or discontinuous. The second process of preservation of new mutations is achieved by natural selection or genetic drift. This view was conceptually somewhat different from Darwin's view of natural selection, which is supposed to act for the "preservation of favored races in the struggle for existence." Darwin's view was developed under the assumption that "favored and unfavored races" always exist in the population. Furthermore, because the genetic entities of favored or unfavored races were not known in Darwin's time, the outcome of the struggle for existence was always vague. Morgan was clearly in a better position than Darwin in the conceptual formulation of natural selection because of the new genetic knowledge. In his early days Morgan (1903) was critical of the efficacy of natural selection and found it somewhat teleological. In his 1916 and 1932 books, however, he presented a clear form of mutation-selection theory.

Some authors stated that Morgan was a typologist and did not have a population concept (e.g. Allen 1978; Mayr 1982). In his book *The Scientific Basis of Evolution* (Morgan 1932), however, he presents a rather incisive discussion of population and quantitative genetics. On page 132, he states: "*If the mutant is dominant…and is bred to the wild type, the mutant character will appear again in half of the progeny, and if advantageous, i.e., if one that increases the chance of survival of the individual and the race, it will gradually spread through the race. If the new mutant is neither more advantageous than the old character, nor less so, it may or may not replace the old character, depending partly on chance; but if the same mutation recurs again and again, it will most probably replace the original character. If the new character is a disadvantageous one, it will soon be eliminated.*" This statement indicates that Morgan had a good grasp of the concept of population genetics.

Morgan's mutationism was quite popular in the first quarter of the twentieth century. Actually, this view was held by most evolutionary geneticists at that time, and Morgan was merely a spokesman for them (Wright 1960). Another leader of mutationism or the mutation-selection theory was Hermann Muller (1929), who was more proficient in mathematics than Morgan. This theory clearly showed that any evolutionary change of phenotypic characters should be studied by examining the mutational change of genes, and this idea has changed the concept of evolution forever though there are still some evolutionists who investigate only phenotypic changes. However, as the study of mutations and polymorphic alleles proceeded, a number of observations which were unfavorable to mutationism were noticed. First, most mutations observed in laboratories were deleterious and did not appear to contribute to evolution at all. Second, morphological characters were mostly quantitative, and their variation was thought to be a product of interaction of a large number of genes and environmental factors. Therefore, the relationship between mutation and selection was unclear. Third, even though some discrete characters showed seemingly Mendelian inheritance, the characters were often controlled by additional modifier genes. For example, the heredity of guinea pig coat color pattern could be explained by a number of loci, but there was a complicated interaction among different loci (Wright 1927).

To explain these observations, Morgan's mutationism appeared too simplistic. Although Morgan had a reasonably good grasp of the population concept as an experimentalist, he was not proficient in mathematics and was gradually left behind as the mathematical theory of population genetics advanced in the 1920s and 1930s. As mentioned in Section 1.5, this development initiated a new era of neo-Darwinism, and the rise of neo-Darwinism caused Morgan's mutationism to decline gradually. At present Morgan's view is often called "mutationism," but this is not really appropriate because he accepted natural selection as an agent of eliminating unfit genotypes, as Darwin did. However, because this terminology is widely accepted and Morgan certainly considered mutation as the primary force of evolution, I will use the word mutationism to represent Morgan's view in this book.

1.5. Neo-Darwinism

The term neo-Darwinism has been used to represent various forms of modified Darwinism since the late nineteenth century. At the present time, however, it usually refers to the evolutionary theory formulated by Fisher (1930), Wright (1931, 1932), and Haldane (1932). Evolution refers to the long-term genetic change of populations or species, so that it is difficult to do experimental studies. However, once Mendelian genetics was established, it was possible to predict the evolutionary changes of populations under simplified assumptions. Although this prediction was very crude, it was much better than intuitive speculations. After conducting extensive mathematical studies in the 1920s and 1930s, the three founders of population genetics, Fisher, Wright, and Haldane, reached the conclusion that natural selection is much more important than mutation. This was opposite to the view of mutationism. Around this time, however, various experimental evolutionists initiated studies of natural selection and obtained results suggesting that the intensity of natural selection for a pair of alleles is much higher than previously thought.

For this reason, the mathematical work was soon accepted by a number of leading experimental geneticists, notably Dobzhansky (1937, 1951), who wrote an influential book on evolution titled *Genetics and the Origin of Species*. Through this book, neo-Darwinism was gradually disseminated among biologists, and various authors (Huxley 1942; Mayr 1942; Simpson 1944, 1953; Stebbins 1950; Ford 1964) made further refinements of the theory. Because of these works, neo-Darwinism was accepted by most evolutionists by 1960. Neo-Darwinism is also called the synthetic theory of evolution, because it is based on a mixture of Darwinism and Mendelism (Huxley 1942). In practice, however, it is a stronger version of selectionism than the original version of Darwinism. A detailed account of neo-Darwinism will be presented in Chapter 2.

The main difference between mutationism and neo-Darwinism is in the relative importance of mutation and selection in evolution. In mutationism, new mutations are classified into beneficial, neutral, and deleterious mutations. Beneficial and neutral mutations may contribute to evolution, whereas deleterious mutations are eliminated from the populations. Here beneficial mutations may be advantageous from the time of their occurrence, though the possibility that the environmental change may affect the fitness of mutations later is not excluded. In mutationism, the driving force of evolution is mutation, and natural selection is merely a sieve to save favorable mutations and eliminate unfavorable ones (Morgan 1916, 1932). Of course, this does not mean that the gene interaction among different loci or the genotype and environmental interaction is unimportant.

Neo-Darwinism also assumes that mutation is the ultimate source of genetic variation, but its effect on gene frequency change is so small that it plays a minor role in evolution. It is also assumed that because of the mutations that have occurred in the past, natural populations contain a sufficient amount of genetic variability to respond to almost any kind of selection. Therefore, evolution is determined mainly by environmental changes and natural selection. Since there is enough genetic variability, no new mutations are required for a population to evolve in response to an environmental change. Mutations are assumed to occur recurrently so that a majority of advantageous mutations have been fixed or have reached their optimum frequencies in the population. Therefore, the genetic composition of a population is at or near its optimum for a given environment. Evolutionary change of a species occurs gradually by environmental changes.

Of course, the above statements characterize the general properties of neo-Darwinism. In practice, there have been considerable differences among the views of different authors such as Fisher, Wright, and Haldane. Fisher (1930, 1958) was essentially a panselectionist and believed that evolution has occurred almost exclusively by natural selection. Wright's (1931, 1932) view was considerably different from Fisher's (1930). He emphasized the importance of gene interaction and random changes of allele frequencies due to genetic drift. Haldane's (1932) view was somewhere between Fisher's and Wright's, but his work was primarily concerned with allele frequency changes in large populations. He was a rationalist and studied various evolutionary problems with common sense and relatively simple mathematical methods.

It should also be noted that the transition of evolutionary theory from mutationism to neo-Darwinism occurred only gradually (Wright 1960, 1977), unlike the opposite statement by Mayr (1963, 1982). Dobzhansky's (1937) book is often regarded as the initiation of neo-Darwinism by experimental evolutionists. However, the general view presented in this book was very similar to that of Morgan (1932), though he certainly presented much empirical data obtained from natural populations and admired Wright's shifting balance theory of evolution. For example, he believed that most genetic polymorphisms including those of inversion chromosomes are more or less neutral, following Darwin (1859) and Morgan (1932). Only in the 1951 edition of *Genetics and the Origin of Species* did he present a strong selectionist view similar to Fisher's (1930). The books written by Huxley (1942) and Mayr (1942) were composed of a heterogeneous mixture of mutationism and selectionism. By around 1960, however, many empirical evolutionists became selectionists.

1.6. Neomutationism or Mutation-Driven Evolution

In the eras of mutationism and neo-Darwinism, the molecular structure of a gene was not known, and genetic variation was studied by using morphological or physiological characters. A pair of alleles that showed Mendelian inheritance was automatically thought to have been generated by a single mutation. However, the mutation was a black box at this time. Only after the molecular structure of a gene was discovered and its role of protein synthesis was clarified could biologists study the mechanism of mutation and evolution in a reasonable way. The basic mechanism mutation and of gene duplication or chromosomal mutations was known, but their evolutionary implications were unclear until genome sequence data became abundant.

In the 1960s and 1970s molecular biologists such as Ingram (1961), Zuckerkandl and Pauling (1962), and Margoliash (1963) initiated evolutionary studies of protein molecules, and these studies showed that the rate of amino acid substitution is roughly proportional to evolutionary time, as will be discussed in Chapter 4. These findings led Kimura (1968b) and King and Jukes (1969) to propose the neutral theory of molecular evolution. Yet, many evolutionists including Motoo Kimura and Jack King believed that phenotypic evolution is caused primarily by natural selection. By contrast, Nei (1975, 1987, 2007) proposed that since phenotypic evolution is ultimately controlled by DNA and RNA molecules, both molecular and phenotypic evolution must be primarily caused by mutation. Nei's view has been based on the new findings in the study of molecular evolution and developmental biology. He considered all kinds of DNA changes (nucleotide substitution, gene duplication, polyploidization, epigenetics, etc.) as mutations and tried to explain phenotypic evolution by mutation.

Previously I called this view neomutationism (Nei 1983, 1984), neoclassical theory (Nei 1987), and the new mutation theory (Nei 2007), but in this book I have decided to call it the theory of mutation-driven evolution or neomutationism depending on convenience. At this point, I would like to indicate that a mutation is defined here as any change of genetic material, i.e. nucleotide sequences, genes, chromosomes, and genomes. During the last five decades, mutation has been studied intensively at the molecular level, and our knowledge of mutational changes of genes has expanded enormously. At the DNA level, the basic process of a mutation is the replacement of a nucleotide (say A) by one of the three remaining nucleotides (T, C, or G) at a given nucleotide site or deletion or insertion of nucleotides. This discovery made it possible to study the pattern and effects of mutations in protein-coding or noncoding regions of DNA, and this ushered a new discipline of molecular evolutionary biology in the 1970s. We now know that a large amount of genetic variation is generated by gene duplication and deletion as well as by nucleotide changes and that the variation of gene copy number contributes to the formation of innovative phenotypic characters. The relationships between mutations and phenotypic evolution in which most evolutionists are interested are quite complicated, because the development of phenotypic characters is controlled by interaction of many genes. Yet, we already know the basic principle of developmental biology, and we are in a position to study phenotypic evolution by examining both processes of mutations and natural selection at the molecular level for the first time in the history of evolutionary biology. The results so far obtained indicate that mutation is the driving force of evolution, as will be discussed in Chapters 4–9. The idea of mutation-driven evolution has already changed the methodology of studying evolution and the interpretation of the mechanisms of evolution.

1.7. Survival of the Fittest and Survival of the Niche-Filling Variants

Darwin's theory of natural selection has been symbolized by Herbert Spencer's phrase "the survival of the fittest." Although this phrase has been criticized for its potential tautology, it captures the essence of natural selection (Gayon 1998). In fact, this idea constitutes the foundation of theoretical population genetics in the neo-Darwinian era. In population genetics, it is customary to assign different fitness values (number of offspring per individual) for all the genotypes in a population and then examine the evolutionary change of the mean fitness ($\overline{w}$) of the population. As long as there are

(A) Darwinian theory of evolution

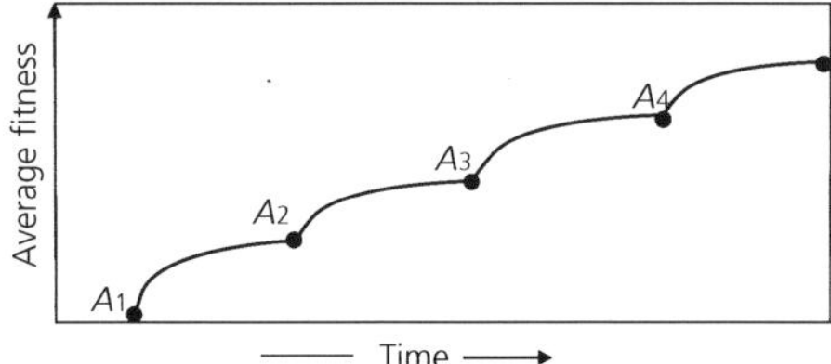

(B) Niche-filling mutation theory

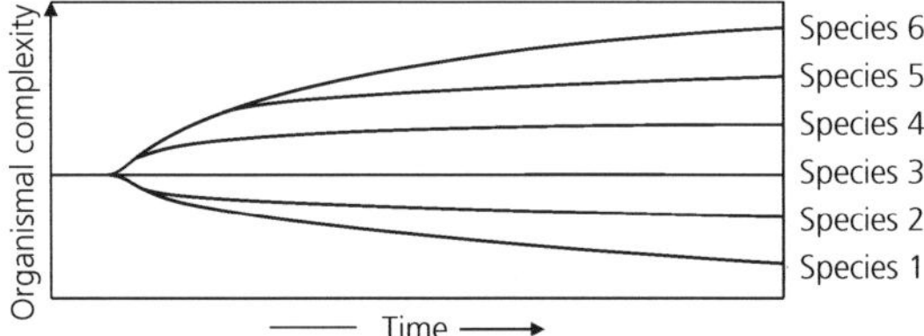

Fig. 1.1. (A) Darwinian theory of evolution as illustrated by the development of drug resistant strains of bacteria through mutation and selection. A_1, A_2, A_3, and A_4 refer to the occurrence of resistant mutations. Here only the mutations fixed in the population are considered. (B) Niche-filling mutation theory. Here only species or species groups which show large morphological differences (e.g. different orders of mammals) are considered. In some cases organismal complexity may decline as in the case of cave animals. Niche-filling evolution is expected to enhance biodiversity.

some advantageous mutations segregating in the population and the environmental condition remains constant, ($\overline{w}$) always increases. Thus, Fisher's (1930) fundamental theorem of natural selection (see Chapter 3) states that the rate of increase of mean fitness is equal to the additive variance (V_A) of fitnesses among different individuals. Because V_A is a non-negative quantity, ($\overline{w}$) always increases.

However, Fisher's theorem works only for limited cases of short-term evolution, and in the case of long-term evolution the biological meaning of the theorem is unclear (see Chapter 3 for details). Let us consider a simple case of evolution of a drug-resistant strain (population) of bacteria. In bacteria many antibiotic-resistant mutations occurred after World War II, because antibiotics such as penicillin and streptomycin were widely used in hospitals and in the livestock industry. Figure 1.1A shows a schematic picture of the successive replacement of the wild-type allele by the drug-resistant mutant at four different loci. In this case the mean fitness of the population is expected to increase gradually as the allelic replacement proceeds, so that the population with the four mutant alleles has the highest fitness. This is in agreement with the Darwinian view of evolution.

In practice, however, this scenario does not always work because the environmental condition never stays constant and there are many different niches or localities in which different strains are adapted. Actually, some bacterial species like *Salmonella enterica* and *Escherichia coli* are distributed worldwide and consist of many different strains (e.g. Tindall et al. 2005; Sims and Kim 2011). Therefore, there must be strains which do not have any antibiotic-resistant mutations, and some are likely to have one or two. However, because there are many different niches, strains with one or two mutations or even no mutation may survive without trouble. Actually, drug resistance is merely one of many characters that determine bacterial survival. Therefore, it is quite reasonable to see a large number of *Salmonella* strains, which are adapted to different niches and survive simultaneously. This example can easily be extended to the evolution of different strains within any species or the evolution of different species, genera, families, etc.

These observations suggest that evolution does not necessarily occur by the struggle for existence. In the presence of changeable environments, Fisher's theorem does not work (see Chapter 3 for details). Note also that the mean fitness is not a good quantity for measuring the extent of evolutionary change because we cannot compare the mean fitness of different species. Is there any way to define evolution in a reasonable way when long-term evolution is considered? In my view evolution should be defined as a process of increase of organismal complexity, as is generally believed by many biologists. Strictly speaking, this definition does not always work well, because there are organisms that simplify their complexity to adapt to a particular environment. If we consider this possibility, we may define evolution as a process of increase or decrease of organismal complexity. However, how should we measure the extent of organismal complexity? This is not a simple problem, but there is a crude way to measure the extent. It is to use the number of cell types in the organism (Vogel and Chothia 2006). This measure is often used by evolutionists who are interested in the evolution of complex organisms (see

Chapter 5), though it is not always easy to compute the number of cell types.

In this connection, we can introduce another concept of evolution. It is the survival of the niche-filling variants. This concept is for describing the tendency for new mutant organisms to occupy new niches and propagate there (Fig. 1.1B). The new group of organisms may not produce more offspring in the new niche than in the previous one, yet they may generate a new evolutionary lineage with a different degree of organismal complexity because of the different environmental factors. Many marine mammals such as whales and sea lions probably originated in this way. In Darwinian or neo-Darwinian theory of evolution, survival of the fittest has been emphasized. In the theory of mutation-driven evolution, however, emphasis is given to the generation of different mutant forms with a higher or lower degree of organismal complexity. For this reason, survival of the niche-filling variants symbolizes the importance of mutation in evolution. This phrase may not be as attractive as survival of the fittest, yet it conveys the basic principle of mutation-driven evolution. Of course, this principle can be used for Darwinism or neo-Darwinism as well (Darwin 1859), but it is more appropriate for mutation-driven evolution, because it does not require the struggle for existence, though this may happen. It should also be noted that we are here using the word *niche* very loosely because the strict definition of a niche is impossible. In this view, evolution is not for increasing fitness but for enhancing the diversity of organismal complexity or biodiversity. We shall return to this issue again in Chapter 9.

CHAPTER 2

Neo-Darwinism and Panselectionism

2.1. Backgrounds

As mentioned in the previous chapter, the theoretical foundation of neo-Darwinism was laid out by three pioneers of theoretical population genetics, R. A. Fisher, Sewall Wright, and J. B. S. Haldane. In the 1920s and 1930s, these authors conducted extensive mathematical studies on the changes of allele and genotype frequencies due to mutation, selection, and random genetic drift, and reached the conclusion that natural selection is much more important in evolution than mutation. This theoretical conclusion was gradually accepted by leading empirical evolutionists such as Theodosius Dobzhansky, G.G. Simpson, Ernst Mayr, Julian Huxley, and E.B. Ford by 1960.

When Mendel's principle of inheritance was rediscovered in 1900, it was believed to support the Huxley-Galton view of evolution by discontinuous variation rather than Darwin's view of evolution by continuous variation. William Bateson (1902) championed as a Mendelian and severely attacked the biometricians such as Weldon and Pearson, who supported the Darwinian view of gradual evolution. Because Mendelism dealt with discrete characters, the camp of discontinuous evolution appeared to have won the battle with the biometricians at this time.

Later, Johannsen (1909) clarified the genetic nature of the variation of quantitative characters. Using the self-fertilizing plant *Phaseolus vulgaris* (beans), he isolated purelines and showed that variation within purelines is caused by environmental factors and therefore it is not heritable. Only when different purelines were mixed in a population was part of the variation heritable. He coined the words phenotype and genotype as well as gene. Another important step of progress that occurred with respect to quantitative characters was the experimental demonstration that these characters are usually controlled by multiple genetic loci apart from environmental factors (Nilsson-Ehle 1909; East 1910; Emerson and East 1913). The alleles at these multiple loci were then shown to follow Mendelian inheritance. These studies now showed that the inheritance of discrete characters as well as continuous characters can be explained by Mendelism. This finding resolved the controversy between Mendelians and biometricians and provided a general approach for studying the evolution of discontinuous and continuous characters. Weinberg (1910) and Fisher (1918) also showed that the correlation of close relatives (e.g. parents and offspring or siblings) discovered by Galton and Pearson can be explained by Mendelian inheritance. Weinberg's work on this subject was not known to English-speaking scientists for a long time until Stern (1962) wrote a commentary on Weinberg's work in English. This occurred apparently because his paper was difficult to understand even if one could read German (Crow 1999). This paper is now translated into English (Weinberg 1984).

Once it was shown that the inheritance of almost all characters is controlled by Mendelian genes, it was possible to study the evolution of both discontinuous (qualitative) and continuous (quantitative) characters in terms of allele frequency changes at least theoretically. In practice, however, the genetic study of evolution was confined within species, because it was almost impossible to identify orthologous genes between different species by Mendelian methods and without knowing orthologous genes we could not study the extent of genetic variation between species. For this reason, the genetic

Mutation-Driven Evolution. First Edition. Masatoshi Nei.

study of evolution in this period was done almost exclusively by studying allele frequency changes within species rather than between species. Furthermore, since it was usually difficult to identify the allelic differences for morphological or physiological variations even within a species, the evolutionary changes of populations were studied by using mathematical models under various assumptions of gene effects, population structures, environmental changes, ecological changes, etc. Although these mathematical studies were conducted under many simplified assumptions, they were better than the pre-Mendelian speculative approach of predicting evolution. This was the main reason why neo-Darwinism became popular and dominated the field of evolutionary biology for the last 70 years. Experimentalists who could not study long-term change of genes simply followed the guidelines of research established by theoreticians.

Although neo-Darwinism is characterized by the importance of natural selection, there were considerable differences among different authors with respect to minor aspects. However, the major features of neo-Darwinism may be summarized as follows. (1) Mutation is the primary source of genetic variation, but its effect on gene frequency change is so small that it plays a minor role in evolution. (2) Because of the mutations that have occurred in the past, natural populations contain a sufficient amount of genetic variation to respond to almost any kind of selection. (3) Evolution is determined mainly by environmental changes and natural selection. Since there is enough genetic variation, no new mutations are necessary for a population to evolve in response to an environmental change. There is no relationship between the rate of mutation and the rate of evolutionary change. (4) Natural selection has the power of creating innovative characters in the presence of raw genetic material provided by mutation. (5) Because evolution is determined almost exclusively by natural selection, virtually no characters evolve in a neutral fashion. Even the number of hairs on a part of human body or human facial structures is a product of adaptation to environmental conditions. (6) Generally speaking, the size of a natural population is so large that the allele frequency changes in different generations can be described with a deterministic model without the effects of random errors. Only in very small populations is it necessary to consider the random errors of allele frequency changes due to genetic drift. In this way neo-Darwinism has generated an idea of panselectionism in its extreme form.

2.2. Allele Frequency Changes as the Basic Process of Evolution

In the early twentieth century, Mendelian genetics established that all the genetic variation is ultimately caused by allelic differences at genetic loci. This finding led to the idea that the evolutionary change of populations can be studied by examining the changes in allele frequencies over time. This is the reason why mathematical studies of allele frequency changes were initiated (Fisher 1930, 1958; Wright 1931, 1969; Haldane 1932; Crow and Kimura 1970). In these studies only a pair of alleles was considered at each locus partly because most loci were known to have only two alleles at that time and partly because the mathematical treatment with multiple alleles was quite complicated. To simplify the mathematical treatment, it was also often assumed that the population size is infinitely large and the allele frequency changes in a population can be treated deterministically. In practice, all natural populations are finite, so the allele frequency changes are subject to sampling errors at the time of reproduction. The mathematical model taking into account the sampling errors is called the stochastic model and the allele frequency changes due to sampling errors are called genetic drift. In the following I first use the deterministic model to present the basic idea of population genetics. I will start with the allele frequency changes due to mutation, disregarding the effect of natural selection.

Mutation

One of the early discoveries by population geneticists was that the allele frequency change due to mutation is much smaller than that due to natural selection. Let us consider a pair of alleles A_1 and A_2 at a locus in a large population and denote the relative frequencies of alleles A_1 and A_2 by x and $1 - x$, respectively. We assume that A_1 mutates to A_2 at a rate of u per generation and A_2 mutates to A_1 at a rate of v and that there is no selection. Then, the frequency of allele A_1 in the next generation (x') becomes

$$x' = x(1-u) + (1-x)v \quad (2.1)$$

Therefore, the amount of change of the frequency of allele A_1 per generation ($\Delta x = x' - x$) is

$$\Delta x = v - (u+v)x \quad (2.2)$$

This formula indicates that if v is greater than $(u + v)x$, Δx is positive and therefore x increases. However, as x increases, Δx eventually becomes 0 and no more change in allele frequency occurs. At this point, x is given by

$$\hat{x} = v/(u+v) \quad (2.3)$$

This $\hat{x}$ is called the equilibrium allele frequency.

In classical genetics it was generally believed that the mutation rate is of the order of 10^{-5} per locus per generation in many different organisms. For simplicity, let us assume that there is one-directional mutation with $v = 10^{-5}$ and $u = 0$. In this case Equation (2.2) becomes $\Delta x = v(1 - x)$. This indicates that the allele frequency change per generation is at most $v = 10^{-5}$ when $x = 0$. Therefore, the evolutionary change of allele frequency is very slow. For this reason, early population geneticists concluded that mutation is not an important factor for evolution.

At the present time, we know that mutation generates many different alleles (nucleotide sequences) at a locus and that the above mathematical formulation with two alleles is not very meaningful. In the 1920s and the 1930s, no one knew the molecular structure of genes, and this hampered proper mathematical modeling of allele frequency. Existence of multiple alleles at a locus was known, but few population geneticists took into account this observation in mathematical modeling. One exception was Wright (1939, 1948a), who developed a theory of allele frequency distribution caused by mutation, selection, and genetic drift, assuming that a gene is capable of generating almost infinitely many different alleles. This idea was conceived when he came to know that the locus controlling self-incompatibility in some plants (e.g. *Oenothera*) can have a large number of alleles even in a relatively small population (Wright 1939). He then developed a mathematical model that is capable of generating a large number of alleles at a locus. However, this model was considered to apply only to a special case, and few population geneticists were aware of this work until the molecular structure of genes was clarified. Wright's model is now called the infinite-allele model (Kimura 1983), and it is widely used in molecular evolutionary genetics (see Section 2.6).

Natural Selection with Constant Fitness

In the presence of natural selection, the allele frequency change per generation was considered to be much greater than the change by mutation pressure. In population genetics the effect of natural selection on allele frequency changes was studied by considering the relative fitnesses of genotypes. In a randomly mating diploid population there are three genotypes at a locus with two alleles, A_1 and A_2. They are A_1A_1, A_1A_2, and A_2A_2. The fitness of a genotype is defined as the relative number of adult offspring produced by the genotype. Let w_{11}, w_{12}, and w_{22} be the fitnesses of A_1A_1, A_1A_2, and A_2A_2, respectively. If we assume that A_1 is advantageous over A_2 and the genotype fitnesses for a locus are determined solely by the locus without the effect of other loci, the amount of change in the frequency (x) of allele A_1 is given by

$$\Delta x = xy\left[x(w_{11} - w_{12}) + y(w_{12} - w_{22})\right]/\bar{w} \quad (2.4)$$

where $y = 1 - x$ and w is the mean fitness of the population and is given by $x^2 w_{11} + 2xy\, w_{12} + y^2 w_{22}$ (see Appendix A).

If we write $w_1 = x\, w_{11} + y w_{12}$ and $w_2 = x\, w_{12} + y\, w_{22}$, Equation (2.4) can also be written as follows.

$$\Delta x = xy(w_1 - w_2)/\bar{w} \quad (2.5)$$

In real populations the definition and estimation of fitnesses are very difficult, but the concept of fitness is useful in understanding how the relative frequency of a particular allele changes by selection over evolutionary time. The simplest form of selection is genic selection, in which gene effects are assumed to be additive (semidominant). In this case we may write the fitnesses of A_1A_1, A_1A_2, and A_2A_2 as $w_{11} = 1$, $w_{12} = 1 - s$, and $w_{22} = 1 - 2s$, respectively, where s is called the selection coefficient and represents the extent of reduction of fitness due to the disadvantageous allele A_2. The frequency of the advantageous allele A_1 is expected to increase, and the amount of change per generation is given by

$$\Delta x = sxy/(1-2sy) \quad (2.6)$$

(See Appendix A.)

The equations for Δx for dominant, recessive, and overdominant selection (heterozygote advantage) are presented in Appendix A. Equation (2.6) indicates that Δx is large when s is large and x is close to 0.5. For example, when s = 0.1 and x is 0.5, Δx is approximately 0.028, whereas if s = 0.001 and x = 0.5, Δx becomes 0.00025. These values are much greater than the possible change due to mutation ($\Delta x \approx v = 0.00001$). The pattern of allele frequency change varies considerably depending on whether the advantageous allele is semidominant (genic selection), dominant, or recessive. However, Δx is generally much larger than the change due to mutation unless s is very small. Figure 2.1 shows the allele frequency change of semidominant advantageous alleles for various values of s and indicates that the frequency change is quite rapid unless s is very small. In classical population genetics s was thought to be at least 0.01 and often greater than 0.1. For this reason, population geneticists believed that the effect of natural selection is much more important than that of mutation in evolution. In practice, however, this view was derived from a small number of studies of genes controlling morphological characters. Note also that with morphological characters it is very difficult to measure small values of selection coefficients. Therefore, the idea of large selection coefficients was derived from a biased set of genes studied.

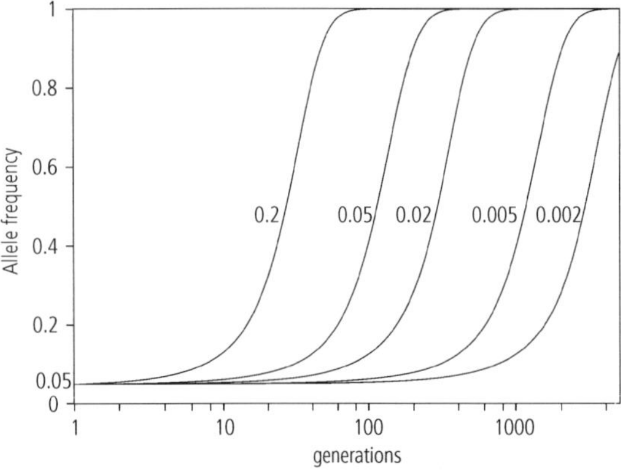

Fig. 2.1. Allele frequency changes of semidominant advantageous genes (genic selection). The number given for each curve shows the selection coefficient (s) for the advantageous allele. The initial frequency of the advantageous allele is assumed to be 0.05. Here s is assumed to be constant throughout the evolutionary process. In reality, this assumption would not hold, and s may change from generation to generation. In this case the frequency change curves would be very different from those given in this figure particularly when s is small (see text and Fig. 2.2).

At this point, it should also be noted that the roles of mutation and natural selection in evolution are fundamentally different. When we discussed the allele frequency change by mutation, we assumed that all mutations are neutral. In reality, this is not the case, and most mutations are known to be deleterious. However, there are some mutations that are advantageous over the pre-existing alleles, and these mutations are important for adaptive evolution. By contrast, natural selection is for shifting allele frequencies in populations, and conceptually they do not create any new genotypes. To understand this issue more clearly, we have to know the molecular basis of mutation and natural selection, and this problem will be discussed in Chapters 3—9.

Note also that the allele frequency changes in Fig. 2.1 are derived under the assumption of a constant selection coefficient (s) for the entire evolutionary process. In the real world this assumption would never hold because of the environmental change and the effect of other genetic loci which are also changing every generation. Therefore, s may increase or decrease with time. It may also fluctuate with generation almost at random (Fisher and Ford 1947; Bell 2010). In these cases the allele frequency change would be very different from those given in Fig. 2.1.

Figure 2.2A shows the frequency change of the *medionigra* gene of the moth *Panaxia dominula* in a seemingly isolated population near Oxford, England. The gene *medionigra* is a dominant allele generating a dark form of the moth, and the study of its frequency change was initiated by Fisher and Ford (1947) and continued by Sheppard and Cook (1962), Clarke et al. (1991), Cook and Jones (1996), O'Hara (2005), and others for over 50 years. The frequency of *medionigra* initially declined gradually and then appears to have reached an equilibrium value. However, the frequency change was never smooth

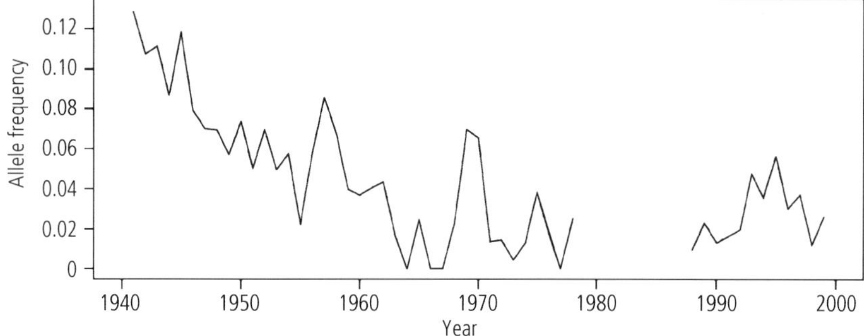

(B) Selection coefficients for the brown shell form of *Cepaea nemoralis*

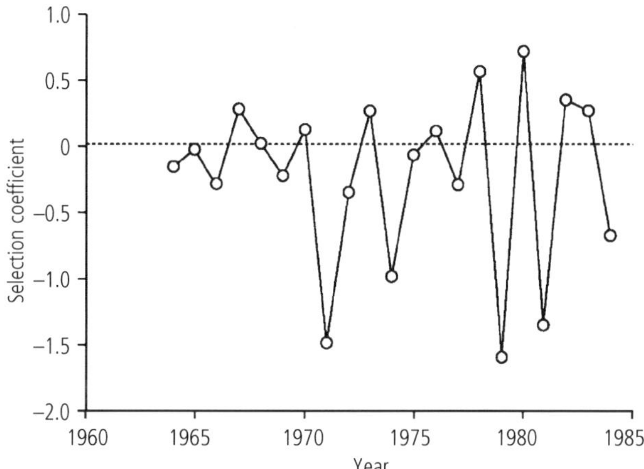

Fig. 2.2. Random changes of allele frequencies and selection coefficients. (A) Frequency changes of the *medionigra* gene in the moth *Panaxia dominula*. The frequency was not studied from 1979 to 1987. From O'Hara (2005). Reproduced with permission from the Royal Society of Biological Sciences. (B) Fluctuation of selection coefficient (s) for the brown phenotype of the snail *Cepaea nemoralis* for a period of 21 generations. This fluctuation includes sampling errors, so that the actual fluctuation of s should be smaller than shown here. From Bell (2010).

and fluctuated from generation to generation even in large populations. This type of fluctuation of allele frequencies occurs because the environmental conditions and the genetic background of the locus change almost every generation. By contrast, the reason for the apparent equilibrium frequency near the end of the experiment is not very clear, but it is possible that the frequency-dependent selection is operating at this frequency level (O'Hara 2005) or the gene migration has occurred from the neighboring populations. The estimate of average selection coefficient for the entire evolutionary process was $s = -0.103$ (Bell 2010).

Figure 2.2B shows that the selection coefficients for the brown phenotype of the land snail *Cepaea nemoralis* also vary extensively across different generations (Cain et al. 1990; Bell 2010). The significance of this variation will be discussed in Section 2.4.

Mutation-Selection Balance

In the 1930s to the 1950s most mutations were shown to be deleterious and occur repeatedly. Therefore, natural populations were thought to contain many unfavorable alleles in low frequency because of the balance between mutation and selection. It was also thought that some of these unfavorable alleles become favorable when the environment changes and therefore they are an important source of evolution (Muller 1950; Dobzhansky 1951; Haldane 1957).

One example was the melanic form of the peppered moth *Biston betularia*, which will be mentioned in Section 2.3. It was therefore important to know the equilibrium frequencies of mutant alleles in populations.

Suppose that A_1 and A_2 are the favorable and unfavorable alleles, respectively, and A_1 mutates to A_2 at a rate of u per generation. Let 1, $1-h$, and $1-s$ be the relative fitnesses of A_1A_1, A_1A_2, and A_2A_2, respectively. If selection coefficients are large and $s >> h >> 0$, the frequency of A_2A_2 is negligibly small, and the equilibrium frequency of allele A_2 is approximately given by

$$\hat{y} = u / h \tag{2.7}$$

(see Appendix A). However, if A_2 is completely recessive to allele A_1 with $h = 0$, $\hat{y}$ becomes

$$\hat{y} = \sqrt{u / s}. \tag{2.8}$$

Therefore, if $u = 10^{-5}$, $h = 0.1$, and $s = 0.5$, the equilibrium frequency of the codominant allele A_2 is given by 0.0001, whereas the frequency of a completely recessive allele A_2 with $h = 0$ and $s = 0.1$ becomes 0.01. Many early Mendelians detected low frequency unfavorable mutations. One example is the recessive gene controlling albinos in humans. The frequency of albinos in human populations is one in 20 000, and the allele frequency is about 0.007. Human populations contain many such unfavorable alleles whose frequency ranges from 0.0001 to 0.02, depending on the population and the gene studied (Haldane 1957; McKusick 1986). The frequencies of these mutations are higher than the average mutation rate of 10^{-5} per generation.

Balanced Polymorphism

It has long been known that a species often contains different morphological and physiological types (polymorphisms). Darwin (1859) was aware of the existence of these polymorphic characters and thought that they are neither advantageous nor disadvantageous and are floating in the population as neutral characters. As the genetic study of such polymorphic characters advanced in the early twentieth century, it became clear that many such polymorphic characters are controlled by single genes or supergenes (tightly linked genes). At the present time, there are three explanations for the occurrence of polymorphic alleles. First, the polymorphism is more or less neutral. Second, the polymorphism may represent the case where one disadvantageous allele is being replaced by the advantageous allele by natural selection. This type of polymorphism is called transient polymorphism. Third, the polymorphism may occur because the heterozygote (A_1A_2) has a higher fitness than the two homozygotes (A_1A_1 and A_2A_2) (overdominant selection). In this case the polymorphism may persist in the population for a long time. For this reason, it is called balanced polymorphism. Balanced polymorphism may occur by frequency-dependent selection as well (e.g. Wright and Dobzhansky 1946), but overdominant selection is generally considered more important than frequency-dependent selection.

The idea of balanced polymorphism was first proposed by Fisher (1922). He showed mathematically that if the fitness of heterozygotes A_1A_2 is higher than the fitnesses of the two homozygotes (A_1 A_1 and A_2 A_2) the allele frequency of A_1 eventually reaches an equilibrium and stays there indefinitely in large populations unless the fitnesses of the three genotypes change due to some environmental change. When the genotype fitnesses are expressed as $w_{11} = 1 - s$, $w_{12} = 1$, and $w_{22} = 1 - t$, the equilibrium frequency of allele A_1 is given by

$$\hat{x} = t / (s + t) \tag{2.9}$$

(Appendix A). Therefore, if the magnitudes of s and t are nearly equal to each other, $\hat{x}$ will be nearly equal to 0.5. In other words, both alleles A_1 and A_2 would exist with a high frequency. After the discovery of this mathematical property, many biologists reported examples of overdominant selection with respect to morphological characters, but these examples have been controversial because the genes responsible were not clearly identified (Dobzhansky 1951; Ford 1964; Lewontin 1974). Only in recent years have some clear cases of overdominant selection been found by molecular studies (Chapter 4). Some polymorphisms seem to have been maintained for nearly 50 million years.

Natural Selection for Multiple Loci

If natural selection occurs for two or more loci simultaneously, the genetic change of a population cannot be described by allele frequencies alone, because in this case the unit of selection will be the chromosome carrying multiple loci rather than single loci. Let us consider the case of two loci each with two alleles A_1, A_2 and B_1, B_2, and let r be the recombination value between the two loci. In this case there will be four different chromosomes, i.e. A_1B_1, A_1B_2, A_2B_1, and A_2B_2, and we denote the frequencies of these chromosomes by X_1, X_2, X_3, and X_4, respectively. The allele frequencies of alleles A_1, A_2, B_1, and B_2 will then be given by $x = X_1 + X_2$, $(1 - x) = X_3 + X_4$, $y = X_1 + X_3$, and $(1 - y) = X_2 + X_4$, respectively. The chromosome frequencies are not necessarily given by the products of allele frequencies involved. Instead, they are given by $X_1 = xy + D$, $X_2 = x(1 - y) - D$, $X_3 = (1 - x)y - D$, and $X_4 = (1 - x)(1 - y) + D$, where D is

$$D = X_1X_4 - X_2X_3 \tag{2.10}$$

This D is called the linkage disequilibrium. Therefore, the chromosome frequencies are given by the products of allele frequencies only when $D = 0$.

In a randomly mating population the genotype frequencies are obtained by expanding $(X_1 + X_2 + X_3 + X_4)^2$. They are given in Table 2.1. This table gives the relative fitnesses of all genotypes as well as the frequencies. We can therefore compute the amount of changes (ΔX_i) of chromosome frequencies (Lewontin and Kojima 1960; see also Nei 1975, pp. 44–47). They become

$$\Delta X_1 = [X_1(w_1 - \overline{w}) - rw_{14}D] / \overline{w} \tag{2.11a}$$

$$\Delta X_2 = [X_2(w_2 - \overline{w}) - rw_{14}D] / \overline{w} \tag{2.11b}$$

$$\Delta X_3 = [X_3(w_3 - \overline{w}) - rw_{14}D] / \overline{w} \tag{2.11c}$$

$$\Delta X_4 = [X_4(w_4 - \overline{w}) - rw_{14}D] / \overline{w} \tag{2.11d}$$

where $w_1 = X_1w_{11} + X_2w_{12} + X_3w_{13} + X_4w_{14}$, $w_2 = X_1w_{21} + X_2w_{22} + X_3w_{23} + X_4w_{24}$, $w_3 = X_1w_{31} + X_2w_{32} + X_3w_{33} + X_4w_{34}$, $w_4 = X_1w_{41} + X_2w_{42} + X_3w_{43} + X_4w_{44}$, and $\overline{w} = X_1^2w_{11} + 2X_1X_2w_{12} + 2X_1X_3w_{13} + 2(X_1X_4 + X_2X_3)w_{14} + X_2^2w_{22} + 2X_2X_4w_{24} + X_3^2w_{33} + 2X_3X_4w_{34} + X_4^2w_{44}$.

Table 2.1. Frequencies and fitnesses of nine possible genotypes for two loci each with two alleles.

		A_1A_1	A_1A_2	A_2A_2
B_1B_1	Frequency	X_2^2	$2X_1X_3$	X_3^2
	Fitness	W_{11}	W_{13}	W_{33}
B_1B_2	Frequency	$2X_1X_2$	$2(X_1X_4 + X_2X_3)^a$	$2X_3X_4$
	Fitness	W_{12}	$W_{14} = W_{23}$	W_{34}
B_2B_2	Frequency	X_2^2	$2X_2X_4$	X_4^2
	Fitness	W_{22}	W_{24}	W_{44}

[a] The double heterozygotes are composed of coupling (A_1B_1/A_2B_2) and repulsion (A_1B_2/A_2B_1) genotypes. The frequencies of A_1B_1/A_2B_2 and A_1B_2/A_2B_1 are $2X_1X_4$ and $2X_2X_3$, respectively.

Therefore, the changes of chromosome frequencies depend on the recombination value (r) and linkage disequilibrium (D) as well as the genotype fitnesses. The allele frequency changes of A_1 and B_1 are then given by

$$\Delta x = \Delta X_1 + \Delta X_2 \tag{2.12a}$$

$$\Delta y = \Delta X_1 + \Delta X_3 \tag{2.12b}$$

Here ΔX_1, ΔX_2, and ΔX_3 are given by Equations (2.11a), (2.11b), and (2.11c), respectively.

Therefore, the prediction of the allele frequency changes in future generations becomes more complicated when the fitness of a locus is affected by another locus. The situation becomes even more complicated when the fitness of a locus is affected by many loci. For example, if we consider three loci, the number of chromosome types in the population becomes eight and the number of genotypes is $3^3 = 27$ when there are two alleles at each locus. Therefore, if a locus interacts with many other loci, the allele frequency change at this locus is affected by all other interacting loci.

Although this problem has not been studied in detail in the neo-Darwinian era, it is worth noting that Δx in Equation (2.12a) is the sum of ΔX_1 and ΔX_2, each of which is a complicated function of the fitnesses and frequencies of chromosome types. If there are more interacting loci, Δx will be affected by a large number of factors. Therefore, it may not be always positive but it may take a positive or

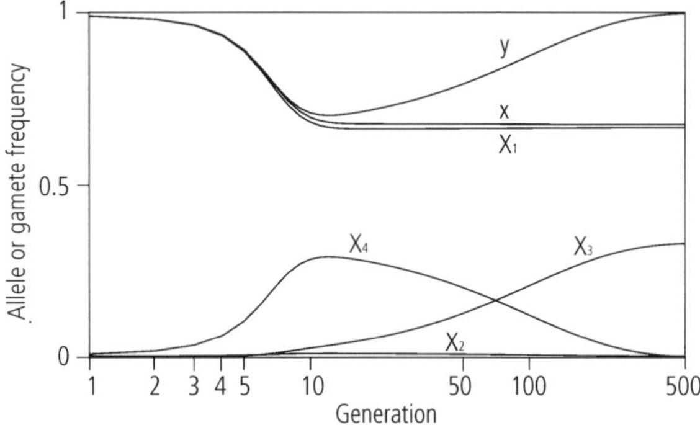

Fig. 2.3. Changes in allele and gamete frequencies of two linked overdominant lethal genes. $w_{11} = w_{12} = w_{14} = w_{23} = 1$, $w_{13} = 2$, $w_{23} = w_{24} = w_{33} = w_{34} = w_{44} = 0$, and $r = 0.01$ were assumed. The initial values of X_1, X_2, X_3, and X_4 were assumed to be 0.99, 0.0, 0.0, and 0.01, respectively. From Nei (1964).

negative value before the allele is fixed in or lost from the population. Figure 2.3 shows the gamete and allele frequency changes for two overdominant lethal genes in a deterministic computer simulation. Around 1960 the idea of overdominant lethal genes was quite popular (e.g. Mukai and Burdick 1959), and a number of experimental studies were conducted with respect to allele frequency changes in population cages of *Drosophila melanogaster*. The results presented in Fig. 2.3 show that changes in the allele (e.g. y) and gamete (e.g. X_4) frequencies can be quite complicated depending on the initial gamete frequencies, relative fitnesses, and the recombination value ($r = 0.01$) between the two loci. That is, the allele frequency (y) may first decrease and then start to increase before reaching fixation.

In natural populations phenotypic polymorphisms, which are seemingly controlled by a pair of Mendelian alleles, are often observed. A good example is the pigment polymorphism of the peppered moth *Biston betularia*, which will be discussed in detail in Section 2.3. However, the fitnesses of these polymorphic characters may be affected by several other genes which are tightly linked with the primary character. In this case the frequency of the advantageous allele for the primary character may temporarily decrease before it becomes fixed in the population as in the case of allele frequency y in Fig. 2.3. Therefore, a temporary decrease of allele frequency does not necessarily mean that the allele is disadvantageous.

2.3. Difficulties of Defining and Estimating Selection Coefficients

Although it is easy to develop mathematical theories of natural selection, it is very difficult to estimate genotype fitnesses or selection coefficients in natural populations. If we have a series of genotype frequencies obtained from a population for many generations, it is theoretically possible to fit a mathematical model of the changes of genotype frequencies and estimate selection coefficients. In practice, such data are rarely obtainable partly because it requires an enormous amount of labor and partly because genotype fitnesses vary from generation to generation due to environmental changes or changes in the genetic background. In the following I first consider a few examples of estimation of selection coefficients and then discuss some general problems.

Estimates of Selection Coefficients and their Reliability

One dataset that is often used to illustrate the intensity of natural selection is that of the increase in frequency of the melanic form of the peppered moth *Biston betularia* in industrial areas of England during the latter half of the nineteenth century. This melanic form, which is dominant over the light-colored wild type, was virtually absent around 1850, but its frequency reached about 98 percent around 1900. This increase in frequency was caused by the darkening of the environment by coal soot during the industrial revolution. Using this information, Haldane (1924) estimated that the selection coefficient against the wild-type form during this period is about 50 percent (see also Clarke and Sheppard 1966). This estimate of selection coefficient was the first to be obtained and gave a strong impression to population geneticists and evolutionists, though it was no more than a guessing game. The frequency of melanic forms is now declining in England after the introduction of clean air legislation around 1960. This decline was of course caused by the elimination of coal-soot pollution in the environment. Grant et al. (1996) studied the frequency of the melanic form every year from 1959 to 1995 and estimated the selection coefficient against the melanic form in a nonpolluted area near Liverpool, England. They obtained $s = 0.153$. This estimate is considerably smaller than Haldane's estimate of s against the wild-type allele in the polluted area.

In these studies we cannot exclude the possibility that the yearly change of the frequency of the melanic form was affected by migration of the nonmelanic wild type into the populations studied because the populations studied were not isolated from other populations. It is known that males of *B. betularia* regularly fly about two miles a night (Ford 1975, p. 317) and the melanic and nonmelanic forms have a tendency to migrate to the locations of black-colored and nonblack-colored areas, respectively (Kettlewell 1955). Therefore, even the estimate of $s = 0.153$ could be an overestimate. Furthermore, no studies have been done on the molecular basis of the melanic and light-colored forms of this moth. Therefore, we still don't know whether this polymorphism is controlled by a single mutation or not.

There are many other examples of natural selection observed in wild populations (Ford 1975; Endler 1986). However, it is very difficult to estimate genotype fitnesses and selection coefficients, because natural populations are widely distributed and it is difficult to separate the effects of selection and migration. Temporal changes of environmental factors also make it difficult to estimate genotype fitnesses accurately.

For this reason, a number of investigators initiated the study of natural selection in artificially controlled populations such as chemostat populations of bacteria and cage populations of *Drosophila*. The most well-known example is the experiment conducted by Dobzhansky and his colleagues with respect to inversion polymorphism in *Drosophila* species. When Sturtevant and Dobzhansky (1936) first discovered the polymorphism of inversion chromosomes in *Drosophila pseudoobscura*, they postulated, following Darwin's (1859) idea, that it represents a case of neutral polymorphism. However, when Wright and Dobzhansky (1946) studied the frequency changes of a pair of inversion chromosomes (ST and CH) in a cage population of *D. pseudoobscura*, they found that the polymorphism is maintained by overdominant selection, the selection coefficients, s and t in Equation (2.9), being 0.3 and 0.4 respectively, and the frequency of chromosome ST reached the approximate equilibrium frequency of 0.7. Furthermore, Dobzhansky and his colleagues found several other inversion chromosomes in *D. pseudoobscura* and its sibling species *D. persimilis* and the polymorphism for these chromosomes varied with geographical area and year of investigation. From these observations, Dobzhansky (1951) concluded that inversion polymorphism is generally maintained by strong overdominant selection.

These studies again gave a powerful message to evolutionists about the rapid genetic change of a population by natural selection. At this time, little attention was given to the fact that a pair of inversion chromosomes contains a large number of genetic loci in the inverted segment and the overdominant selection observed may have been caused by linkage of positively and negatively selected alleles which are located at different genetic loci. In this case, if the positively selected allele is dominant over the negatively selected allele, the inversion heterozygote will have a higher fitness than the inversion homozygotes due to associative overdominance (Haldane 1957; Ohta 1971; Yamazaki 1971; Lewontin 1974). However, if mutation occurs recurrently from the dominant to the recessive alleles for a large number of generations, the selective advantage of inversion heterozygotes will gradually diminish and eventually disappear (Nei et al. 1967). In fact, the inversion polymorphism in *D. pseudoobscura* has not lasted very long because most inversions apparently originated 0.5–1.5 million years ago (Wallace et al. 2011). This suggests that the large values of selection coefficients observed by Wright and Dobzhansky were due to associative overdominance, because the cage population was formed with a small number of individuals (Yamazaki 1971).

Another genetic polymorphism which was intensively studied by the ecological genetics school led by E. B. Ford was that of color and banding types of the snail *Cepaea nemoralis*, which are indigenous in Western Europe including the United Kingdom and France. The shells of this species can be classified into several different types in terms of color and banding patterns of shells. Noting that different polymorphic phenotypes are randomly distributed irrespective of the environmental conditions in French populations, Lamotte (1951, 1959) concluded that the polymorphism is effectively neutral (see Ford 1975). However, examining the frequencies of color and banding types of snails on different habitats, Cain and Sheppard (1950, 1954) noticed that the proportions of different phenotypes are related to the ecological condition of the site and the most common phenotypes are the least conspicuous against the prevailing backgrounds. They then proposed that this correlation is caused by visual predation by birds, particularly the thrush *Turdus erictorum*, which preys on easily distinguishable snails. They presented some evidence supporting this hypothesis, but they also noted that there is some other nonvisual selection operating on the phenotypic characters.

Fluctuation of Selection Coefficients

It should be noted that the study of allele frequency changes in wild populations usually does not produce clear-cut evidence of natural selection. One reason for this is that it is hard to know the population size even if it is a closed population. To overcome this problem, Dowdeswell et al. (1940) invented the so-called capture-recapture method for estimating the population size. In this method, a certain number of individuals are first sampled from the population, and all sampled individuals are marked with some noninvasive markers and then released into the original population. After these marked individuals are mixed well with the rest of the population, a given number of individuals are again sampled, and the size of the population in question is estimated from the proportion of marked individuals in the second sample. In practice, this method may not give a reliable estimate of population size when animals are sedentary and do not mix well or when the mortality of marked individuals is different from the rest of the individuals, etc. However, since rough estimates are still useful, the method has often been used in ecological genetics.

Fisher and Ford (1947) and Ford (1975) used this method for studying the frequency changes of a pair of alleles controlling the wing color patterns of the moth *Panaxia dominula* in a seemingly isolated population near Oxford, England. The estimated population size varied from year to year ranging from about 200 to 18 000. This population had two mutant color types called *medionigra* (heterozygote of the *medionigra* gene) and *bimacula* with respect to the wings. Type *bimacula* is believed to be the homozygote of the mutant gene *medionigra*, and the frequency of the *medionigra* gene was studied for a period of 34 years (1939–1972), or for 34 generations. The frequency of the allele *medionigra* showed a considerable amount of yearly fluctuation when a sample of 117–986 individuals were examined. (Data for 1962–1967 were excluded because of excessively small sample size.)

Fisher and Ford (1947, 1950) then proposed that the yearly fluctuation is caused by the variation of selection coefficient for the *medionigra* gene rather than by random genetic drift. Fisher (1930; p. 10) believed that genetic drift is unimportant whenever the population is greater than 100. Because the population size estimate was much greater than 100 in this species, Fisher and Ford criticized Wright's (1932) shifting balance theory of evolution in which genetic drift plays an important role. However, when Wright (1948a, 1951) reanalyzed the allele frequency changes by estimating the effective population size, he found that a substantial portion of the changes is due to genetic drift (see Section 2.4). He then concluded that the yearly variation of allele frequencies observed was caused by both the variation of selection coefficients and the genetic drift of allele frequencies. A similar conclusion was obtained by O'Hara (2005).

General Considerations

If the population size is large and the environmental conditions are the same for all generations, it is possible to estimate genotype fitnesses and selection coefficients from genotype frequency data. In practice, however, such cases are very rare, and it is difficult to obtain data on selection coefficients in wild populations (Lewontin 1974; Endler 1986). Some of the reasons are as follows. (1) Our lifetime is too short to observe allele frequency changes for many generations in large animals or perennial plants. (2) In microorganisms such as bacteria and fungi, the generation time is sufficiently short for this type of study. However, the definition of a population in these organisms is very difficult in nature and a population may expand rapidly or become almost extinct depending on the environmental conditions. Migration of individuals also occurs so frequently that it is not always clear whether we are studying the same population or not. In the preceding subsection I mentioned that the *Panaxia dominula* population studied by Fisher and Ford was seemingly "isolated." In reality, however, it is not really a closed population, because the individuals of this species are widely distributed in the United Kingdom and Europe and a long-range migration occurs. (3) Variation in morphological characters in natural populations is often controlled by many genes even if the variation is discrete and a pair of major genes is identifiable. In this case the study of natural selection with major genes can be complicated. (4) The fitnesses of different genotypes are affected by many environmental factors such as weather, availability of food, predation, etc. in the wild, and the fitnesses are never constant over evolutionary time. In nature, some genotypes may survive well in wet years while others may do so in dry years. In plants some genotypes are more resistant to fungal infection than others, but fungal infection occurs differently in different years. Note also that natural selection occurs among different individuals rather than among genotypes for a particular locus. These factors make it very difficult to define and measure selection coefficients.

Lewontin (1974, chapter 5) reviewed various methods of estimating genotype fitnesses in wild populations and evaluated the estimates obtained for polymorphic alleles in snails (Lamotte 1951; Cain and Sheppard 1954), humans (Levene 1953), *Drosophila* species (Dobzhansky 1970; Ayala et al. 1971), live-bearing fish *Zoarces viviparus* (Christiansen and Frydenberg 1973), and other organisms. He concluded that although there may be genuine natural selection it is very difficult to estimate genotype fitnesses. Lewontin stated (1974, p. 236): *"Although there is no difficulty in theory in estimating fitnesses, in practice the difficulties are virtually insuperable. To the present moment no one has succeeded in measuring with any accuracy the net fitnesses for any locus in any species in any environment in nature."*

In addition to the difficulties of estimating relative fitnesses of genotypes for a pair of alleles at a locus, there is another problem in the application of population genetics theory to the study of evolution. It is gene interaction among different loci and epigenetic effects (environmental effects). The importance of gene interaction in the development of a specific morphological character is well known from the time of the early stage of Mendelian genetics in the twentieth century. For example, Wright (1916) showed that the coat color of guinea pigs is determined by interaction of many different genes. However, it was difficult to study the relative fitnesses of different genotypes concerned with a specific phenotypic character. One such study was done by Lewontin and White (1960) with respect to the inversion types of two different chromosomes in the grasshopper *Moraba scurra*, though the phenotypic characters affected by these inversion chromosomes were not known.

Table 2.2 shows the relative fitnesses of the nine genotypes with respect to the inversion polymorphism of two different chromosomes (Lewontin and White 1960; Lewontin 1974). The inversion chromosomes involved are the *ST* and *TD* inversion types in chromosome *EF* and the *ST* and *BL* types in chromosome *CD*. One distinctive feature of this dataset is that the relative fitnesses of the three genotypes for one chromosome, say *ST/ST*, *ST/TD*, and *TD/TD*, vary according to the genotype for the other chromosome. Thus, the three genotypes show heterozygote advantage (overdominance) when the genotype for the other chromosome (*CD*) is *BL/BL*, but genotype *ST/TD* does not show the highest fitness when the genotype for chromosome *CD* is *ST/BL* and the

Table 2.2. Estimated fitnesses for the nine genotypes with respect to two polymorphic inversion systems, Blundell (*BL*)/Standard (*ST*) on chromosome (*CD*) and Tidbinbilla (*TD*)/Standard (*ST*) on chromosome *EF* in *Moraba scurra.* Data for the Wombat population are presented. From Lewontin and White (1960).

		Chromosome EF		
Year	**Chromosome CD**	**ST/ST**	**ST/TD**	**TD/TD**
1956	*ST/ST*	0.789	0.801	0.000
1958	*ST/ST*	1.353	0.000	0.000
1959	*ST/ST*	0.970	1.282	0.000
1956	*ST/BL*	1.000	0.876	1.308
1958	*ST/BL*	1.000	0.919	0.272
1959	*ST/BL*	1.000	0.672	1.506
1956	*BL/BL*	0.922	1.004	0.645
1958	*BL/BL*	0.924	1.113	0.564
1959	*BL/BL*	0.917	1.029	0.645

genotype *ST/ST* tends to show a higher fitness than *ST/TD* when the genotype for chromosome *CD* is *ST/BL*. Although the estimates of these fitnesses were subject to large estimation errors, it seems obvious that the fitnesses of genotypes for a locus depend on the genotypes of other loci (Lewontin and White 1960). If this is true, the mathematical theory of allele frequency change available now is not appropriate. It is also important to note that the relative fitness of genotypes vary considerably in different years.

Here we have considered chromosome genotypes, because there seem to be no equivalent data for two polymorphic gene loci. However, this type of gene interaction in fitness has been known for genes that control morphological characters for a long time. Wright (1932) developed the shifting balance theory of evolution considering these gene interactions. Yet, most population genetics theories about natural selection are based on single loci without gene interaction mainly because of mathematical difficulties. In the 1950s to 1970s there were some attempts to rectify this situation by considering two or more loci (Kimura 1956; Lewontin and Kojima 1960; Franklin and Lewontin 1970; Lewontin 1974), but they are primarily mathematical and insufficient for understanding the general effect of natural selection. Recent studies in developmental biology have shown that the formation of morphological characters is controlled by many different genes that regulate gene expression, as will be discussed in Chapter 6. Here I would like to indicate that in the presence of complex gene interaction the selection coefficient for a particular pair of alleles may fluctuate from time to time depending on the allelic combination with other loci and the average fitness differences for a pair of loci should be very small (Wagner 2008).

2.4. Stochastic Changes of Allele Frequencies

In the 1950s and 1960s it was commonly believed that since natural populations are very large and contain a large amount of genetic variation, evolution occurs primarily by changes of the frequencies of pre-existing alleles in the population. However, several authors such as Fisher, Haldane, and Wright were interested in the process of introduction of genetic variation by new mutations. Conducting detailed mathematical studies, they showed that the allele frequency change in a population is affected considerably by genetic drift when the frequency is low. In fact, if a new mutation occurs in a population, the initial survival of the mutant gene largely depends on chance, whether it is advantageous or not or whether the population size is large or small, and there is a high probability that the mutant allele will be lost in the first several generations (see Nei 1987, pp. 352–353).

In a large population (e.g. 10^9 individuals), however, once the frequency of advantageous mutant alleles reaches a certain level, say, five percent, the mutant allele will almost certainly be fixed in the population unless the mutant allele has a very small selective advantage over the wild-type allele. In this case the process of allele frequency change can be expressed by a formula equivalent to Equation (2.4), and the allele frequency change by chance effects becomes negligible as long as the selection coefficient remains the same. In practice, however, all natural populations are finite, and the environmental conditions never stay the same. For this reason, it is more appropriate to use a stochastic theory to understand the evolutionary changes of allele frequencies. In this section we will consider this problem without the derivation of mathematical formulae.

We will first consider the probability of fixation of a new mutant allele considering just one locus.

Probability of Fixation of Mutant Alleles

For the reason mentioned in the previous subsection, even advantageous mutations may be lost from the population in the early generations, and the probability of fixation of a new mutant allele is generally quite small. This problem was first studied by Haldane (1927), and later Fisher (1930) derived a more general formula. However, the most general formula is due to Malecot (1948, 1969) and Kimura (1957). When the fitnesses of A_1A_1, A_1A_2 and A_2A_2 are 1, 1 + s, and 1 + 2s, respectively, the probability of fixation of the advantageous allele A_2 is given by

$$U = (1 - e^{-4Nsp}) / (1 - e^{-4Ns}) \qquad (2.13)$$

where N and p are the effective population size and the initial allele frequency, respectively.

Here the effective population size means the number of individuals that participate in reproduction in each generation. Pre-adult or post-reproductive individuals are excluded, because they do not contribute genes to the next generation. With this definition of N, the initial frequency of a new mutant gene (A_2) is given by $p = 1/(2N)$, and therefore U becomes $2s/(1 - e^{-4Ns})$ or $2s$ approximately when s is small but $4Ns$ is large (>>1) (Haldane 1927; Fisher 1930). This indicates that the mutant gene will be lost from the population with a probability of $1 - 2s$, which is very high when s is of the order of 0.02. In the case of neutral mutations with $s = 0$, Equation (2.13) becomes equal to p or $1/(2N)$ (Fisher 1930).

Examining the fixation probability of a new mutation $p = 1/(2N)$ given by $U = 2s/(1 - e^{-4Ns})$, Fisher (1930, p. 94) noted that U increases 50 times when Ns changes from –1 to +1 and therefore even small values of s have significant effects on U. This means that if $N = 10^9$, mutant genes with even $s = \pm 10^{-9}$ cannot be regarded as strictly neutral. Because Fisher believed that the population size is very large (10^4–10^{12}) in many species, he denied the importance of genetic drift in most natural populations. (Fisher did not have the concept of effective population size.) For this reason, he became a panselectionist, and there was no need for him to consider the possibility of neutral mutations. However, Fisher's argument is not very meaningful from the biological point of view, because the absolute values of U are very small when $Ns = \pm 1$ and N is large. For example, when $N = 10^9$ and $Ns = 1$, s becomes 10^{-9}. Therefore, the absolute value of U is 2.04×10^{-9} for $Ns = 1$ and 3.73×10^{-11} for $Ns = -1$. The ratio of U for $Ns = 1$ to that for $Ns = -1$ is 54.6. However, how significant is the ratio of these two very small values? This is particularly so if the selection coefficient varies from generation to generation.

At the present time, Equation (2.13) is widely used. However, this equation is based on the assumption that the selection coefficient s remains constant for the entire process of gene substitution, and this assumption is unlikely to be correct in the presence of gene interaction and the fluctuation of environmental factors. What should we do then? One crude way to handle this problem is to assume that the selection coefficients of a pair of alleles at a locus fluctuate randomly every generation and compute the probability of fixation of a mutant allele under this condition. This approach can be justified to take care of the random fluctuation of selection coefficients due to environmental factors. The effects of gene interaction on the selection coefficient surely change over time, because the gene pools (allele frequencies) of other loci will also change in the course of evolution and the latter change is generally unpredictable. Although the change in gene pool at a locus could be directional, the overall effects of a large number of loci on the selection coefficients for the locus under consideration would be unpredictable and may be treated as random effects. A mathematical formulation of this problem will be presented later in this section.

Equilibrium Distribution of Allele Frequencies

In Section 2.2 we considered the equilibrium frequency of a mutant allele under the assumption that the population size is infinitely large. In finite populations, however, the equilibrium frequency is not uniquely determined but shows a certain distribution. For example, when allele A_1 mutates to A_2 at a rate of u per generation and A_2 mutates to A_1 with a rate of v, and overdominant selection operates, the equilibrium distribution of the frequency (x) of allele A_1 is given by

$$f(x) = ce^{-2N(s+t)(x-\hat{x})^2} x^{4Nv-1}(1-x)^{4Nu-1} \quad (2.14)$$

where N is the effective population size, and s and t are the selection coefficients for genotypes A_1A_1 and A_2A_2, respectively. Therefore, the relative fitnesses of A_1A_1, A_1A_2, and A_2A_2 are given by $1 - s$, 1, $1 - t$, respectively, and $\hat{x} = t/(s+t)$. Here c is a constant such that $\int_0^1 f(x)dx = 1$ (Wright 1937).

Figure 2.4 shows the frequency distributions of an overdominant allele A_1 in a population of effective size $N = 1000$. In this figure it is assumed that the mutation rates u and v are equal to 10^{-6}. Four cases of symmetric overdominant selection are shown in Fig. 2.4A. $s = 0.10$ indicates the case where s and t are both 0.1. In this distribution the mean of x is 0.5, which agrees with the deterministic solution of $\hat{x} = 0.1/(0.1+0.1) = 0.5$. However, x takes some other values too, because of the stochastic variation. When selection coefficients are smaller, the variation of x becomes greater. When $s = t = 0.003$ or 0.001, allele A_1 may be fixed ($x = 1$) or lost ($x = 0$) with a certain probability because of chance effects. When $\hat{x} = 0.66$, the distribution of x is distorted toward $x = 1$, because the selection is asymmetric (Fig. 2.4B), but otherwise the property of the distribution is similar to that of Fig. 2.4A.

The study of allele frequency distributions in finite populations for a pair of alleles is conceptually important for understanding the possible mechanisms of maintenance of genetic variability within populations. In practice, however, there are no ways to prove or disprove the mathematical distributions experimentally. The population size (N) also varies drastically from generation to generation, and it is difficult to estimate it. Therefore, the mathematical distributions have remained largely unused in population genetics. Furthermore, the stochastic changes of allele frequencies were thought to be important only in small populations, and for this reason many leading evolutionists such as Fisher (1930), Haldane (1932), Mayr (1963), and Ford (1964) ignored the stochastic changes of allele frequencies. Wright, who developed the shifting balance theory of evolution, also thought that while genetic drift is important within subpopulations of a species the evolutionary change of the entire species is determined primarily

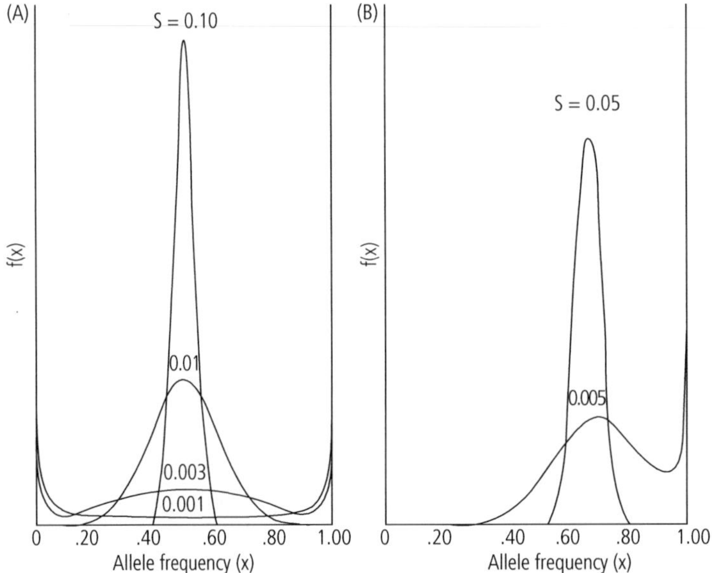

Fig. 2.4. Frequency distributions of an overdominant allele (A_1) in a population of $N = 1000$. The mutation rate $u = v = 10^{-6}$ is assumed. (A) Four cases of symmetric selection ($s = t = 0.001$, 0.003, 0.01, and 0.1) are considered. (B) Two cases of asymmetric selection ($s = 0.005$ and $t = 0.01$; $s = 0.05$ and $t = 0.1$) are considered. The $\hat{x} = t/(s + t)$ value is 0.66. From Wright (1948a).

by natural selection. In the neo-Darwinian era most evolutionists were essentially panselectionists and believed that natural selection is responsible for almost all evolutionary changes of phenotypic characters. This view was challenged only after molecular study of evolution was initiated in the 1960s (Chapter 4). Molecular data clearly showed the importance of stochastic changes of allele frequencies in evolution.

Furthermore, some types of allele frequency distributions have become useful for studying the effect of natural selection on polymorphic alleles at the molecular level. One of them is the distribution of mutant allele frequencies under the irreversible mutation model (Wright 1938a). In this model the wild-type allele *A* and a mutant allele *a* are considered at each nucleotide site, and the equilibrium distribution of frequencies of allele *a* is derived under the assumption that mutation occurs only from *A* to *a*. In this case allele *a* will eventually be lost from the population, but if mutation occurs every generation at monomorphic sites, the frequency distribution of allele *a* reaches an equilibrium when only the polymorphic sites are considered.

Because the mutation rate at a nucleotide site is very low (order of 10^{-9}), this irreversible mutation model can be used for studying the effect of natural selection by comparing the theoretical distribution with the observed distribution. The model of allele frequency distribution under irreversible mutation is now called the infinite-site model (Kimura 1983). Wright (1938a, 1968) derived this distribution for various types of natural selection as well as for neutral mutations. Here the distribution for overdominant selection is of special interest, because many neo-Darwinian evolutionists believed in the importance of this type of selection, as will be discussed in Section 2.6.

Figure 2.5 shows the theoretical distributions for overdominant mutant allele *a* for various values of $2Ns$ under the assumption that the fitnesses of *AA*, *Aa*, and *aa* are given by $w_{AA} = 1 - s$, $w_{Aa} = 1$, and $w_{aa} = 1 - s$, respectively. These distributions were obtained by the method given in Appendix B. The number given for each curve in the figure represents the $2Ns$ value, $2Ns = 0$ indicating neutral mutations. It is clear that overdominant selection enhances the number of nucleotide sites

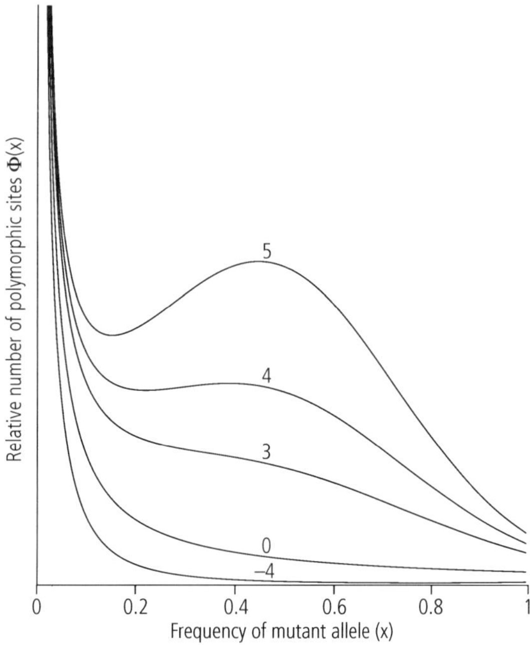

Fig. 2.5. Frequency distributions of overdominant mutations under the infinite-site model. The number given for each curve is the $2Ns$ value. $2Ns = -4$ represents a case of underdominant selection, where the heterozygote has a lower fitness than the two homozygotes.

with middle allele frequencies (mean $x = 0.5$) compared with the case of neutral mutation even when $2Ns = 5$ or $s = 5/(2N)$. When w_{Aa} is lower than w_{AA} and w_{aa} (underdominance), the relative number of polymorphic sites declines as expected but not to a great extent (see the case of $2Ns = -4$).

It is also known that when $w_{AA} = 1, w_{Aa} = 1 + s$, and $w_{aa} = 1 + 2s$, the number of polymorphic sites with x close to 1 increases but the extent of the increase is not as high as in the case of overdominant mutations (Wright 1968). Therefore, it is relatively easy to detect overdominant mutations by studying the distribution of allele frequencies.

Effective Population Size and Sampling Errors of Allele Frequencies

Earlier in this section I mentioned that the effective size of a population is equal to the number of individuals that participate in reproduction in a generation. In practice, it is convenient to consider a dioecious diploid population of N adult individuals, in which all individuals mate at random simultaneously. Let us consider a locus with two alleles A and a and designate the frequency of A and a by p and $1 - p$, respectively. Now we assume that all individuals mate at random and consider the frequency of allele A in the next adult population. As long as N is finite, this frequency will not be equal to p but take some value (x) with a binomial distribution. Therefore, the variance (V_x) of the distribution of x is given by

$$V_x = x(1-x)/(2N) \quad (2.15)$$

where $2N$ appears because we are considering $2N$ alleles in a diploid population.

In real populations this idealized situation rarely occurs, and the variance V_x often becomes larger than the value given by Equation (2.15). For example, if there are N_m males and N_f females in the adult population, V_x is given by

$$V_x = x(1-x)/(2N_e) \quad (2.16)$$

where

$$N_e = 4N_m N_f/(N_m + N_f) \quad (2.17)$$

(Wright 1931; Hedrick 2000). This N_e is called the effective population size.

In this case if the sex ratio is 1:1, we have $N_m = N_f$ and $N_e = (N_m + N_f)$, which is the total population size. Therefore, the effective population size is the same as that for a monoecious population. However, if a population consists of one male ($N_m = 1$) and 19 females ($N_f = 19$), then N_e will be $4 \times 1 \times 19/20 = 3.8 \approx 4$. Therefore, N_e is much smaller than the total adult size (20). In this case, V_x becomes very large, and allele A may be fixed in the next generation with a probability of p^{2N_e}. This probability is $0.9^8 = 0.43$ if $p = 0.9$ and $N_e = 4$. Even if $p = 0.5$, p^8 becomes 0.004. Therefore, if this unusual sex ratio continues for many generations, either allele A or a will be fixed in the population. Actually, a situation similar to this one may happen in some small mammals, as will be discussed in Chapter 8.

Another situation in which the effective size can be much lower than the actual population size is the case where the population size shows a seasonal cyclical change. In some insect species (e.g. fruitflies), there are several generations in a year, and the population size (N) often changes seasonally. Suppose that there are six generations a year and the N for the six generations (spring to winter) is $N_1 = 10^4$, $N_2 = 10^6$, $N_3 = 10^8$, $N_4 = 10^6$, $N_5 = 10^4$, and $N_6 = 10^2$. Then, N_e is given by the following harmonic mean.

$$N_e = n / \sum_{i=1}^{n} \frac{1}{N_i} = 594$$

where n is the number of generations per year (Wright 1938b).

Therefore, even if N is very large in summer, N_e can be much smaller than the average size (17 003 350) for the entire year if the winter size is very small.

Another situation in which N_e can be much smaller than total population size occurs when a population is divided into many colonies and these colonies experience expansion and extinction continuously. This situation would occur often in microorganisms. For this reason, the effective size in *E. coli* can be as small as 10^9 instead of trillions (Maruyama and Kimura 1980; Nei and Graur 1984).

In the above discussion we considered some important factors that would make the effective population size much smaller than the actual size. In practice, there are several other factors that reduce the effective size, including overlapping generation and large progeny size variation. It is therefore important to realize that the deterministic models of population genetic theories that are often used by neo-Darwinians are only crude approximations to reality and may give misleading conclusions. In the following subsection I would like to discuss another type of random errors of allele frequency changes.

Random Errors Caused by Fluctuation of Selection Coefficients

As mentioned in Section 2.2, the random fluctuation of selection intensity was observed in the evolutionary change of the *medionigra* gene of the moth *Panaxia dominula*. Actually, such a fluctuation of allele or genotype frequencies is quite common as documented by Bell (2010) in various species. Figure 2.2B shows variation of selection coefficient (s) for the brown phenotype of the snail *Cepaea nemoralis* for a period of 21 generations. In this case the variation of s is essentially random. This type of variation is expected to reduce the extent of genetic variation.

The mathematical formulation of this problem was initiated by Wright (1948a, 1951) and Kimura (1954). Using the diffusion method, these authors showed that random fluctuation of selection coefficient is essentially a force to reduce genetic variation and its effect is often greater than that of genetic drift due to the sampling error of allele frequencies.

In their mathematical formulation Wright and Kimura considered only the case of the mean selection coefficient equal to 0 and the random change of allele frequency caused by the fluctuation of selection coefficients. A more realistic formulation, in which the mean ($\bar{s}$) and variance (V_s) of selection coefficient s and the random fluctuation of allele frequencies due to finite population size (N) are taken into account, was given by Ohta (1972). In her formulation in terms of the diffusion method (see Appendix B), the mean ($M_{\delta x}$) and variance ($V_{\delta x}$) of allele frequency change per generation were given by

$$M_{\delta x} = \bar{s}x(1-x) \qquad (2.18)$$

$$V_{\delta x} = V_{\delta}x^2(1-x)^2 + x(1-x)/(2N) \qquad (2.19)$$

These equations have been criticized by a number of theoretical population geneticists (Gillespie 1973, 1991; Jensen 1973; Avery 1977; Ewens 2000). However, these criticisms are based on rather trivial problems concerning the mathematical formulation as mentioned in Appendix C. If we use the competitive selection model incorporating population size regulation (Mather 1969; Nei 1971), Equations (2.18) and (2.19) are biologically more realistic than formulas based on Gillespie's stabilizing selection model. Furthermore, a recent study of the frequency distribution of mutant nucleotides in human populations indicates that the criticisms by Gillespie and others are not important (see Fig. 2.9).

One of the important quantities in population genetics is the probability of fixation of a mutant allele discussed in the preceding section. This probability (U) can be computed by using Kolmogorov's backward equation if we know $M_{\delta x}$ and $V_{\delta x}$ (Kimura 1962). In the present case it is difficult to obtain a simple analytical formula for U. For this reason, Ohta (1972) evaluated this probability numerically. Figure 2.6 shows some of the results obtained by using her approach. In this figure the initial allele frequency is assumed to be $p = 0.001$, and the fixation probability is expressed as a function of NV_s. The numbers given for each curve represent the value of $N\bar{s}$. When V_s or NV_s is very small, the effect of random fluctuation of s is negligible. Therefore, U is given by Equation (2.13). In Fig. 2.6, the U values for small $N\bar{s}$ are indicated by a, b, c, d, and e at $NV_s = 0.001$. However, as NV_s increases the U value gradually converges to $U = 0.001$, i.e. the expected U value for a neutral allele. In other words, the random fluctuation of s has the same effect as that of genetic drift and makes allele frequency change similar to that of neutral alleles.

Bell (2010) obtained $\bar{s} = -0.103$ and $V_s = 0.0075$ for the data of the *medionigra* gene in *Panaxia dominula* (Fig. 2.2A). Therefore, we have $N\bar{s} \approx -0.1N$ and $NV_s = 0.008N$ approximately. Figure 2.6 indicates that the *medionigra* allele is expected to become extinct with a high probability if N is larger than 100. However, if $\bar{s}$ were -0.0001 with $V_s = 0.008$ and $N = 10^5$, the allele would have behaved just like a neutral allele and fixed in the population with a probability of 0.001. Bell gives many other examples of fluctuating selection, in which V_s takes various values. Lynch (1987) estimated the mean and variance of s for many isozyme genes in large poplulations of

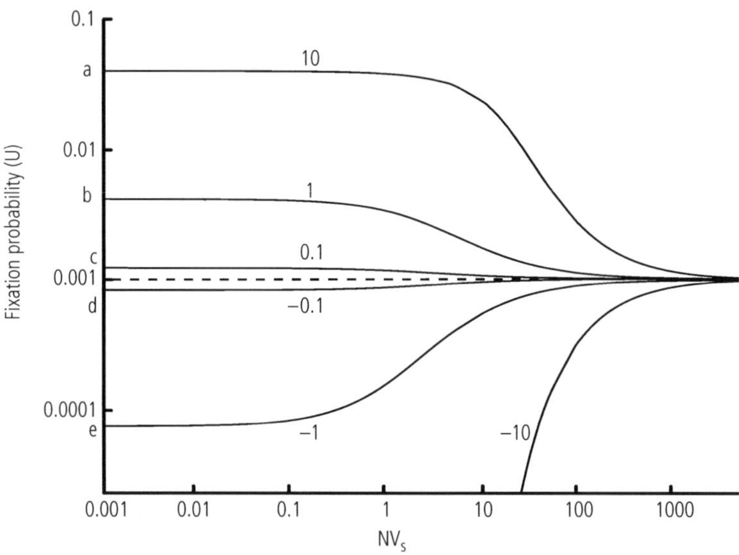

Fig. 2.6. Probabilities of fixation (U) of a mutant gene with random fluctuation of selection coefficient (s). N is the effective population size, whereas $\bar{s}$ and V_s are the mean and variance of s, respectively. The numbers beside the curves represent the values of $N\bar{s}$. The initial frequency of the mutant gene is assumed to be 0.001. The values of a, b, c, d, and e represent the probabilities for the case of $V_s = 0$. The broken line is for neutral alleles. U and NV_s are given on a logarithmic scale.

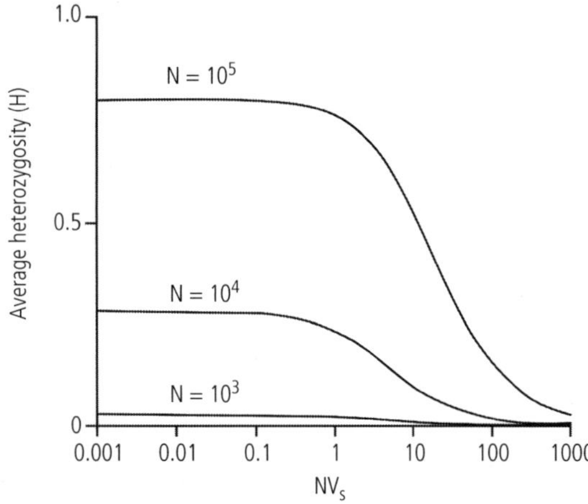

Fig. 2.7. Relationships between the average heterozygosity per locus (H) and NV_s. The mutation rate is assumed to be 10^{-5}. NV_s is given on a logarithmic scale. From Nei and Yokoyama (1976).

Daphnia. In his study the mean of s was close to 0 and the variance was of the order of 0.01 to 0.05 in most genes.

Nei and Yokoyama (1976) studied the expected heterozygosity of a locus (H) using Equation (2.18) and (2.19) for the infinite-allele model with a mutation rate of $v = 10^{-5}$. Figure 2.7 shows the relationships between the heterozygosity (H) and NV_s for various values of effective population size (N). As expected, the expected heterozygosity decreases as V_s or NV_s increases for a given population size. However, if NV_s is smaller than 0.1, H is nearly the same as that for $NV_s = 0$, i.e. the case of no fluctuation of s. When NV_s is larger than 1, H is considerably reduced particularly in large populations. These results indicate that in the presence of random fluctuation of s even alleles under selection tend to behave as though they are neutral. This seems to be one factor that is responsible for the fact that the observed heterozygosity of electrophoretic protein loci is almost always lower than the level expected from the actual population size and the mutation rate (Nei and Graur 1984). Using a slightly different mathematical model, Takahata and Kimura (1979) studied the same problem, but their results are essentially the same as the above.

In recent years a large number of authors have investigated single nucleotide polymorphisms (SNPs) in human populations (International HapMap Consortium 2005, 2007, 2010). Because these datasets were obtained by examining hundreds of genomes, they can be used for studying the effect of natural selection on the frequency distribution of mutant nucleotides. For this reason, Miura et al. (2013) studied the effects of fluctuation of selection intensity on the frequency distribution. They considered two models of fluctuating selection: competitive selection model $[M_{\delta x} = \bar{s}x(1-x)]$ and stabilizing selection model $[M_{\delta x} = \bar{s}x(1-x) + V_s x(1-x)(1-2x)]$ (Appendix C). The relative distributions of mutant allele frequency x for these two fluctuating selection and the neutral mutation models are presented in Fig. 2.8. Here the NV_s value is given for each curve except for the case of neutral mutation, where $NV_s = 0$. It is clear that the competitive selection model reduces the number of polymorphic nucleotide sites whereas the stabilizing selection model enhances the number, as expected.

Miura et al. now examined which of the three models fits the observed polymorphism data best. For this purpose, they used SNP data for 689 genomes from a central African population. However, because SNP data do not show which of the two alleles at a site is ancestral or a new mutant nucleotide, they first examined the nucleotide sequences of the chimpanzee and macaque genomes. If the nucleotide of a given nucleotide site of human SNPs was identical with the nucleotide of the homologous site of both

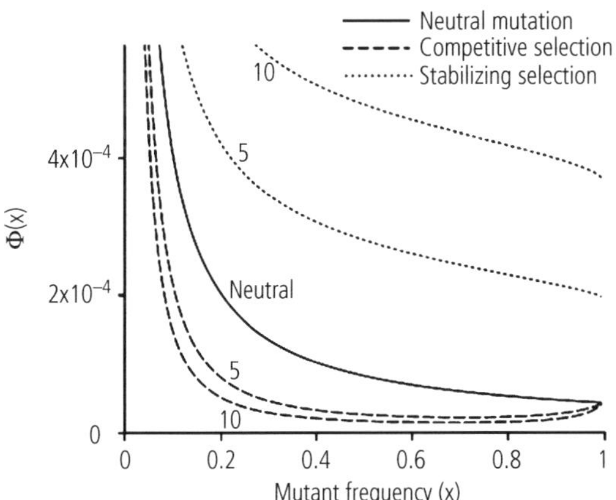

Fig. 2.8. Equilibrium distributions of the frequency of mutant nucleotides for the neutral, competitive selection, and stabilizing selection models. The number for each curve represents the NV_s value. $4Nv = 4\times10^{-5}$ is assumed. $\Phi(x)$ is the relative number of nucleotide sites with mutant allele frequency x. From Miura et al. (2013).

chimpanzee and macaque sequences, they assumed that this nucleotide is ancestral and the other nucleotide is a mutant allele. In this way they could compute the frequency (x) of this allele in the human population. In this case there was one problem for data analysis, because SNP data did not include low-frequency and high-frequency alleles for a technical reason. However, as far as the alleles with a frequency between $x = 0.2$ and $x = 0.8$ were concerned, the accuracy of the estimates of allele frequencies appeared to be sufficiently high.

Figure 2.9 shows the observed frequency distribution in comparison with the expected distributions under the neutral and the competitive selection models. The distributions for the two theoretical models were fitted to the observed distribution by using the least squares method. The results given in Fig. 2.9 clearly show that both models fit the observed data quite well and consequently the NV_s value was very close to 0. Actually, the stabilizing selection model was also shown to fit the SNP data when NV_s is very small. This was true for both protein-coding and intergenic regions of the genomes. Therefore, Miura et al. concluded that the frequency distribution of mutant nucleotides can be explained by the neutral mutation model and there is no need to consider fluctuating selection in the present case. A similar conclusion was obtained by analysis of DNA sequence data for about 4000 genomes from an African American population (Miura et al. unpublished).

In the past there have been rather strong debates between the supporters of the two different models of fluctuating selection (Nei and Yokoyama 1976; Takahata and Kimura 1979; Gillespie 1980, 1991; Nei 1980a), because there were no empirical data to test the two hypotheses. It is interesting to see that this long-standing controversy can be solved by empirical data in a simple way. This issue will be discussed again in Chapter 4.

2.5. Mutation and Standing Genetic Variation

Artificial and Natural Selection in Quantitative Characters

As is well known, the effectiveness of artificial selection with quantitative characters played an important role

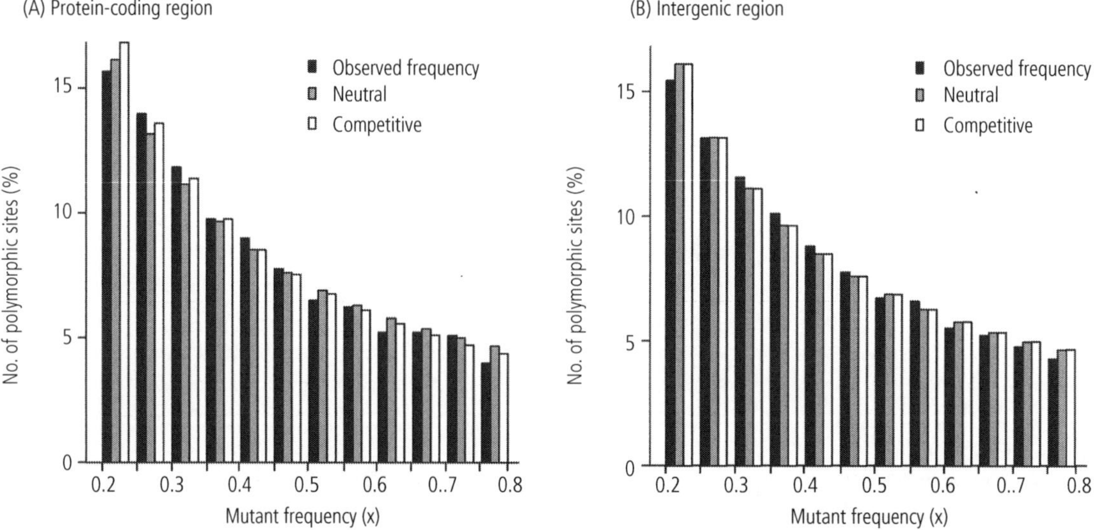

Fig. 2.9. Observed and expected frequency distributions of mutant nucleotides for a central African population of humans. The expected frequency distribution for the stabilizing selection model is virtually the same as that for the competitive selection model. The number of polymorphic nucleotide sites used was 6924 for the protein-coding region and 269 323 for the intergenic region. The number of genomes sampled was 689. The least squares residual was nearly the same for the three models. From Miura et al. (2013).

for the development of Darwin's idea of natural selection. Once the applicability of Mendelian inheritance to quantitative characters was shown, artificial selection was conducted with many quantitative characters in various organisms. One of the best known examples is the upward and downward selection of sternopleural or abdominal bristle number in *Drosophila* species (e.g. Macdowell 1917; Payne 1918; Mather 1948; Falconer 1960). It was soon discovered that artificial selection is generally effective for the first dozen generations but its effectiveness gradually declines. When upward selection was applied for the bristle number of *Drosophila* and was stopped after a plateau was reached, the original gain by selection usually declined in the following generations, suggesting that the plateau occurred not because of the lack of genetic variation but by a balance of artificial selection and some form of counteractive natural selection (Mather 1948; Falconer 1960; Lynch and Walsh 1998).

Similar results were obtained with many different characters in various organisms (Dobzhansky 1970). In fact, whenever artificial selection was conducted with any quantitative characters, the response to selection almost always occurred. These results supported the idea that natural populations contain almost all kinds of genetic variation (standing variation) and no new mutations are necessary for evolutionary changes, and that natural selection for advantageous alleles is triggered by environmental changes. (See Jones et al. 2012 for a recent example of the standing variation observed in the genomes of marine stickleback fish.) However, this view is not entirely correct, and several experiments have shown that mutation is actually necessary for a long-term response to artificial selection (Clayton and Robertson 1955; Hill 1982; Mackay 2010).

As is well known, artificial selection is very different from natural selection with respect to quantitative characters. In artificial selection, a particular character, say abdominal bristle number in *Drosophila*, is measured at the adult stage, and a few percent of males and females with the highest bristle numbers are chosen. These males and females are then mated to raise the next generation. If the heritable component of variation of the character is high, the response to selection or genetic gain (ΔG) is high (Appendix D and Fig. 2.10). The proportion of heritable component of variation is called heritability (h^2). If artificial selection is continued, the

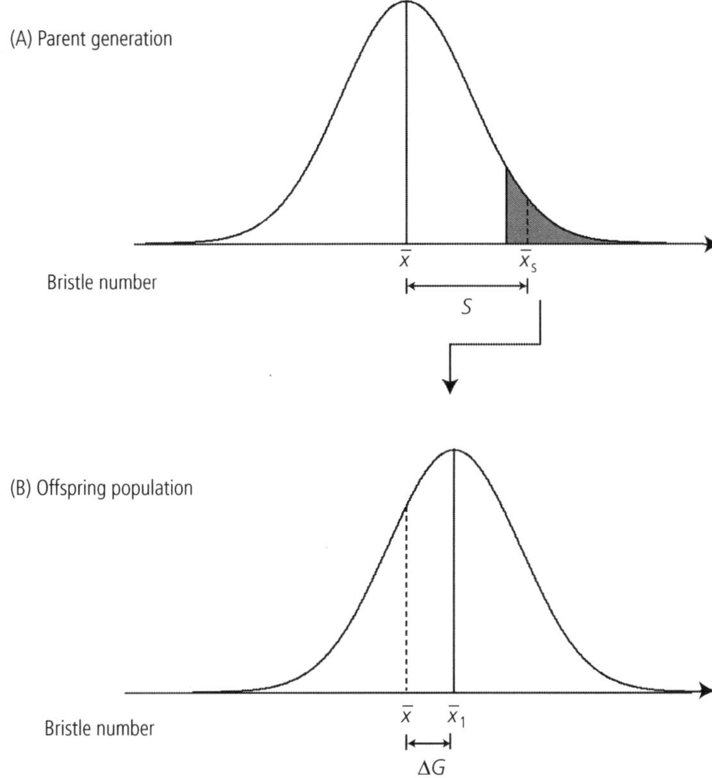

Fig. 2.10. Diagram showing the artificial selection applied to a quantitative character and their effect that is revealed in the next generation. S represents the selection differential, which is defined as the difference between the population mean ($\bar{x}$) of the quantitative character and the mean ($\bar{x}_s$) of the selected individuals. In the next generation the mean value ($\bar{x}_1$) of the offspring is expected to be higher than the mean ($\bar{x}$) of the parental population. The difference (ΔG) between $\bar{x}_1$ and $\bar{x}$ is called the genetic gain. This genetic gain can be expressed as $\Delta G = Sh^2$, where h^2 is the heritability of the character. Therefore, ΔG is high when h^2 is high.

initial genetic variation is gradually dissipated, and the genetic gain gradually becomes 0 unless new variation is supplied by mutation. The results of artificial selection conducted by early Mendelians appeared to confirm this expectation as long as the environmental conditions remained the same, and this observation was used as another factor to support Darwinism or neo-Darwinism.

However, natural selection for a quantitative character occurs in a quite different way. It occurs in every stage of development from eggs to adults rather than just in the adult stage. Therefore, if there are a large number of genes involved in a character, the contribution of each gene to the evolutionary change of the character, particularly such a character as viability, must be very small and largely independent of that of other genes. Therefore, natural selection is less effective than artificial selection for changing a particular character for a given direction.

If a character is subject only to weak selection and mutation occurs with a reasonably high rate, a large amount of genetic variation can be maintained in a population. Fingerprint patterns of humans are known to have a heritability of nearly 100 percent, and the variation in this character is considered to be virtually neutral (Cavalli-Sforza and Bodmer 1971). For this reason, the amount of genetic variation is very large, so that this character can be used for human identification. If the above view is right, a quick response to artificial selection in a character suggests that the natural selection operating on the character is weak. By contrast, a slow or small magnitude of response to artificial selection suggests that strong purifying selection is operating with the character. This is opposite to Darwin's view that natural selection is more or less analogous to artificial selection.

In recent years a number of authors (Santiago et al. 1992; Mackay et al. 2005; Mackay 2010) have conducted mutation accumulation experiments using sternopleural and abdominal bristle number, wing length, etc., in fruitflies and showed that a significant number of spontaneous mutations occur every generation. These studies clearly indicate that standing genetic variation as well as new mutations are responsible for the effectiveness of artificial selection for increasing and decreasing the mean measurements of various quantitative characters. Of course, because quantitative characters are controlled by hundreds or thousands of genetic loci, it is difficult to determine the mutation rate for individual loci. Another conclusion obtained from these experiments is that there is an extensive amount of gene interaction among different loci controlling a specific character.

Evolution of Drug Resistance

It is often stated that one of the strongest pieces of evidence supporting neo-Darwinism is the quick development of drug resistance in insects and bacteria when insecticides or antibiotics are applied. During and after World War II, various insecticides (e.g. DDT) and antibiotics (e.g. penicillin) were used to control pests or bacterial diseases. Initially, these drugs were very effective, but later various resistant strains appeared and the drugs were no longer so effective.

In bacteria the resistance to drugs such as penicillin is sometimes introduced by the lateral transfer of plasmids carrying resistance genes (Watanabe 1963), but the general mechanism appears to be natural selection of mutant strains with drug resistance. According to neo-Darwinism, these mutant strains have existed in the population in low frequency even before the application of antibiotics, and the resistant strains were selected because nonresistant strains were quickly eliminated by the drug. This interpretation is similar to that of artificial selection of quantitative characters mentioned above.

When antibiotic-resistant bacterial strains surfaced in the 1940s, there were authors who claimed that resistant mutations were induced by the application of antibiotics (post-adaptation or Lamarckism) rather than by selection of resistant mutations. This problem was studied by a number of investigators such as Luria and Delbruck (1943) and Lederberg and Lederberg (1952). Among others, the replica plating method used by Lederberg and Lederberg (see Chapter 1) clearly showed that the antibiotic resistance was caused by new mutations rather than post-adaptation. However, it is possible that natural populations contain many drug-resistant mutations that have occurred in the past. If these mutations are kept in low frequency in the population, they may quickly generate drug-resistant strains when the population is exposed to antibiotics, as expected from neo-Darwinism. In recent years, however, the molecular basis of drug resistance has been clarified for many antibiotics such as penicillin and streptomycin. This study has shown that drug resistance was generated primarily by mutation of enzymes involved in the biosynthesis of antibiotics (e.g. Furuya and Lowy 2006; Morar and Wright 2010; Sykes 2010). These results indicate that drug resistance is essentially controlled by mutation. It appears that the persistence of drug resistance genes in a bacterial species is caused by the fact that bacteria live in highly isolated niches and there are always some niches where resistant strains can survive.

2.6. Classical and Balance Theories of Maintenance of Genetic Variation

Another issue which was debated by neo-Darwinians is the mechanism of maintenance of genetic variation within species. This issue was important for knowing the mechanism of microevolution within species to infer long-term evolution. It was also important for developing new methods of animal and plant breeding and for understanding the hereditary mechanism of genetic diseases.

Dobzhansky (1955) classified the hypotheses about the maintenance of genetic variation into two major groups: classical theory and balance theory. The classical theory refers to the view that genetic variation within species is primarily caused by mutation-selection balance or transient polymorphism that occurs when the original allele at a locus is replaced by a new advantageous allele (Morgan 1932; Muller 1950). At this time, most mutations were thought to be deleterious, so that the extent of genetic variation generated by this mechanism was considered to be relatively small. Yet, it was thought that if there are a large number of loci and occasionally there occur transient polymorphisms caused by gene substitution, the classical theory would be sufficient to explain the extent of genetic variability. As mentioned earlier, gene substitution was often assumed to be triggered by environment changes and a previously rare allele replaces the high-frequency allele. By contrast, the balance theory assumed that most genetic variations are caused by overdominant or frequency-dependent selection (balancing selection). There were many examples of genetic polymorphisms at that time, and it was clear that if there were a large number of balanced polymorphisms any large population would contain a large amount of genetic variation.

In the 1950s and the 1960s, population geneticists were polarized into two camps, one supporting the classical theory and the other supporting the balance theory. The major figures of the classical camp were Hermann Muller, James Crow and Motoo Kimura, whereas the balance camp was led by Theodosius Dobzhansky, Bruce Wallace and E. B. Ford. The controversy between the two camps was quite intense and often confrontational, and it was not resolved until the molecular study of DNA polymorphism was initiated (Chapter 4).

Genetic Load

One of the important issues considered in relation to this controversy was the genetic load that is generated by the maintenance of genetic polymorphism. To maintain a polymorphism by selection, some genotypes have to produce more offspring than others. For this reason, the average individuals must have a fertility higher than 1. However, to quantify the fertility excess required, it is easier to consider the extent of fitness reduction caused by natural selection by using the standard mathematical theory of population genetics. We can then know how much fertility excess is necessary to offset the amount of fitness reduction or genetic load. In population genetics, genetic load is defined as

$$L = (w_{max} - \bar{w}) / w_{max} \qquad (2.20)$$

where w_{max} is the fitness of the best genotype and $\bar{w}$ is the mean population fitness.

In the classical theory of maintenance of genetic polymorphism mutation-selection balance is considered to be important. In this case the mutant genotypes are less fit than the original wild genotype. Therefore, it reduces the mean population fitness. However, since a highly deleterious mutation is quickly eliminated, the equilibrium frequency is low. By contrast, a recessive mutation with mildly deleterious effects is expected to be in appreciably high frequency (see Equation 2.8). For this reason, the genetic load or mutation load becomes approximately equal to the mutation rate u (see Appendix E). Because the mutation rate was thought to be of the order of 10^{-5} per locus per generation, this indicates that genetic polymorphism can be maintained at a large number of loci though the frequency of unfavorable alleles may be generally quite low.

For example, if the polymorphism by mutation-selection balance is maintained for 10 000 loci with a mutation rate of $u = 10^{-5}$ in a population, the total genetic load will be $L = 0.1$ and the population fitness will be $\bar{w} = (1 - u)^{10,000} \approx e^{-0.1} = e^{-L} = 0.90$. Therefore, the average fertility required ($\bar{F}$) is $1/\bar{w} = 1.11$. Since mammalian species apparently have a fertility excess higher than this requirement (Haldane 1957), there will be no problem for the survival of the population. Actually, the genetic load cannot be very high. For example, the mammalian DNA consists of about 3×10^9 nucleotide pairs. Around 1960, it was thought that the human genome may contain about 3 million genes if one gene was composed of 1000 nucleotide pairs on average. If the mutation load per locus is 10^{-5}, the total load for 3 million genes would be $L = 30$, and $\bar{w}$ becomes $e^{-30} = 9 \times 10^{-14}$. The average fertility required per individual will then be $\bar{F} = 1.1 \times 10^{13}$, which is absurdly high. For this reason, Muller (1950, 1967) suggested that this would bring too much genetic load and that the number of human genes is probably no more than 30 000, which would give $L = 0.3$ and $\bar{w} = 0.72$. Interestingly, his estimate of gene number was found to be approximately correct when the DNA sequencing was completed (International Human Genome Sequencing Consortium 2004).

By contrast, the genetic load created by an overdominant locus, which is often called segregation load, can be quite high. When the fitnesses of genotypes A_1A_1, A_1A_2, and A_2A_2 are $1 - s$, 1, and $1 - t$, the genetic load is given by

$$L = st/(s+t) \tag{2.21}$$

(Appendix E). Therefore, if $s = 0.3$ and $t = 0.7$, it will be 0.21. This means that if there are 50 such loci, L and $\overline{w}$ will be 10.5 and $e^{-10.5} = 2.8 \times 10^{-5}$, respectively. Therefore, the genetic load is outrageously high, and it is unlikely that a large number of polymorphic loci can be maintained by overdominant selection with a small number of alleles at each locus. This is true even if s and t are one order smaller than the above values.

In the neo-Darwinian era, supporters of the balance theory believed that a large number of polymorphic loci such as shell polymorphism, inversion chromosomes, blood group loci, etc., were maintained by overdominant selection, but the actual values of s and t in natural populations were not known. For this reason, supporters of the balance theory had a hard time convincing other investigators. For example, Lewontin and Hubby (1966) studied protein polymorphism by using the electrophoretic technique and obtained data suggesting that about 30 percent of protein loci in *Drosophila pseudoobscura* are polymorphic. Therefore, if this species has about 15 000 protein-coding loci, the estimate suggests that about 5000 protein-coding genes are polymorphic. Lewontin and Hubby considered various possible explanations but could not reach any definitive conclusion. They stated that if these loci were to be maintained by overdominant selection, an absurdly high genetic load would be imposed.

Sved et al. (1967), King (1967), and Milkman (1967) questioned the assumption of independent action of genes at different loci in the genetic load theory. Arguing that natural selection must be largely competitive because population size remains more or less constant and competitive ability must be controlled by a large number of loci, they developed a model of truncation selection, in which only individuals whose competitive ability is higher than a certain threshold can survive to adulthood. This model is a direct application of the model used for artificial selection (Fig. 2.10). In the case of artificial selection, the character of interest is measured after the character is fully developed, and the upper few percent of individuals is selected to produce the next generation (Appendix D). In natural populations this type of selection cannot occur because natural selection operates at every developmental stage of life. Therefore, a model of independent gene action is more realistic than the model of truncation selection. Furthermore, Felsenstein (1971) and Nei (1971) have shown that the fertility excess required under competitive selection is essentially the same as that for the standard model of viability selection.

At this point, however, it should be indicated that the genetic load for overdominant alleles can be small when there are many alleles at a locus as in the case of MHC loci. In a large population the genetic load for overdominant selection with m alleles can be written as

$$L = \tilde{s}/m, \tag{2.22}$$

where $\tilde{s}$ is the harmonic mean of selection coefficient (Crow and Kimura 1970). In human populations HLA (MHC) polymorphism is believed to be maintained by overdominant selection, and the average $\tilde{s}$ has been estimated to be 0.01 (Satta et al. 1994). In the HLA-B locus there seem to be about 25 alleles in each population (Roychoudhury and Nei 1988). Therefore, the genetic load required for maintenance of this polymorphism is $0.01/25 = 0.0004$ per locus. This genetic load is quite low because most individuals are heterozygous at this locus. Note that if every individual is heterozygous with the same selection coefficient there will be no load. In the case of self-incompatibility S alleles in the plant *Oenothera organensis*, Emerson (1939) counted 37 alleles in a population of about 500 individuals. Because the homozygotes for S alleles cannot be produced, all the 500 individuals must be heterozygotes. Therefore, no genetic load is generated by this locus even though homozygotes are effectively inviable.

Number of Alleles that can be Maintained in Finite Populations

While the controversy on the classical and balance theories was going on, Kimura and Crow (1964) published an influential paper. They first considered the possibility that at each locus a large number of alleles (theoretically infinite numbers of alleles) can be generated. Because an allele at a locus represents one nucleotide sequence, the number of possible alleles can be equated to the number of different nucleotide sequences that can be generated by random nucleotide substitution. Since there are four possible nucleotides (A, T, C, G) at each nucleotide site, the number of possible alleles for a locus consisting of 100 nucleotides will then be given by $4^{100} = 1.6 \times 10^{60}$. This number is practically infinite. Even if a large proportion of alleles are inviable because of the functional constraints of the gene, the possible number of alleles can still be very large. This model is now called the infinite-allele model (Kimura 1983). Interestingly, the same model had been used by Wright (1939, 1948b), but this was not noticed until the molecular nature of alleles was clarified.

In this model a new mutation is assumed to be always different from the pre-existing alleles in the population. Therefore, the number of alleles increases whenever a new mutation occurs. However, the number may decrease by the effect of genetic drift in a finite population or by natural selection. The number of alleles at a locus will then be determined by the mutation rate per locus (v) and the effective population size (N) in the absence of natural selection.

Kimura and Crow did not compute this expected number of alleles per locus but studied the effective number of alleles (m_e) given by $1/J$, where J is the expected homozygosity defined by $J = 1/(4Nv + 1)$ (see Nei 1987). They showed that the effective number can be very large when Nv is large, but there will be no genetic load because no selection is operating. Incidentally, the expected frequency of heterozygotes (heterozygosity), which is often used to measure the extent of polymorphism, is given by

$$H = 1 - J = 4Nv/(1 + 4Nv) \qquad (2.23)$$

In practice, the heterozygosity for a locus is computed by $1 - \sum_i x_i^2$, where x_i is the frequency of the ith allele in the population and Σ_i stands for the summation for all alleles.

Kimura and Crow then studied the heterozygosity for overdominant alleles under the assumption that all homozygotes have fitness 1 and the fitness of all heterozygotes is $1 + s$. Mutation was assumed to occur at a rate of v per locus per generation creating always new alleles. The results showed that when s is greater than 0.001 the heterozygosity for overdominant alleles at a locus is much higher than that for neutral alleles. For example, when $s = 0.1$ and $Nv = 0.1$, H becomes 0.9 for overdominant alleles, whereas it is 0.285 for neutral alleles ($s = 0$). This indicates that overdominant selection is very powerful for increasing genetic variability. At the same time, the amount of genetic load generated by overdominant selection can be quite high.

It should be noted that the theory of genetic load is a very crude concept and it is for getting a rough idea about whether the extent of natural selection proposed is appropriate or not. If genetic load is outrageously high, the proposed selection model is questionable. However, if the average fertility required is only slightly lower than the biologically possible value, we cannot draw a definite conclusion. Note also that the number of offspring born to a female is much larger in invertebrates than in mammals. This means that invertebrates can bear a higher genetic load than mammals. In practice, however, a large proportion of larvae in invertebrate species are killed by non-genetic factors such as bad weather, food shortage, and predation, and the average number of individuals surviving to adulthood per mating pair is not much different from 2 except when population size is expanding for some reason. Therefore, the extent of genetic load bearable in these species cannot be very high.

It should also be noted that the genetic load was originally defined deterministically for an infinitely large population. In finite populations, the extent of genetic load is expected to decline considerably even for the same selection model, because genetic drift may reduce the number of polymorphic alleles at each locus. Therefore, caution is necessary when the concept of genetic load is applied to real populations.

2.7. Natural Selection as a Creative Force

As mentioned in Section 2.1, neo-Darwinians developed the idea that the role of mutation in evolution is merely to provide raw genetic material and it is natural selection that has the power of creating innovative characters. Arguing that natural selection is for enhancing reproductive success of a population rather than for eliminating unfit genotypes, Mayr (1963, pp. 201–202), stated: "*Is not a sculptor creative, even though he discards chips of marble? As soon as selection is defined as differential reproduction, its creative aspects become evident. Characters are the developmental product of an intricate interaction of genes and since it is selection that supervises the bringing together of these genes, one is justified in asserting that selection creates superior new gene combinations. This viewpoint has been ably presented by Muller (1929), Simpson (1949), Fisher (1958), Dobzhansky (1951), and virtually every recent writer familiar with population genetics.*"

Similarly, Dobzhansky (1970) wrote "*Evolution is a creative process, in precisely the same sense in which composing a poem or symphony, carving a statue, or painting a picture are creative acts. An artwork is novel, unique, and unrepeatable. The evolution of every phylectic line yields a novelty that never existed before.*" This kind of statement about the creativity of natural selection is opposite to Darwin's view mentioned earlier but is still quite common (Strickberger 1996; Gould 2002; Futuyma 2005). (In the later years of his life, Mayr (1997, p. 2093) changed his view and stated that natural selection is an elimination process.)

These rhetorical statements are impressive, but it is not easy to find their logical basis from Mendelian inheritance. One argument which is often used was provided by Fisher (1930), Haldane (1932), and Muller (1932). According to this argument, natural selection rapidly enhances the frequencies of advantageous alleles at each locus, and therefore this makes it easy to combine two or more advantageous alleles at different loci into a single individual by recombination. In this way a novel character may be generated more rapidly when natural selection occurs simultaneously at different loci than when selection occurs sequentially one by one at different loci. This is certainly a legitimate argument, but recombination is not natural selection but a genetic process that produces new genotypes. In this sense the new genotypes produced should be called "mutations." However, a more serious problem with this argument was that there were no empirical data to support it. It is true that in complex genetic loci like the HOX gene cluster in animals there is gene interaction between different component loci and therefore a better combination of alleles at different loci may have occurred by simultaneous allelic replacement. However, this does not mean that simultaneous allelic replacement occurs more often than sequential replacement of alleles. Evolution is a slow process, and there is no reason to believe that fast evolution is more beneficial than slow evolution.

Another argument that was used for this purpose was Wright's (1932) shifting balance theory of evolution (see Chapter 3). In this theory, it is assumed that one species is composed of many partially isolated populations and allelic interaction exists among different loci. For this reason, the average fitness of the species is expected to show multiple peaks in a hyper-dimensional space of allele frequencies, and evolution can be assumed to occur by moving from one peak to another by means of selection and random genetic drift. Therefore, if the environmental conditions change, a new genotype which is better adapted may be created by selection and drift (Dobzhansky 1951, 1970; Simpson 1953). However, this is again a rhetorical argument, and there are no empirical data to support this argument. Furthermore, there are a number of theoretical problems in this argument as will be discussed in Chapter 3. For these reasons, the argument that natural selection has a creative power is weak despite the fact that there are many investigators and textbooks supporting this view.

The third argument was that there are many examples of supergenes of which the component genes are polymorphic (e.g. Rh blood group genes in humans, heterostyly in primulas, snails' shell types, etc.) (Ford 1975). In the absence of knowledge of the molecular structures of these gene complexes, they were thought to be products of natural selection. It was commonly believed that there are many modifier genes controlling such gene complexes. Fisher (1928) proposed that even the phenomenon of dominance observed in many crossing experiments was generated by natural selection of modifier genes. This problem will be discussed in detail in Chapter 3.

2.8. Summary

The idea of neo-Darwinian evolution was developed when Mendelian geneticists showed that most new mutations are deleterious and therefore do not appear to contribute to evolution significantly. However, it was observed that some disadvantageous mutations become advantageous when the environmental conditions change. This appeared to happen particularly when recombination occurs between different genetic loci. These observations led to the idea that natural populations contain a large amount of genetic variability due to mutations that have occurred in the past but advantageous alleles are created when the environmental conditions change or when recombinations occur between different loci. In this view mutation is considered to generate raw genetic materials for evolution, but the driving force of evolution is natural selection that picks up advantageous alleles in a given environment and drives them to fixation.

In the neo-Darwinian era various mathematical theories of allele frequency changes in populations were developed. These theories were very useful for visualizing how the genetic structures of populations would change during evolution. Because evolution was believed to occur mostly by allele frequency changes due to natural selection, most of the theories were deterministic under the assumption that the population size is infinitely large. Unfortunately, a large portion of these theories remain unused because the validity of the biological assumptions made is unclear.

One of the important issues in the neo-Darwinian era was to obtain empirical evidence of natural selection and show that natural selection is the major force of evolution. This was necessary because Charles Darwin did not present any empirical evidence for natural selection. Darwin's proposal of natural selection was based on the effectiveness of artificial selection in domestic animals and plants and his deductive argument about the occurrence of natural selection when the population size grows rapidly according to Malthusian rule. However, natural selection is very different from artificial selection because there is no designer and his argument of natural selection based on Malthusian population growth was a mere speculation. In practice, however, it turned out a difficult job to prove natural selection in wild populations. One reason was that natural populations are widely distributed and it is difficult to find an isolated population which is not affected by gene migration from the surrounding populations. Furthermore, because allele frequency changes are often affected strongly by environmental factors as well as other genetic loci, it is difficult to obtain reliable estimates of selection coefficients. For this reason, the natural selection inferred is always very crude.

Some theoreticians were fully aware that all natural populations are finite and some are quite small and that the allele frequency changes in finite populations occur stochastically and are subject to random errors. Therefore, many different theories of stochastic changes of allele frequencies were developed during the middle twentieth century, but these theories were rarely used by empirical investigators, mainly because it was difficult to collect proper data for studying this problem empirically and the species population size was thought to be large enough even if it included many small local populations. Only recently, we have come to realize that this problem can be studied by using molecular polymorphism data.

Another issue that was debated extensively in the neo-Darwinian era was the classical and balanced theories of maintenance of genetic variation. This debate was not resolved until genetic variation was studied at the molecular level, and molecular data suggest that very high polymorphisms are restricted to a small proportion of genetic loci.

As mentioned above, neo-Darwinism has suggested that most natural populations contain a large amount of hidden genetic variation and this standing variation can be used for future evolution. This is certainly true with quantitative characters under artificial selection. However, recent studies indicate that new mutations occur with significant frequency in quantitative characters and this mutation plays important roles in responding to artificial selection. Similarly, for generating antibiotics resistance in bacteria a significant portion of variation appears to be caused by fresh spontaneous mutations. However, because bacteria inhabit various different niches, different mutant strains can be maintained in a species and these strains may reappear when antibiotics are administered.

CHAPTER 3

Evolutionary Theories in the Neo-Darwinian Era

As discussed in the preceding chapter, neo-Darwinians developed sophisticated mathematical theories for predicting the evolutionary change of gene or genotype frequencies and for understanding the evolution of intricate characters such as sexual reproduction and altruism. These theories were immensely useful for visualizing the process of long-term evolution and provided the guiding principles for studying evolution. However, they were dependent on many simplifying assumptions such as the large population size, the same selection coefficients throughout the evolutionary process, absence of the interaction between genotypes and environments, etc. For these reasons, the mathematical predictions easily could go wrong. Unfortunately, the experimental verification of these predictions was extremely difficult because human life is too short to observe long-term genetic changes of populations.

Furthermore, it was difficult for most biologists to understand the mathematical theories, and therefore they simply accepted or rejected the theories on an intuitive basis. Of course, there were biologists who conducted experiments or data analyses to test the mathematical predictions. Well-known examples are the collaboration between Theodosius Dobzhansky and Sewall Wright (Dobzhansky 1951, 1970) and that between E. B. Ford and R. A. Fisher (Ford 1964, 1975). In these collaborations experimental data were obtained by biologists, but the interpretation of data was provided by theoreticians. For this reason, evolutionary studies in the twentieth century were dominated by theoreticians. Yet, there were vocal biologists who were critical of mathematical theories. These criticisms occurred mainly because some mathematical theories were too abstract and could not explain the evolution of any concrete characters such as bird wings and mammalian brains. Furthermore, mathematical population geneticists themselves often disagreed with one another about the theories, and this disagreement sometimes accelerated or decelerated the progress of evolutionary biology.

In this chapter I would like to discuss various evolutionary theories developed by neo-Darwinians. However, because it is impossible to cover all the theories developed, I have decided to focus on the theories that are directly related to the subject of this book and discuss their current status in the light of recent molecular biology. My discussion will be primarily concerned with the theories developed by the major figures of neo-Darwinism. I have chosen subjects that have been widely accepted by neo-Darwinians and tried to present a critical examination of the subjects. Some more specific theories such as kin selection (Hamilton 1964) and selfish gene theory of evolution (Williams 1966; Dawkins 1976) will be discussed in Chapter 8, because some knowledge of molecular evolution is necessary for critical evaluation of these theories.

3.1. Modifier Genes

As mentioned earlier, neo-Darwinism asserts that natural populations contain various kinds of genetic variations and these variations can be used for adapting to new environments or for creating new phenotypic characters. For this reason, it has been thought that even the mutation rate, recombination values, the degree of dominance, etc. can be modified to the optimal values by natural selection. Therefore, various mathematical theories have been

Mutation-Driven Evolution. First Edition. Masatoshi Nei.

developed to predict the evolutionary change of these genetic events. In the following I would like to consider two examples, which have been studied in the neo-Darwinian era.

Evolution of Dominance

The first example is Fisher's (1928) theory of evolution of dominance. Although this subject is somewhat outdated, it represents a good example of neo-Darwinian theory of evolution (see Ewens 2004, pp. 221–224). In his hybridization experiment with garden peas, Gregor Mendel discovered the three laws of inheritance, and one of them was the law of dominance. This law states that the F_1 hybrid shows the same phenotype as that of the homozygote of the dominant allele. Initially, this was thought to be a general rule, but later experiments with other organisms revealed that this was not the case and there were many exceptions. Fisher noticed that in general a new mutant allele (*a*) at a locus is partially recessive and deleterious compared with the wild-type allele (*A*) and the recessive mutation occurs recurrently. He then proposed that if there is another locus that modifies the degree of dominance and the allele *M* at this locus enhances the degree of dominance at the *A* locus compared with its allelic gene *m*, allele *M* would be selected for because the higher degree of dominance of the wild-type allele *A* at the primary locus reduces the harmfulness of the mutant allele *a* in heterozygous condition. He then argued that if the mutation $A \rightarrow a$ occurs repeatedly for a large number of generations the mutant allele *a* may become fully recessive by natural selection occurring at the *M* locus.

This idea is typically neo-Darwinian, because in this theory almost any kind of genetic variation is assumed to exist in natural populations and therefore even the degree of dominance at the *A* locus can be modified by natural selection at the *M* locus. If there are no modifier genes (polymorphism of alleles *M* and *m*) controlling the degree of dominance between alleles *A* and *a*, there will be no change of the degree of dominance of *A* over *a*. In fact, there was no empirical data suggesting the existence of modifier genes at that time. Wright (1929a) criticized Fisher's evolutionary theory for three reasons. First, he questioned the effectiveness of natural selection for changing the degree of dominance even if the modifier gene *M* exists in the population. Fisher assumed that the degree of dominance at the primary *A* locus occurs when the frequency of allele *M* at the modifier locus increases relative to the frequency of allelic gene *m* by natural selection. Wright found that the selection coefficient for allele *M* is at most of the order of the mutation rate from *A* to *a* per generation. Because this magnitude of selection coefficient is so small, he questioned the effectiveness of natural selection at the *M* locus. A more detailed analysis of this problem was conducted by Ewens (1967), and it was shown that Wright's conclusion is essentially correct. Wright also indicated that because any mutation would affect many characters simultaneously the biological significance of the selective advantage of allele *M* over *m* is unclear.

Second, Wright (1929b) later argued that, because the selective advantage of allele *M* over *m* is so small, allele *M* may not be fixed in the population because of genetic drift. For this argument, he briefly presented his equation for the equilibrium distribution of allele frequencies under the effect of mutation, selection, and genetic drift (see Wright 1931). He also argued that because the potential number of alleles at a locus is likely to be infinite instead of two the probability of fixation of allele *M* would be even smaller. (Incidentally, the idea of the infinite-allele model, which later became popular in the study of molecular evolution, was presented for the first time in this paper.) Third, the dominance relationships between alleles should be considered in terms of the biochemical genetics of gene expression. If the expression level of allele *A* in the heterozygote *Aa* is as high as that of two copies of *A* in the genotype *AA*, we will have complete dominance. If the expression level of genotype *Aa* is half that of *AA*, we will have semidominance. For this reason, Wright did not think that the mathematical study of this problem is meaningful.

For Fisher, however, Wright's criticism was not the end of the study. He continued to write papers on the subject and never gave up his theory (Fisher 1931). A detailed account of the controversy between Fisher and Wright is presented by Provine (1986). At the present time, we know that there are several genetic mechanisms that control the expression level of a gene and these mechanisms can change during evolution (see Chapter 6). Therefore, it may not be meaningless to study the effect of natural selection on gene expression levels. However, because such mechanisms must have evolved in the early stage of evolution and are maintained by strong purifying selection, it is not clear whether dominance can be modified easily (see Chapter 9).

Modification of Linkage Intensity

Another genetic quantity of which the evolutionary modification has been studied is the linkage intensity of genes or the recombination value between different genetic loci. In this case there is a large amount of evidence that linkage intensity is under genetic control (see Bodmer and Parsons 1962). In *Drosophila ananassae*, Moriwaki (1940) discovered a dominant gene, *En-2*, located on the right arm of chromosome 2, which enhances the recombination in males between almost every pair of genes on the same chromosome, whereas Kikkawa (1937) found another gene on chromosome 3, which induces recombination in males. (*Drosophila* males usually do not show any recombination.) Furthermore, recent molecular studies have shown that both prokaryotic and eukaryotic genomes contain a large number of recombination-controlling genes such as RecA/RAD genes (Lin et al. 2006) and DNA mismatch repair genes (Lin et al. 2007).

Fisher (1930) argued that the recombination value between genetic loci may decrease under certain types of natural selection, but he never studied the problem mathematically. Nei (1967) appears to be the first to examine this problem considering the allele frequency change at a modifier locus. For simplicity, let us consider two linked loci (A and B loci) each with two alleles, A_1 and A_2 and B_1 and B_2 respectively, in haploid organisms. We then have four different haplotypes A_1B_1, A_1B_2, A_2B_1, and A_2B_2, as shown in Fig. 3.1. Let us designate their relative fitnesses by w_1, w_2, w_3, and w_4, respectively. In this case the extent of gene interaction or epistasis can be measured by $E = w_1 - w_2 - w_3 + w_4$. When gene action is additive, E becomes 0. Otherwise, E is positive or negative. We denote the frequencies of haplotypes A_1B_1, A_1B_2, A_2B_1, and A_2B_2 by P_1, P_2, P_3 and P_4, respectively. The linkage disequilibrium is then given by $D = P_1P_4 - P_2P_3$ (see Equation 2.10). Let us now consider recombination modifier alleles M and m at another locus (M locus) and denote their frequencies by x and $1 - x$, respectively. We also denote the average recombination value between loci A and B by r_M when the modifier allele M is present and by r_m when allele m is present. Nei then showed that the evolutionary change of x per generation is given by

$$\Delta x = \frac{x(1-x)(r_m - r_M)DE}{\bar{w}} \tag{3.1}$$

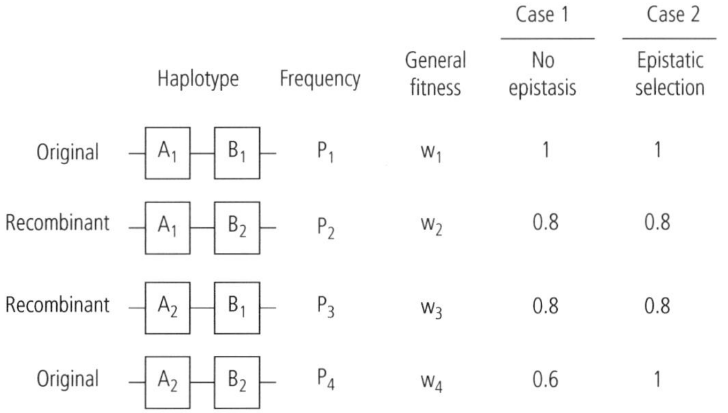

	Haplotype	Frequency	General fitness	Case 1: No epistasis	Case 2: Epistatic selection
Original	A_1 — B_1	P_1	w_1	1	1
Recombinant	A_1 — B_2	P_2	w_2	0.8	0.8
Recombinant	A_2 — B_1	P_3	w_3	0.8	0.8
Original	A_2 — B_2	P_4	w_4	0.6	1

Fig. 3.1. Diagram showing the importance of epistatic selection for reducing recombination value. Loci A and B are linked and form four different haplotypes with alleles A_1, A_2, B_1 and B_2. w_1, w_2, w_3, and w_4 are the relative fitnesses of the four haplotypes. Gene interaction or epistatic parameter is defined by $E = w_1 - w_2 - w_3 + w_4$. If $E = 0$, there is no epistasis. In case 1, $E = 0$, so there is no advantage in reducing recombination. In case 2, $E = 0.4$, and haplotypes A_1B_2 and A_2B_1 are less fit than A_1B_1 and A_2B_2. Therefore, reduction in recombination is advantageous. In this haploid model, all haplotypes mate at random after selection, and offspring haplotypes are produced through meiosis and recombination. The recombination value is controlled by another recombination-modifying locus, which is polymorphic.

where $\bar{w}$ is the average fitness determined by the A and B loci.

Nei also showed that D tends to be positive when E is positive but tends to be negative when E is negative. Therefore, if M is a new mutant allele and reduces the recombination value, Equation (3.1) shows that the frequency of allele M always increases and therefore the allele will be fixed in the population. This means that the recombination value always tends to decrease by natural selection if there is epistasis between the loci A and B. Nei showed that essentially the same equation holds for diploid organisms. A more detailed study of this problem was conducted by Feldman (1972), but the conclusion was the same. Note that in this case the same modifier gene may control the recombination for many loci and therefore the selection pressure for the modifier gene can be quite strong.

After obtaining the above result, Nei (1968) examined the average recombination value per genome in various groups of organisms and showed that complex multicellular organisms generally have a lower recombination value per unit length of DNA than single-cell organisms. He suggested that this represents the results of natural selection. This suggestion now seems to be supported by extensive genomic data on the recombination value and genome size (Lynch 2007, p. 87). However, recent genomic data indicate that functionally interacting genes are often closely located on a chromosome. A good example is the vertebrate major histocompatibility complex (MHC) comprising 20–100 functional genes. These genes are located in the same chromosomal region and are tightly linked in most classes of vertebrates (Kulski et al. 2002). Another example is the HOX gene cluster, in which even the same order of genes with different functions are maintained in diverse groups of animal species despite the frequent occurrence of chromosomal rearrangements. Similarly, groups of histone genes or globin genes have been maintained as clusters.

Recently Pal and Hurst (2003) studied the relationship between the cluster of genes essential for viability and the level of recombination within the cluster in the genome of baker's yeast. They divided the genome into a large number of contiguous non-overlapping blocks of 10 genes and examined the relationship between the number of essential genes per block and the recombination value as measured by meiotic doublestrand breaks. The results were surprising. Whenever the number of essential genes per block was high, the recombination value in the region was low, and this negative correlation was highly significant. If we consider the possibility that the E value is likely to be greater for essential genes than for weakly selected genes, Pal and Hurst's finding supports the linkage modification theory.

In this connection it should be mentioned that the initial formation of a cluster of functionally related genes probably occurred by repeated tandem duplication and only clusters of strongly interacting genes have been maintained unchanged in the evolutionary process. Other genes were probably scattered to different chromosomes and evolved into genes with different functions. Only strongly interacting genes appear to have remained as part of the same cluster.

3.2. Fisher's Fundamental Theorem of Natural Selection

In the early neo-Darwinian era, the primary concern was the study of allele frequency changes within populations, and the important problem was how to explain the evolutionary relationships of genetic variability within and between species. Although there were the classical and balance schools of population geneticists, their views about long-term evolution were nearly the same. Because mutations were thought to provide merely raw genetic materials with which natural selection creates innovative characters, the occurrence of a new mutation was not a starting point for a gene substitution (see Fig. 1.1A). Rather it was believed that new mutations are almost always deleterious and are retained at low frequencies in the population but some of them become advantageous when the environmental condition changes (Muller 1950; Haldane 1957). However, because the environmental change for a species occurs haphazardly without any regular pattern, we cannot expect any clear relationship of phenotypic change with evolutionary time. For this reason, few neo-Darwinians studied the long-term evolution of phenotypic characters mathematically.

In neo-Darwinism it was customary to measure the extent of evolutionary change of a population in terms of the increase of mean fitness of the population, as mentioned earlier. Without any mathematical proof, Fisher (1930, 1941) introduced the so-called fundamental theorem of natural selection, which stated that the rate of increase in the average fitness of a population is equal to the genetic variance of fitness of that population. This means that the mean fitness always increases by natural selection because the variance is a nonnegative quantity. Mathematically it is an appealing theorem because it supports our intuitive expectation about the effects of natural selection and is similar to the increase of entropy in the second law of thermodynamics. Later a number of authors (e.g. Kimura 1958; Price 1972; Ewens 1989, 2004; Lessard 1997) attempted to prove the theorem mathematically under various assumptions.

However, the real problem with this theorem is not mathematical but biological. First, for the theorem to be biologically meaningful the fitnesses of all genotypes must remain the same in all generations though allele frequencies may change, but this assumption obviously does not hold, because genotype fitnesses depend on the environmental condition and this condition varies with generation. Without knowing the genotype fitnesses in consecutive generations, it is impossible to compute the rate of increase of population fitness. Second, this theorem is for short-term evolution even if the above assumption is satisfied, and it cannot work for long-term evolution. If we consider long-term evolution, we cannot avoid drastic changes of environmental conditions due to geological or meteorological factors. For example, if an asteroid hits the earth, many organisms may be wiped out as in the case of the Cretaceous mass extinction and the remaining organisms would have to survive in entirely different environments. This indicates that the survival of organisms depends on many unpredictable events and Fisher's fundamental theorem has no biological meaning in this case.

Third, if any species is well adapted to a given environment (say an isolated island) because of the favorable mutations that have accumulated in the past, no new species then would be able to invade the environment. In reality, the opposite is often true. It is well known that when a new species from a continent is introduced into an island or another continent, the species often proliferates rapidly and indigenous species are decimated. This indicates that the mean fitness of a population is a complicated function of genotypes and environments and the prediction of evolutionary change is very difficult.

At this point, one might argue that the above comments are not appropriate because Fisher's theorem is intended to be applied to a given constant environment rather than to changing environments and that it is a conceptual theorem rather than a practical one. This argument is also questionable if we consider Felsenstein's (1971) study about the variance of fitness in relation to the cost of natural selection, which will be discussed below. He indicated that conceptually the variance of relative fitnesses can be computed whether the population size is increasing, decreasing or remains constant. If we note that the variance is positive even in a decreasing population, the meaning of Fisher's theorem becomes obscure. Note also that Fisher's theorem does not apply in small populations because it is based on the deterministic model and ignores the effects of genetic drift and mutation.

Fourth, biologists are generally interested in understanding the evolutionary change of a particular character such as an elephant's trunk or a whale's body structure. For this purpose, Fisher's theorem is powerless, because it is impossible to relate the evolutionary change of a specific character to the change in mean fitness of the population.

In fact, Fisher's theorem is an abstract concept based on the principle of survival of the fittest, and it has never been shown to be useful for explaining the evolution of a particular character or for predicting the success of a particular population. In real worlds, even a small population with small genetic variability may succeed in occupying a given territory or a wide geographical area. Note also that many biologists are interested in explaining the evolution of complex organisms from simple ones rather than in the enhancement of population fitness.

The above arguments suggest that the important problem in the study of evolution is to understand the molecular basis of formation of phenotypic characters rather than the mathematical basis of increase of population fitness.

3.3. Cost of Natural Selection and Fertility Excess Required

Haldane's (1957) theory of the cost of natural selection is based on a more concrete concept of natural selection than Fisher's fundamental theorem. Haldane first noticed that in animal or plant breeding artificial selection is more effective when the best animals or plants (fixed number) are chosen from a large group of individuals than when they are chosen from a small group. He therefore thought that natural selection would be more effective when the number of offspring from a pair of parents is large than when it is small. He then argued that the number of genes that can be substituted simultaneously in a population depends on the fertility of the organism. In the process of gene substitution a less fit gene generates a reduction in mean fitness, and if there are many gene substitutions occurring in the same population, the total amount of reduction in fitness is so large that the species may not be able to survive. The total amount of fitness reduction or genetic deaths in the process of gene substitution is called the cost of natural selection.

For simplicity, let us consider a haploid organism and assume that the frequencies of alleles A_1 and A_2 are given by x and $y = 1 - x$, respectively. We also assume that the fitnesses of A_1 and A_2 are $w_1 = 1$ and $w_2 = 1 - s$, respectively. In this case, the mean fitness $\overline{w}$ is given by $x + (1 - x)(1 - s) = 1 - sy$, and the allele frequency change per generation becomes

$$\Delta x = \frac{dx}{dt} = \frac{sxy}{1-sy} \tag{3.2}$$

The cost of natural selection for one generation is then defined as the amount of genetic deaths due to allele A_2 (sy) relative to the mean fitness ($1 - sy$), i.e., $sy/(1 - sy)$. This means that the mean fertility of the population must be higher than 1 at least by this amount to sustain the same or a larger population size. If the average fertility is not high enough, the population size will decline. In other words, Haldane's cost of natural selection is the fertility excess required for natural selection to operate (Crow 1970; Felsenstein 1971; Nei 1971). However, this fertility excess required is just for one generation, and the total amount of excess required or the total cost of natural selection for the entire process of gene substitution when x changes from its initial frequency x_0 to 1 is given by

$$C = \int_0^\infty \frac{sy}{1-sy}\,dt \tag{3.3}$$

Since $dt = \dfrac{1-sy}{sxy}\,dx$ from Equation (3.2), we have

$$C = \int_{x_0}^1 \frac{dx}{x} = -\log_e x_0 \tag{3.4}$$

where x_0 is the initial allele frequency.

It is interesting to note that C is independent of s and is solely determined by x_0. It becomes 14 for $x_0 = 10^{-6}$ and 9 for $x_0 = 10^{-5}$. Therefore, the total amount of genetic deaths or fertility excess required for the entire process of gene substitution is 14 times the total population size for $x_0 = 10^{-6}$ and 9 times the total size for $x_0 = 10^{-5}$.

In the above formulation we considered a simple haploid model. In diploid organisms computation of C is somewhat complicated because the cost depends on the degree of dominance of the advantageous allele (Haldane 1957). Considering various degrees of dominance and initial allele frequencies, Haldane concluded that the average cost of one gene substitution would be about C = 30 in diploid organisms. In other words, the total amount of genetic deaths required for one gene substitution is about 30 times the population size. He also thought that the fertility excess that can be used for gene substitution in mammalian species is about 10 percent of the population size if we consider accidental or unrelated deaths. Therefore, the upper limit of the number of gene substitutions (n) in mammalian organisms is about one in 300 generations ($n = 0.1/30 = 1/300$).

Haldane's original paper has been misunderstood by a number of authors. For example, Brues (1969) stated that because gene substitution occurs when the population adapts to a new environment natural selection should not incur any cost but it should be beneficial. As stated by Crow (1970), however, "*the cost is the excess in survival and fertility that the favored genotype must have in order to carry out gene substitution at a specific rate, while the entire population stays roughly constant.*" Later Nei (1971) developed a mathematical model of natural selection under the regulation of population size and showed that the fertility excess required is the same as that of Haldane's cost. He also showed that the number of possible gene substitutions per generation is given by

$$n = \log_e k / C, \quad (3.5)$$

where k is the average fertility of the population and C is about 30 in diploid organisms. Using a slightly different model, Felsenstein (1971) derived a similar formula independently. Equation (3.5) indicates that when $k = 1.1$ and $C = 30$, n becomes approximately $(k - 1)/C = 0.1/30 = 1/300$, which agrees with the number obtained by Haldane.

Ewens (1970, 1972) attempted to compute n considering the standard deviation of genotype fitnesses rather than the mean fitness. He concluded that the expected standard deviation of fitness among different individuals is so small that the fitness value of an individual belonging to an extremely high fitness group is only slightly higher than the mean fitness. For this reason, he concluded that the cost of natural selection cannot be as high as the one computed by Haldane. However, because Ewens computed the standard deviation by using the relative fitness values rather than the absolute fitness values, his results are not very meaningful. His results would remain the same whether the population size is increasing or decreasing. In fact, as indicated by Felsenstein (1971), Ewens' conclusion would not change even if the population is becoming extinct. In the computation of the cost of natural selection, we must consider the absolute fitness values of individuals. This is intuitively obvious if we consider a special case where all individuals cannot produce more than one offspring in a haploid population. In this case natural selection still could operate if some individuals do not produce any offspring, but the population size will decline every generation and the population will become extinct eventually.

Let us consider an example to clarify the above argument. Suppose that there are 10 loci at which the favorable (A_1) and unfavorable (A_2) alleles exist at each locus with a frequency of 0.5 and A_1 and A_2 have the absolute fitnesses of 1 and $1 - s$, respectively, where $s = 0.01$. Under the assumption of multiplicative fitness, the mean fitness of this haploid population becomes $\bar{w} = (1 - 0.5s)^{10} = 0.995^{10} = 0.95$. Therefore, the population size declines to 95 percent of the original size. If this type of selection continues for many generations, the population eventually becomes extinct. Therefore, to prevent the extinction of the population, each individual must be able to produce more than one offspring on the average. That is, there must be a fertility excess to offset the population size reduction.

The above example is certainly artificial. However, if we consider many deaths due to nongenetic factors, the fertility excess available for gene substitution could be about 10 percent in mammals as argued by Haldane. Therefore, as a rough approximation as implied by him, it seems to be reasonable to accept his cost of natural selection. As is well known, Kimura (1968a) used the argument of the cost of natural selection to propose his neutral theory of molecular evolution. Ewens (1993, 2004) criticized Kimura's argument by using his estimate of n based on the standard deviation of fitnesses. In my view his criticism is not justified because his estimate of n is based on incorrect arguments.

In the above formulation of the cost of natural selection we used the deterministic model of selection, assuming that the population size is effectively infinite. In reality, of course, the population size is always finite, and the effective population size can be quite small. In this case the allele frequency change is affected by both natural selection and genetic drift. The allele frequency change due to genetic drift does not cause any genetic deaths. Therefore, if we consider a small population, the cost of natural selection is expected to be reduced substantially. In fact, Kimura and Maruyama (1969) showed mathematically that this is exactly what would happen. If we note that natural populations are always finite, the actual upper limit of gene substitutions per generation can be much higher than Haldane's calculation. Nevertheless, Haldane's upper limit is useful for knowing what natural selection can do without the effect of genetic drift. It should also be noted that when there is no selection there will be no upper limit for the number of gene substitutions, as will be discussed in the next chapter.

As is clear from the above description, there are some ambiguities in the concept of the cost of natural selection. First, the idea that only 10 percent of the fertility excess can be used for gene substitutions is certainly arbitrary. The average human fertility around 1900 without birth control (half the number of adult children for a married couple) was 2.5 in a Japanese population (Imaizumi et al. 1970). This means that the fertility excess available (2.5 – 1 = 1.5) is quite high. However, if we consider the human lineage before the occurrence of modern *Homo sapiens*, it is likely that the average fertility was quite low because of diseases, natural disasters, wars, etc., and the fertility excess may not have been very high. Therefore, Haldane's use of fertility excess may not be so unrealistic, though it is a very crude conjecture.

Second, Haldane assumed that the initial allele frequency (x_0) is about 0.001 because he thought advantageous alleles arise from previously disadvantageous mutations when the environment changes. Actually, there must be some mutations that are advantageous from the time of their occurrence. For these mutations, one can argue that x_0 should be $1/(2N)$, where N is the effective population size. The cost of natural selection could then be higher. In reality, however, when the frequency of a new mutant allele is low, its fate is determined largely by chance, as discussed in Chapter 2, and selection becomes effective only when the frequency reaches a certain critical value. This critical value would depend on the values of N and s, but in the absence of clear knowledge $x_0 = 0.001$ may be used for these mutations.

If we consider various factors mentioned above, it is certainly difficult to know the real extent of the cost of natural selection. However, this theory still gives a crude estimate of the number of gene substitutions that are admissible and this crude estimate is useful for determining the cause of gene substitution, as will be discussed in Chapter 4. In recent studies of natural selection the cost of gene substitution is often neglected partly because the definition of the cost is somewhat ambiguous as mentioned earlier in this section. For example, analyzing the pattern of linkage disequilibrium of 3.6 million single nucleotide polymorphisms (SNPs) in human populations, Hawks et al. (2007) concluded that the rate of nucleotide substitution has been accelerated enormously by natural selection during the last 40 000 years and the current rate of nucleotide substitution per genome is 13.25 per generation or 0.53 per year. This estimate of substitution rate is enormously high compared with Haldane's upper limit, and it is quite likely that their estimate of the rate of nucleotide substitution is too high because of the unrealistic assumptions made in the computation (Nei et al. 2010).

3.4. Shifting Balance Theory of Evolution

Sewall Wright was a physiological geneticist as well as a population geneticist. He conducted an extensive study of the genetics of morphological characters in guinea pigs before he became a mathematical population geneticist. Probably for this reason, his view of evolution is somewhat different from that of Fisher and Haldane. For this reason, his mathematical theories are still widely used in the study of population genetics. However, the most well-known evolutionary theory he developed is the shifting balance theory first published by Wright (1931, 1932). This theory is to show how a phenotypic character evolves in the presence of natural selection, gene interaction, and genetic drift. It is an abstract theory and is based on three premises (see Nei 1987, pp. 419–422). (1) Populations are subdivided into a large number of subpopulations or demes, among which gene migration occurs

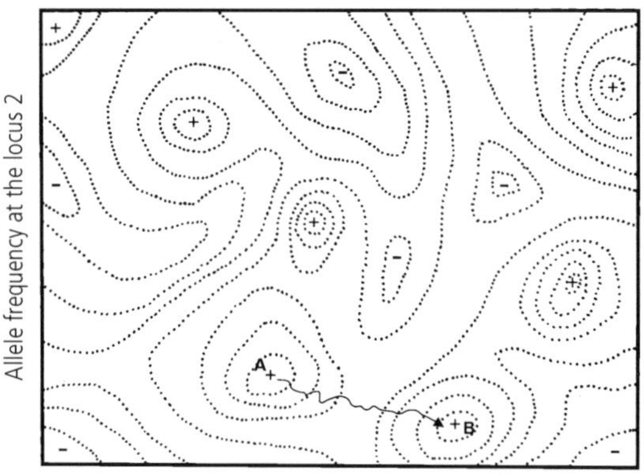

Fig. 3.2. Diagrammatic representation of the adaptive surface ($\overline{w}$) in the two-dimensional space instead of the multi-dimensional. Dotted lines represent contours with respect to fitness. "+" and "–" represent adaptive peaks and adaptive basins, respectively. Evolution is regarded as a process of movement of population fitness $\overline{w}_p$ from one peak (A) to a higher peak (B). Adapted from Wright (1932).

with some probability, and this population structure remains the same for a long evolutionary time. (2) Most loci are polymorphic, and evolution occurs mainly by allele frequency shift rather than by complete gene substitution. Genetic polymorphism is maintained by some sort of balancing selection (overdominance or frequency dependent selection). (Wright (1977, pp. 455–460) considered an example of homoallelic (monomorphic) multiple peaks, but in this case the evolutionary rate depends primarily on the mutation rate.) (3) There are epistatic gene interactions and pleiotropy among many loci. In general, there is no selective advantage or disadvantage for any particular allele, but a particular combination of alleles at different loci determines the fitness of the population.

With these three premises, Wright argued that the rate of evolution in large substructured populations is substantially higher than that in small populations or large nonstructured populations. His argument was as follows. Because of the assumption of gene interaction and pleiotropy for many loci, the mean fitness ($\bar{w}$) of a population is expected to show numerous peaks in a hyperdimensional space (Fig. 3.2), and the evolutionary change of a population can be described as a process of the population mean fitness, $\bar{w}_p$ to move from one peak (A) of the adaptive ($\bar{w}$) surface to another higher peak (B) (Fig. 3.2). He argued that when the population size is small $\bar{w}_p$ would not easily reach a higher peak because of genetic drift, or even if it reaches the peak temporarily, it would not stay there for a long time. By contrast, the $\bar{w}_p$ of a large population may reach a local peak, which is relatively low, and stay there almost infinitely. The reason for the difficulty of this population mean fitness to reach a higher peak is that there is little chance for the population fitness $\bar{w}_p$ to move around the current peak because the extent of allele frequency change by genetic drift is small. Wright assumed that for a population to move up to a new peak $\bar{w}_p$ has to cross a "valley" or "saddle" on the adaptive surface by the aid of genetic drift. However, the story is different in a population which is subdivided into many subpopulations. The effective size of a subpopulation is small so that genetic drift operates in each subpopulation even after the entire population has reached a local peak, and it is possible for a particular subpopulation to cross a "valley" or "saddle" of the $\bar{w}$ surface and reach a higher peak. Once this happens, the allelic combination of this subpopulation may migrate into its neighboring subpopulations and eventually spread through the entire population. In this way, the mean fitness $\bar{w}_p$ of a large subdivided population may increase gradually. For this reason, Wright believed that large subdivided populations would give a favorable condition for the evolution of morphological characters.

This theory was criticized by Nei (1980b, 1987, pp. 419–422) for several reasons. First, the first premise mentioned above rarely holds in nature. The population structure is all temporary. For example, many plant populations in North America and Europe retreated in the ice age, and some of them barely existed as fragmented populations in the southern area. After the ice age was over, the populations again moved up to the north. This has occurred almost every 100,000 years. Actually, even at the present time many natural populations are known to vary from generation to generation (Nei and Graur 1984). For example, Crumpacker and Williams (1973) and Jones et al. (1981) have shown that the population structure of *Drosophila pseudoobscura* varies extensively from year to year and that the populations in Colorado and California rapidly colonize new territories in some seasons or years but quickly disappear in others and therefore the seasonal or yearly fluctuation of population size is enormous.

The second premise is also questionable. If we consider all unimportant nucleotide polymorphism including those in intron regions, genetic loci are often polymorphic, but the nucleotide substitutions that affect phenotypes are not necessarily polymorphic. Gene interactions are also prevalent, but recent studies of developmental biology indicate that most gene interactions occur among homozygous genotypes, and there is no need to consider the adaptive landscape. In this case what are needed for evolution are mutations rather than natural selection. As we will see in Chapter 6, the pattern and extent of gene interaction or the mechanism of gene expression is quite complicated, and it is not always clear how natural selection and genetic drift operate in Wright's abstract multidimensional space. Actually, to understand the importance of gene interaction, we should study various forms of molecular interactions one by one, as is done in current developmental biology (Carroll et al. 2005; Davidson 2006; Gilbert 2006).

As mentioned above, Wright presented the shifting balance theory as the result of his attempt to find the best population structure required for rapid evolution. However, it should be noted that there is no need for any population to evolve rapidly. The evolutionary change of an organism occurs as a passive consequence of the development of a proper genomic structure for a given environment. A particular genomic structure may not be the best for adaptation to a given environment, but if there is no better one, any genomic structure may fill an open niche. After all, evolution does not occur teleologically.

The shifting balance theory has been used as a guiding principle for neo-Darwinian evolution by Dobzhansky (1951, 1970) and Simpson (1949, 1953), but there is little empirical support available. For example, Wright claimed that evolution occurs faster in substructured populations, but this has never been shown mathematically. Rather empirical data suggests that the evolutionary change of morphological characters occurs faster in small populations than in large structured populations (Mayr 1942, 1970). Many highly evolved species such as primates and carnivores have small population sizes compared with other organisms such as fish and invertebrates.

After my criticism on the shifting balance theory was published, Wade and Goodnight (1991) defended the theory by presenting experimental data that were obtained from artificial interdemic selection with the flour beetle *Tribolium castaneum*. However, this experiment was concerned with only one aspect of the shifting balance theory and did not really support the theory, which is based on an abstract and very complex concept. Detailed criticisms of Wade and Goodnight's experiment were presented by Coyne et al. (1997, 2000). These authors were also critical of many aspects of the shifting balance theory from the point of view of population genetics.

3.5. Accumulation of Nonfunctional and Deleterious Mutations

Y Chromosomes

Most of the mutations Thomas Morgan and his colleagues found in their *Drosophila* experiments were deleterious or homozygous lethal. It was also known that although the sex chromosome X contains many functional genes the Y chromosome lacks most of them. It was Muller (1914, 1932), who first provided the evolutionary explanation of the inertness of the Y chromosome. He argued that because the gene loci on the Y are always kept heterozygous, any lethal mutations occurring on the Y are sheltered by the wild-type alleles on the X and therefore they may be fixed in the population of Y chromosomes. However, the lethal mutations on the X may become homozygous in the homomorphic sex (females) and therefore they are eliminated. However, his argument was intuitive and based on no mathematical study. Fisher (1935) therefore examined the probability of accumulation of lethal genes on the Y chromosome considering an infinitely large population and concluded that the probability is virtually 0. For this reason, he rejected Muller's idea. The reason why Fisher obtained this conclusion was that in a large population the frequency of lethal heterozygotes can be rather high even in females and therefore the lethal mutations on the Y are eliminated when they happen to be on both the X and Y chromosomes (X^aY^a and in Table 3.1).

In reality, however, the population size is always finite and can be quite small. For this reason, Nei (1970) re-examined the probability of fixation of lethal genes in a finite population under the effect of genetic drift. He showed that the probability of fixation of lethal genes on the Y chromosome (P) is virtually 0 if the population size is infinitely large as Fisher showed but that when the effective population size (N) is less than 10,000, P is appreciably high, particularly for recessive-lethal genes. Here N refers to the size of a local population rather that of the entire species. However, since many lethal mutations occur recurrently with an appreciable frequency (Crow and Temin 1964), the same mutations may accumulate independently even in different

Table 3.1. Frequencies and fitnesses of genotypes in males and females

Males			
Genotype	X^AY^A	X^AY^a	X^aY^a
Frequency	$(1-x_f)(1-y)$	$(1-x_f)y + x_f(1-y)$	x_fy
Fitness	1	$1-h$	$1-s$
Females			
Genotype	X^AX^A	X^AX^a	X^aX^a
Frequency	$(1-x_f)(1-x_m)$	$(1-x_f)x_m + x_f(1-x_m)$	x_fx_m
Fitness	1	$1-h$	$1-s$

A: Wild-type allele. a: Mutant allele. x_f: Frequency of the mutant allele on the X chromosome in females. x_m: Frequency of the mutant allele on the X chromosome in males. h, s: Selection coefficients.

subpopulations. Therefore, the lethal or nonfunctional mutations may be fixed in the entire population of Y chromosomes. This is similar to the case of the Mexican cavefish *Astyanax mexicanus*, where albino mutations are fixed in most of the isolated cave populations (Avise and Selander 1972).

Charlesworth (1978) raised questions on the above theory on three grounds. First, he stated that the species population size is generally greater than 10,000 and therefore Nei's mathematical formulation cannot explain the inactivation of Y chromosomes. Second, Mukai et al.'s (1972) experimental study showed that the fitness of lethal heterozygotes could be reduced by more than 1–2 percent on average. If this result applies to natural populations, the *P* value would be lowered substantially. Third, Nei's (1970) model cannot explain the evolution of dosage compensation of the X-linked genes that is often observed in many different organisms.

In my view these criticisms are not really justifiable. His first comment has ignored the fact that the population dynamics of lethal mutations are different from neutral mutations because lethal mutations occur recurrently at the same loci. Note also that a new species is often generated from a subpopulation of a species and the effective size of a subpopulation is generally quite small, as mentioned above. Charlesworth's second comment is heavily dependent on Mukai's experimental results, but are these results general and really applicable to natural populations?

Certainly, Crow and Temin (1964) also showed that lethal mutations are not completely recessive on average but reduce the fitness of heterozygotes by 1–2%. However, Wallace (1966), Dobzhansky and Spassky (1968), and Maruyama and Crow (1975) reported that they have slightly beneficial effects in heterozygous condition. Nei (1970) showed that if lethal genes enhance the heterozygote fitness, the *P* value will increase substantially. Furthermore, as discussed in Chapter 2, the selection coefficient (s) of a genotype varies considerably with environmental condition and genetic background, and therefore s is expected to vary from generation to generation randomly. If we consider this factor, the s value of a Y-linked lethal mutation sheltered by the X may behave as though it were a neutral mutation even in relatively large populations. It should also be noted that when sexual selection occurs and there are dominant males to produce a much larger number of offspring than other males, the effective population size of Y chromosomes would be reduced substantially (Charlesworth and Charlesworth 2000). Therefore, the probability of fixation of Y-linked lethal mutations may be reasonably high. Charlesworth's third comment is somewhat complicated, and this will be discussed in Chapter 8.

Before leaving this subject, I should mention that the first step of evolution of the Y chromosome is the establishment of linkage of the sex-determining gene and other sex related genes with no recombination in the proto-Y chromosome. There are many species in which sex is determined by a single gene but no sex chromosomes have been established (Schartl 2004). As long as sex is determined by a single locus, there is no need to form a set of dimorphic X and Y chromosomes. However, if a system evolves in which the sex-determining gene and the genes controlling other sex-related characters (e.g. fertility genes) are linked and inherited together, the system would be advantageous compared with the system where these component genes are inherited independently. In fact, Nei (1969b) studied this problem mathematically and showed that the recombination value between the proto-X and proto-Y chromosomes would be reduced if all the functionally related genes are linked. The reduced recombination may be achieved by natural selection of modifier genes or by development of a genetic system which prohibits male recombination completely as in the case of *Drosophila*. The teleost fish medaka has a primitive sex determination system, and this system seems to represent an early stage of recombination reduction (Schartl 2004).

Previously we considered the evolution of inertness of the Y chromosome. However, the accumulation of nonfunctional genes on the Y chromosome occurs only after the establishment of reduced recombination between the X and Y chromosomes (Nei 1969b, Nei 1969a). Different sex related genes are probably assembled in the same proto-Y chromosome by translocation of genes.

Nonfunctional Mutations in Duplicate Genes

Haldane (1933) suggested that if a gene is duplicated by chromosomal doubling, gene duplication, etc., one of the two gene copies may become nonfunctional because of the fixation of deleterious mutations. Examining the pattern of genome size increase in prokaryotes and eukaryotes, Nei (1969a) postulated that the genomes of complex organisms contain a large number of nonfunctional genes (pseudogenes) because of accumulation of lethal mutations. In this case recessive lethal mutations are effectively harmless because there are extra copies of genes, and therefore they would accumulate as though they are neutral.

However, Fisher (1935) again had shown that lethal mutations would not be fixed in large populations. Nei and Roychoudhury (1973) studied the probability of fixation of recessive lethal mutations in one of two duplicate loci and showed that the probability is appreciably high if $N \leq 4000$. For this reason, they supported Nei's (1969b) conjecture about the accumulation of nonfunctional mutations. Recent genome sequence data indicate that a mammalian genome contains unexpectedly large numbers of duplicate genes and nonfunctional genes (pseudogenes). In fact, Torrents et al. (2003) reported that the human genome contains over 20 000 pseudogenes. This is partly because the number of duplicate gene copies is often very large and therefore pseudogenes can accumulate as though they are neutral. I shall return to this problem in Chapters 5 and 6.

Deleterious Mutations and Muller's Ratchet

Muller (1964) argued that, in the absence of backward mutation, deleterious mutations accumulate more rapidly in asexual organisms than in sexual organisms. The reason is that in asexual organisms all deleterious mutations in a genome are inherited together from the parent to the offspring and the number of deleterious mutations would never decrease in the absence of recombination. Deleterious mutations can also accumulate in sexual organisms, but the rate of accumulation is lower than that of asexual organisms because some of the mutations may be eliminated when recombination occurs between different genomes (individuals). This form of accumulation of deleterious mutations in asexual populations is called the ratchet effect or Muller's ratchet.

Muller's argument was later confirmed by computer simulations (e.g. Felsenstein 1974; Haigh 1978; Takahata 1982) and analytical study (Pamilo et al. 1987), and these studies provided a theoretical basis for the advantage of sexual reproduction over asexual reproduction. Felsenstein (1974) argued that the ratchet effect is caused by linkage disequilibrium generated by the interaction of selection and genetic drift or the Hill-Robertson effect (Hill and Robertson 1966). However, Pamilo et al. (1987) later showed that linkage disequilibrium is not essential and the ratchet effect is generated even in the absence of linkage disequilibrium. This occurs because the ratchet effect merely represents the difficulty of elimination of deleterious mutations in the absence of recombination. This process may generate linkage disequilibrium, but this is a consequence of mutation accumulation rather than the cause. Pamilo et al. (1987) also showed that the ratchet effect is unlikely to occur if the selection coefficient (s) against deleterious mutations is higher than about 0.01, and that the accumulation of deleterious mutations stops when the frequency of deleterious alleles reaches its equilibrium value, v/s, where v is the mutation rate. Furthermore, they showed that the ratchet effect occurs even in selfing organisms but the extent of the effect is smaller than that in asexual organisms.

A number of authors have argued that since slightly deleterious mutations are fixed with a higher probability in small asexual populations than in large outbreeding populations, a higher rate of amino acid substitution observed in sheltered chromosomes such as animal mitochondrial genes (Lynch 1996) and the genomes of the parasitic bacteria *Buchnera* species in aphids (e.g., Moran 1996; Clark et al. 1999) is caused by the ratchet effect. It is true that slightly deleterious mutations may be fixed in small populations even if they are eliminated in large populations. However, the continuous accumulation of deleterious mutations will deteriorate the gene function irrespective of population size (see Fig. 4.1B). Since the symbiosis of *Buchnera* and aphids apparently started about 200 million years ago (Moran et al. 1993) and the mitochondrial genome in eukaryotes appears to have originated by infection of an α-proteobacterial species about 1.5 billion years ago (Javaux et al. 2001), the functional genes remaining in these genomes must have been maintained by strong purifying selection and occasional backward mutation. This suggests that the ratchet effect is not the proper explanation of the enhanced rate of amino acid substitution which has been observed in *Buchnera* genomes. Note that the rate of amino acid substitution due to ratchet effect is always lower than the mutation rate (Pamilo et al. 1987).

Comparing the rates of amino acid substitution of a *Buchnera* species with those of their closely related species of free-living bacteria, Itoh et al. (2002) suggested that the higher rate in *Buchnera* is caused by either enhanced mutation rate or relaxation of selective constraints in small populations. The first hypothesis of enhanced mutation rate was supported by the lack of several DNA repair enzymes in the *Buchnera* genome, and the second hypothesis of relaxed selection is likely to apply because of the change of metabolism in symbiotic bacteria. This explanation is more reasonable than Muller's ratchet. Of course, these genomes have lost many original genes either because they were no longer needed under the condition of symbiosis or because they were transferred to the host nuclear genome (Martin et al. 2002).

3.6. Bottleneck Effects and Genetic Variability

According to Darwin (1859), a new species arises when a group of individuals become competitively stronger than other groups by mutation and selection and eventually drive pre-existing groups to extinction (Fig. 1.1A). He assumed that this process occurs gradually and slowly. Mayr (1942) proposed that a new species often arises when a small number of individuals move to a new isolated location. He conceived this idea by observing that the population living in the main territorial area (e.g. a continent) is usually homogeneous whereas populations living in the peripheral areas (e.g. small islands) are often conspicuously different. For example, Mayr (1942, 1963) noticed that the populations of the common paradise-kingfisher *Tanysiptera galatea* are morphologically uniform throughout New Guinea but the populations living in nearby small islands have distinct morphological characters. He then thought that this occurred because the island populations were derived from a small number of New Guinean birds.

This led him to propose that speciation often occurs through the bottleneck effect or the founder principle. According to this hypothesis, the genetic composition of an isolated population derived in this way would be quite different from that of the parental population and therefore a genetic revolution may occur. He stated (Mayr 1963, p. 538):

"The genetic changes that occur in an isolated population have manifold consequences. With the cohesion of the parental gene pool disrupted, conditions are favorable for a departure in new directions. The direction of such departure is largely unpredictable, since chance enters the picture at many separate levels, gametic, zygotic, developmental, behavioral, and environmental, even though the results of these chance events are continuously guided by natural selection. These results concern the genotype as well as every aspect of the phenotype. The passing of the population through a bottleneck permits rather drastic shifts which would be resisted by the homeostatic system of the well-integrated continental parent population."

Here Mayr proposed the hypothesis that the genetic composition of a population may change rapidly by going through bottlenecks and this change may generate innovative morphological characters. He called this event genetic revolution or the founder effect. This view has been accepted by many biologists, but some population geneticists are quite critical. One of the arguments raised against Mayr's theory is that the number of favorable mutations occurring in a population should be higher when population size is large than when population size is small and selection should be more effective in large populations than in small populations (e.g. Barton and Charlesworth 1984). If coadaptive evolution occurs in small populations, there is no reason to believe that it does not occur in large populations (Coyne and Orr 2004).

However, the controversies concerning Mayr's founder principle are caused largely by ambiguity of the definition of the principle. For example, what is the cohesion of the parental gene pool? What is the definition of genetic revolution? What is the expected amount of loss of genetic variability due to bottleneck effects? In Fig. 3.3, he sketched his idea of the effect of a bottleneck on the change of genetic variability intuitively, but it is not clear how he measured the extent of genetic variability and the evolutionary time. Because there was no standard way of measuring the extent of genetic variability at his time, the ordinate of Fig. 3.3 must represent an intuitive concept of variability, but theoretically it can be the variance of quantitative characters, the number of alleles per locus or average heterozygosity per locus. Therefore, it was not possible for different authors to argue or study Mayr's theory against the same background, and this has made the controversy even more confusing. This was particularly so when speciation based on bottlenecks was discussed.

For this reason, Nei et al. (1975) and Chakraborty and Nei (1977) attempted to clarify the nature of bottleneck effects on genetic variability and genetic differentiation of populations mathematically. They used the simple infinite-allele model of neutral mutations, but they could obtain the basic information about the bottleneck effect on genetic variation and population differentiation. Later Nei et al. (1983) studied the effect of population size on the rate of development of reproductive isolation, but the results of this study will be discussed in Chapter 7. In the following I would like to present a summary of these studies.

As we discussed in Chapter 2, the extent of genetic variability is generally measured by average heterozygosity or gene diversity, which is defined as the average of

$$h = 1 - \Sigma_i x_i^2 \quad (3.6)$$

over all genetic loci. Here x_i represents the frequency of the ith allele at a locus. If we consider the infinite-allele model of neutral mutations and assume that the population is in mutation-drift equilibrium, the expected heterozygosity for a locus (H) is given by $4Nv/(1+4Nv)$ in Equation (2.23), where N and v represent the effective population size and the mutation rate per locus per generation, respectively. If N declines suddenly because of a bottleneck, the expected heterozygosity decreases rapidly.

The most extreme case is that a new population starts from a single fertilized female which migrated from the parental population. Carson (1971) suggested that this situation often occurred when new species of Hawaiian *Drosophila* were formed by migration of individuals from one island to another. In this case the bottleneck size is 2 because the fertilized female carries the genome of the male partner as well, but the population size gradually increases

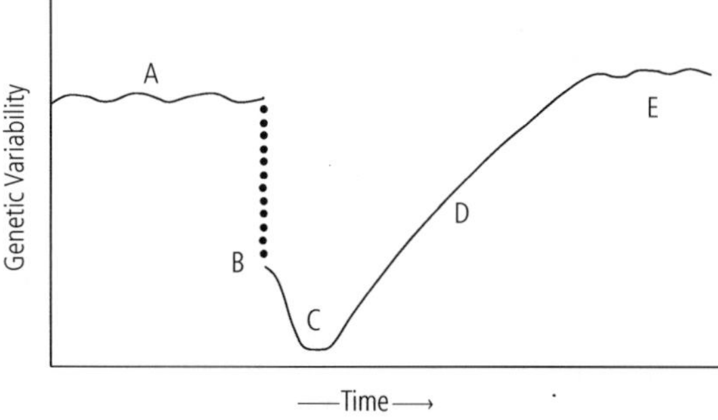

Fig. 3.3. Loss and gradual recovery of genetic variation in a founder population. The founder (*B*) has only a fraction of the genetic variation of the parental population (*A*) and further genes are lost during the ensuing genetic revolution (*B* to *C*). Variation is gradually recovered (*D*) if the population can find a niche until a new level (*E*) is reached. Redrawn from Mayr (1954).

to the size of the parental population. Nei et al. (1975) assumed that the population size increases following a sigmoid curve with a growth rate of r per generation. Considering allozyme data, they assumed that the average heterozygosity in the original population was $H_0 = 0.138$ and examined the changes of H when the population goes through a bottleneck.

The results obtained are presented in Fig. 3.4. They show that the average heterozygosity (H) declines rapidly by the bottleneck effect when the bottleneck size (N_b) and v are small, as predicted by Mayr (1963). However, the increase of H after the bottleneck is very slow, unlike Mayr's prediction. Note that the evolutionary time in this figure is measured by a logarithmic scale rather than by the absolute number of generations. This slow increase of H occurs because new variations have to be created by new mutations. Therefore, it takes the order of $1/v$ generations for H to return to the original level. This makes it rather difficult for a migrant population to undergo genetic revolution after a bottleneck. However, when r is high or when the bottleneck size is about 10, the decline of H is not so extensive, though it again takes a long time for H to recover to the original level. Note that in these cases population size N quickly recovers to the original size N_0 depending on the r value but it still takes a long time for H to recover because the increase of H depends on the mutation rate.

Another quantity that is important in this case is the genetic divergence of a bottlenecked population from the original population. The extent of this genetic divergence can be measured by Nei's (1972) genetic distance defined by

$$D = -\log_e \frac{J_{XY}}{\sqrt{J_X J_Y}}, \tag{3.7}$$

where J_X, J_Y, and J_{XY} are the averages of $\sum_i x_i^2$, $\sum_i y_i^2$, $\sum_i x_i y_i$ over all loci, respectively. Here x_i and y_i represent the frequency of the ith allele at a locus in populations X and Y, respectively. When populations X and Y diverged t generations ago and the population size of the two populations remains the same, D will be given by $2vt$ (Nei 1972). In other words, the genetic distance D is expected to increase linearly as t increases (Fig. 3.5 a).

However, if one of the two populations goes through a bottleneck, the genetic distance rapidly increases in the early generations because of genetic drift (Fig. 3.5 b, c). This early gain of genetic distance gradually diminishes if the population size of the bottlenecked population recovers to the original size N_1 and D eventually converges to the case of linear

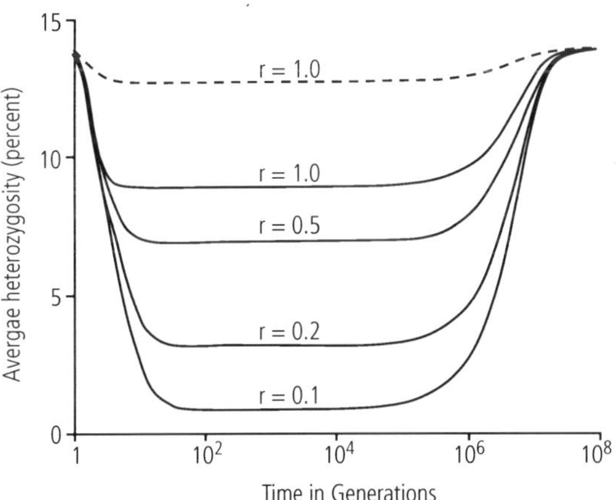

Fig. 3.4. Changes in average heterozygosity when a population goes through a bottleneck. The full lines refer to the case where the bottleneck size (N_b) is 2, while the broken lines refer to the case of N_b = 10. The population growth is assumed to be logistic, and r stands for the intrinsic rate of growth. The original and eventual heterozygosity is 0.138. Generations are given in the logarithmic scale, starting from the original population as generation 1. See the text for further information. From Nei et al. (1975).

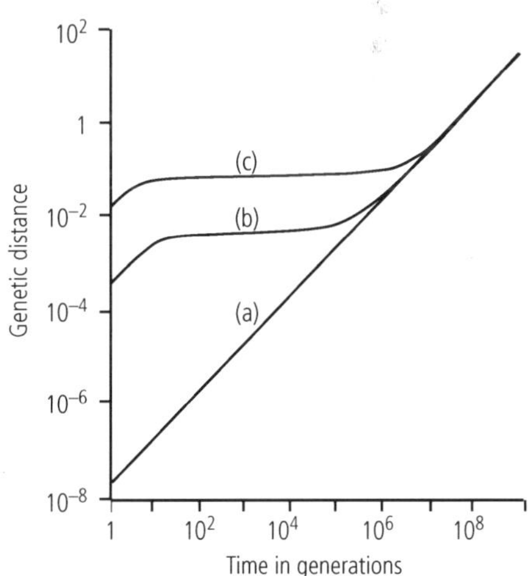

Fig. 3.5. Increase of genetic distance with time under the bottleneck effect. (A) stands for the case of no bottleneck effect, (B) the case of N_b = 100, and (C) the case of N_b = 10, where N_b is the bottleneck size. From Chakraborty and Nei (1977).

increase. However, if the size of the bottlenecked population does not increase substantially and remains small, the initial gain of D is retained and D would increase linearly with time (Chakraborty and Nei 1977). These results would not change very much even if some genes are under selection, because the bottleneck effect influences the genetic variability enormously when N_b is small.

In the above discussion, we considered the change of average heterozygosity or average genetic distance. If we consider the number of alleles per locus as the criterion of genetic variability, the story changes to some extent. The number of alleles per locus is more sensitive to the bottleneck effect, and the number declines more sharply but recovers to the original level more quickly than average heterozygosity (Nei et al. 1975).

Comparison of Fig. 3.3 and Fig. 3.4 suggests that Mayr's conjecture about the bottleneck effect on genetic variability is conceptually similar to the results obtained mathematically, though Fig. 3.3 is scaleless and therefore it can be interpreted in many different ways. Fig. 3.4 gives a much concrete idea about the bottleneck effect, but it was obtained under some simplified assumptions. Figs 3.3 and 3.4 support some of Mayr's conjectures. First, the bottleneck effect certainly produces various forms of genetic variants in the derived population, and therefore the new population may start a new direction of evolution. Second, as will be discussed later in Chapter 7, every population is a group of organisms which retain mating compatibility, and as this group of individuals is developed easily when population size is small, the bottleneck effect may enhance the probability of formation of new species. Genetic revolution is a vague terminology, but if we interpret it as the formation of a group of organisms which have a substantially different genetic constitution, his conjecture appears to be right. It is now important to prove it at the genetic level.

3.7. Beanbag Genetics and Evolution

Population genetics started with an extension of Mendelian genetics to study the correlations of close relatives with respect to quantitative characters, as mentioned in the preceding chapter. The evolutionary change of allele frequencies within populations was then studied by considering various forms of natural selection. These studies gave a theoretical basis for speculating long-term genetic changes of populations or species. In this case many simplifying assumptions had to be made for developing the mathematical formulation. Some of them are large population sizes, constant selection coefficients, no gene interaction, and no environmental effects. These assumptions certainly would not hold in real, natural populations.

Many biologists were unhappy with these unrealistic assumptions, and Mayr (1959, 1963) criticized population genetics emphasizing the lack of attention to the gene interaction in the process of development. He stated "*The Mendelian was apt to compare the genetic contents of a population to a bag full of colored beans. Mutation was the exchange of one kind of beans for another. This conceptualization has been referred to as 'beanbag genetics'... Work in population and developmental genetics has shown, however, that the thinking of beanbag genetics is in many ways quite misleading. To consider genes as independent units is meaningless from the physiological as well as the evolutionary viewpoint*" (*Mayr* 1963, p. 263.) He also stated that "*Fisher, Wright, and Haldane have worked out an impressive mathematical theory of genetical variation and evolutionary change. But what, precisely, has been the contribution of this mathematical school to evolutionary theory?*" (Mayr 1959.)

Reading these comments, Haldane (1964) quickly wrote a spirited rejoinder with a title of "A defense of beanbag genetics." Many population geneticists including myself were elated with this article. Several papers have then been written on this issue mostly by population geneticists (e.g. De Winter 1997; Borges 2008; Crow 2008), and most authors have approved Haldane's defense, emphasizing the importance of simplifying assumptions for extracting essential elements of the problem at issue. However, Mayr never changed his view (Provine 2004). As I have come to understand evolution more and more, I have gradually realized that Haldane evaded Mayr's main point that the evolutionary change of allele frequencies does not explain the evolution of important morphological or physiological characters such as birds' wings and the elephant's trunk. Particularly because population genetics deals with short-term evolutionary changes within species, it is not clear how useful it is for explaining long-term evolution spanning tens of millions of years. Actually, even for explaining relatively short evolutionary events such as geographical differentiation of human pigmentation and apparently rapid speciation of cichlid fishes in Lake Victoria of Africa, population genetics has been practically powerless. For these reasons, Mayr (1963) was skeptical of the utility of population genetics. Yet, population geneticists have endeavored to develop more and more sophisticated mathematical theories under various simplified biological assumptions.

In his defense Haldane indicated how useful population genetics is for estimating the selection coefficient for the dark form of the moth *Biston betularia* in British industrial districts and for estimating the mutation rate for the human hemophilia gene. These were certainly impressive achievements by population geneticists at that time. He also mentioned his estimates of the rate of evolutionary changes of morphological characters in fossilized horses, but this had nothing to do with beanbag genetics. Haldane gave several more interesting results of beanbag genetics, but he did not give proper answers to Mayr's questions. He stated "*Beanbag genetics do not explain the physiological interaction of genes and the interaction of genotype and environment. If they did so, they would not be a branch of biology. They would be a biology. The beanbag geneticist need not know how a particular gene determines resistance of wheat to a particular type of rust or hydrocephalus in mice...*" Here we can see the difference in scope of evolutionary biology between Haldane and Mayr. Haldane was apparently happy if population genetics could explain some aspects of evolution in terms of Mendelian genetics, and he was not interested in explaining the evolution of complex morphological or physiological characters. By contrast, Mayr was opposite and interested in the mechanism of evolution of individual phenotypic characters.

Mayr also criticized Haldane's cost of natural selection, stating that the upper limit of the number of gene substitutions computed by Haldane (one substitution every 300 generations) is too low compared with the actual number of substitutions observed in some natural populations (Mayr 1963, pp. 259–261). At that time there was no way of computing the exact number of gene substitutions in natural populations, but he presented several examples that suggested fast evolution. One of them was Zimmerman's (1960) investigation of the rapid speciation of five species of the pyraustid moth genus *Hedylepta* in Hawaii. These five species were restricted to the banana plants introduced by Polynesians to Hawaii only about 1000 years ago. Zimmerman argued that these five allopatric species evolved from a species living in a palm tree. Because the five species were morphologically quite different, Mayr believed that a large number of gene substitutions had occurred in each species. He also presented several cases in which new species apparently arose from small founder populations and argued that a large number of gene substitutions occurred in these populations.

One problem with Mayr's argument is that he was apparently unaware of Haldane's assumption of large population size for which the cost of natural selection was computed. As mentioned earlier, a large number of gene substitutions may occur in small populations generated through bottlenecks whether natural selection is involved or not, but this does not refute Haldane's theory, which was intended to measure the maximum number of gene substitutions due to natural selection in large constant populations. Nevertheless, Mayr's contention is important because the effective size of natural populations now appears to be considerably smaller than thought by many neo-Darwinians, particularly if we consider the bottleneck effects in speciation. We will discuss this issue again in Chapter 7.

As was discussed in Chapter 2, mathematical theories of population genetics have been very useful for predicting the evolutionary change of populations as long as the theories are based on solid biological assumptions. They are much better than intuitive speculations and have solved many controversies about evolution that existed in the pre-Mendelian era. However, it is true that the current population genetics theory is not powerful enough to explain long-term evolution of morphological characters. So-called naturalists like Mayr and Dobzhansky criticized this deficiency. However, these naturalists also could not explain it very well. To explain this long-term evolution of morphological characters, we had to wait for the establishment of the molecular basis of development and morphogenesis, as will be discussed lin Chapter 6. During the twentieth century, an impressive amount of mathematical theories has been developed, but a large portion of them, including some of mine, have remained unused because they are not realistic. In the future it would be important to develop theories that are based on solid biological principles. It should also be noted that we do not necessarily need mathematical models for understanding the evolution of complex morphological characters. In this case molecular study alone is often sufficient, as will be discussed in the following chapters.

3.8. Summary

In the neo-Darwinian era many theoreticians, particularly three founders of theoretical population genetics, proposed several evolutionary theories, some of which are still considered to be of paramount importance by some traditional evolutionists. However, these theories are generally abstract and are based on the assumption that the population size is effectively infinite. I have examined the adequacy and utility of the theories in the light of recent information on molecular biology.

In neo-Darwinism almost all kinds of genetic variability are assumed to exist in the population because of the mutations in the past, and therefore many genetic parameters such as mutation rate and recombination rate can be modified by natural selection. There are many theories about the modification of genetic parameters, but I have discussed two topics in this chapter: modification of dominance and modification of linkage intensity. The evolutionary change of the extent of dominance for a pair of alleles was initiated by R. A. Fisher. This idea was immediately criticized by Sewall Wright, but Fisher never gave up his view. Recent molecular data, however, suggest that the degree of dominance is determined by the pattern of gene expression at the molecular level and the effect of natural selection is negligibly small. The idea that the recombination value between two genetic loci may be changed by natural selection was also initially proposed by Fisher. In this case there are substantial amounts of molecular data showing that recombination values are under genetic control, and therefore he was right.

Fisher's most well-known theory is his fundamental theorem of natural selection regarding the increase of population fitness. This theorem has been shown to be mathematically correct under the assumptions made. However, its biological significance is unclear and quite misleading. His formulation was based on the assumption that (1) the population size is infinitely large, (2) the allele frequency changes occur only by natural selection, and (3) the relative fitnesses of genotypes remain the same for all the generations. Because many of these assumptions do not hold in nature, the biological meaning of the theorem is unclear. Furthermore, since it is an abstract theory concerning population fitness, it cannot explain the evolution of specific phenotypic characters in which most biologists are interested. Similarly, Sewall Wright's shifting balance theory is an abstract concept, and its utility for understanding the evolution of specific characters is unclear. By contrast, J. B. S. Haldane's theory of cost of natural selection is more concrete, and it is useful for determining a crude upper limit for the number of allele substitutions by natural selection in large populations.

At the time of neo-Darwinism the mathematical approach dominated the field, and some empirical evolutionists were critical of the mathematical studies which did not answer the questions with which they were concerned. In particular, Ernst Mayr labeled population genetics as beanbag genetics and criticized its inability of explaining various specific evolutionary problems such as morphological evolution and speciation. At the same time, he developed several important evolutionary concepts such as the founder principle and the modes of speciation. The founder principle appears to be important in explaining the frequent generation of new species in island populations. This also emphasizes that speciation occurs more frequently when population size is small than when it is large.

The neo-Darwinian theory of evolution was developed primarily by the study of allele frequency changes within species, and therefore it had little power for understanding long-term evolution of organisms beyond the species level.

CHAPTER 4

Molecular Evolution

The discovery of DNA (or RNA) as the genetic material and the subsequent development of molecular biology have revolutionized the study of evolution more significantly than the rediscovery of Mendelian inheritance. The molecular study of evolution can be divided into two types. The first type is to study the evolutionary change of macromolecules such as proteins, DNAs, and RNAs, whereas the second is to understand the molecular basis of phenotypic evolution. Obviously, the second type of study is more complex than the first, and it is still in its infancy. In this chapter I would like to discuss the first type of study, leaving the second type to later chapters.

4.1. Early Studies of Molecular Evolution

Before the molecular study of evolution was introduced in the late 1950s, most studies of the mechanism of evolution were conducted using the Mendelian approach. Because this approach depended on crossing experiments to identify homologous genes, the studies were confined to within-species genetic changes as mentioned earlier. Partly for this reason, evolutionary study was primarily concerned with the change of allelic frequencies within species.

In the molecular approach, however, the evolutionary change of genes can be studied between any pair of species as long as the homologous genes can be identified. This removal of the species barrier introduced new knowledge about long-term evolution of genes. Furthermore, since proteins are the direct products of transcription and translation of genes, we can study the evolutionary change of genes by examining the amino acid sequences of proteins. For this reason, early molecular evolutionists compared the amino acid sequences of hemoglobins, cytochrome c, fibropeptides, etc. from a wide variety of organisms.

In the 100th anniversary year of the publication of Darwin's book *Origin of Species*, Anfinsen (1959) published the first comprehensive book on molecular evolution. In this book he compared the amino acid sequences of proteins such as ribonuclease, insulin, cytochrome c, and hemoglobins from diverse species of animals. In this comparison, he noticed that the rate of amino acid substitution varies considerably among different regions of the proteins and that the substitution rate is lower in important parts of the proteins such as the active center than in functionally less important parts. His interpretation of this finding was that the mutations occurring in important parts of proteins often lead to lethal or deleterious effects while those occurring in other parts may not be deleterious and, therefore, may be fixed in the population with a higher probability. He assumed that the fixation of a new mutation occurs primarily by natural selection. As a protein chemist, he was apparently unaware of the possibility that a mutation can be fixed by chance alone. Ingram (1961) later studied the amino acid differences between myoglobin and four types of hemoglobins (hemoglobin chains α, β, δ, and ε) from humans and again stated that the fixation of a new mutation occurs by selective advantage of the mutation.

The idea that the nucleotide frequencies can be changed by mutation and random genetic drift was first proposed by Freese (1962) and Sueoka (1962). They noticed that the frequency of guanine (G) and cytosine (C) (GC content) or the frequency of adenine (A) and thymine (T) (AT content) in the DNA sequence varies extensively among different species

Mutation-Driven Evolution. First Edition. Masatoshi Nei.

of bacteria. They then argued that the effect of changes of GC content on phenotypic characters is so small that the evolutionary change of GC content can be treated by considering mutation and genetic drift. Under this assumption, the equilibrium frequency of GC content in a DNA sequence is given by Equation (2.3), i.e. $u/(u+v)$, where u is the mutation rate from the AT pair to the GC pair and v is the mutation rate from the GC pair to the AT pair. This indicates that the large variation of GC content in bacterial species can be explained by the variation of u and v. In fact, Cox and Yanofsky (1967) showed the GC content in *Escherichia coli* can be increased dramatically by introducing a mutator gene in the genome.

Freese (1962, p. 85) also noticed that only a small proportion of amino acid changes affect protein function between different species and the remaining amino acids can be altered without any functional change. This implies that most amino acid substitutions occur in a neutral fashion, and this interpretation was opposite to the view of Anfinsen (1959) and Ingram (1961). However, Freese's observation was based on studies of a few proteins from a small number of species, and therefore little attention was given to his view.

More detailed studies on this issue were soon conducted by many investigators, and the results were presented in the symposium volume of Bryson and Vogel (1965), *Evolving genes and proteins*. These studies revealed several interesting properties of molecular evolution. First, the number of amino acid substitutions between two species was shown to increase in proportion to the time since divergence of the species (Zuckerkandl and Pauling 1962, 1965; Margoliash 1963; Doolittle and Blombaeck 1964). Second, amino acid substitutions occurred less frequently in the functionally important proteins or important regions of proteins than in less important proteins or protein regions (Margoliash and Smith 1965; Zuckerkandl and Pauling 1965). Thus, the rate of amino acid substitution was much higher in less important fibrinopeptides than in essential proteins such as hemoglobins and cytochrome c, and the active sites of hemoglobins and cytochrome c showed a much lower rate of evolution than other regions of the proteins. A simple interpretation of these observations was to assume that amino acid substitutions in the nonconserved regions of proteins are nearly neutral or slightly positively selected and that the amino acids in functionally important sites do not change easily because of the functional importance of amino acid sequences involved (Freese and Yoshida 1965; Margoliash and Smith 1965; Zuckerkandl and Pauling 1965, p. 148–149). Third, the evolutionary change of GC content can be explained approximately by the birth-and-death model of stochastic process. These observations suggested the importance of random factors in molecular evolution.

4.2. Neutral Evolution at the Protein Level

Cost of Natural Selection and Neutral Theory

At this stage, Kimura (1968b) and King and Jukes (1969) formally proposed the neutral theory of molecular evolution. Kimura first estimated the average rate of nucleotide substitution per mammalian genome per year from data on amino acid substitutions in hemoglobins and a few other proteins and showed that the rate is about one substitution every two years. He then noted that this rate is very high compared with Haldane's (1957) estimate of the upper limit of the rate of gene substitution by natural selection (one substitution every 300 generations or every 1200 years if the average generation time is 4 years in mammals). Haldane's estimate was based on the cost of natural selection that is tolerable for the average fertility of mammalian organisms (Chapter 3). If we accept Haldane's estimate, such a high rate of nucleotide substitution (one substitution every two years) cannot occur by natural selection alone, but if we assume that most substitutions are neutral or nearly neutral and are fixed by random genetic drift, any number of substitutions may occur. For this reason, Kimura concluded that most nucleotide substitutions must be neutral or nearly neutral.

This paper was immediately criticized by Maynard Smith (1968) and Sved (1968). These authors indicated that if natural selection occurs in a form of truncation selection as in the case of artificial selection of quantitative characters discussed in Chapter 2 (see Fig. 2.6) the cost of natural selection should be much lower than Haldane's estimate. They therefore argued that the mammalian rate of nucleotide substitution can be explained by natural selection. Maynard Smith and Sved's argument, however, does not seem to be justifiable because truncation selection almost never occurs in nature. For truncation selection to occur, the number of advantageous genes in each individual must be identifiable at the time of occurrence of selection so that natural selection allows the best group of individuals to reproduce for the next generation. In practice, natural selection does not occur in this way, as discussed in Chapter 3. It operates in various stages of development for different characters. Therefore, selection must be more or less independent for different genetic loci as a first approximation (Nei 1971). This justifies Haldane's theory of cost of natural selection and supports Kimura's argument for the neutral theory of molecular evolution. Ewens (2004) recently criticized Haldane's theory, but he seems to have misunderstandings about the theory (Chapter 3).

However, Kimura's paper had some problems. First, in the computation of the cost of natural selection, he assumed that all nucleotides in the genome were subject to natural selection. In practice, the unit of selection should be a gene or an amino acid, because noncoding regions of DNA are largely irrelevant to the evolution of proteins and organisms except those in regulatory regions of the genome. As mentioned in Chapter 2, Muller (1967) had examined the extent of mutation load tolerable for mammalian species and estimated that the number of genes in the human genome is probably no more than 30 000. This number is much smaller than the number of nucleotides (3.3×10^9) which Kimura used for his computation. If we consider a gene as the unit of selection, as Haldane did, and assume that the mammalian genome contains 30 000 genes, the average rate of gene substitution now becomes one substitution every 220 000 ($= 2/(3 \times 10^4/3.3 \times 10^9)$) years. (Here one nucleotide substitution every two years was assumed.) This is far less than Haldane's upper limit (one substitution every 1200 years). In contrast, if we consider an amino acid as the unit of selection and each gene encodes 450 amino acid sites on average (Zhang 2000), the average rate of amino acid substitution will be one substitution every 636 ($=2.86 \times 10^5/450$) years. This rate is about two times higher than Haldane's upper limit.

The above computation was done under the assumption of an infinite population size, and it is known that in finite populations the cost of natural selection is reduced considerably because of genetic drift (Kimura and Maruyama 1969). This will further weaken Kimura's original argument. However, at each locus deleterious mutations occur every generation and are expected to impose another kind of genetic load, i.e. mutation load. Therefore, if we consider both the cost of natural selection and the mutation load, Kimura's computation may not be so outrageous. Furthermore, what is important is the fact that Kimura initiated the study of population dynamics of neutral mutations at this stage and that he later became the strongest defender of the neutral theory and provided much evidence for it.

Definition of Neutral Mutations

The second deficiency of Kimura's 1968 paper is concerned with the overly strict definition of selective neutrality. According to him, mutations with $|2Ns| < 1$ or $|s| < 1/(2N)$ are defined as neutral, where N is the effective population size and s is the selective advantage of the mutant heterozygotes (A_1A_2) over the homozygote for the wild-type alleles (A_1A_1). Here the fitnesses of the wild type homozygote (A_1A_1), the mutant heterozygote (A_1A_2), and the mutant homozygote (A_2A_2) are given by 1, $1 + s$, and $1 + 2s$, respectively. Kimura never explained how he derived his definition of neutrality, though the definition has been used by many investigators. (It is possible that he used a method similar to Fisher's; see Section 2.4.) Actually, this definition does not have much biological meaning. For example, if a slightly advantageous mutation with $s = 0.0001$ occurs in a population of $N = 10^6$, this s is much greater than $1/(2N) = 5 \times 10^{-7}$. Therefore, this mutation will not be called "neutral." In this case, the fitness of mutant homozygotes will be higher than that of wild-type homozygotes only by 0.0002. Is this small magnitude of fitness difference biologically significant? In reality, it will have little effect on the survival of mutant homozygotes or heterozygotes because this magnitude of fitness difference is easily swamped by the well-known random variation in the number of offspring. It is known that the distribution of the number of children (progeny size) of a

Table 4.1. Fitness difference between the original (A_1A_1) and the mutant (A_2A_2) populations and the definition of a neutral mutation

	Original population		Population fixed with mutant allele A_2
Genotype	A_1A_1		A_2A_2
Average fitness (w)	1		$1 + 2s$
Standard error of w	$1/\sqrt{N}$		$1/\sqrt{N}$
Average fitness difference		$2s$	
SE of the difference ($2s$)		$\sqrt{2/N}$	
Normal deviate (Z) of the difference		$s\sqrt{2/N}$	
Definition of neutrality			
For $Z = 2$		$s < \sqrt{2/N}$	
For $Z = 1$		$s < 1/\sqrt{2N}$	

N: Effective population size. The mean and variance of progeny size per individual are both 1 under the assumption that the progeny size follows the Poisson distribution and the population size is stable. The standard error (SE) of the average fitness of a population is $1/\sqrt{N}$, and the SE of fitness difference $2s$ is $(1/N+1/N)^{1/2} = \sqrt{2/N}$ (see Nei 2005).

female parent often follows the Poisson distribution (Imaizumi et al. 1970). In this case the mean and variance of progeny size per individual both become 1 in a stable population. (*Drosophila* data have shown that the variance is several times higher than the mean, Crow and Morton 1955).

Considering these problems, Nei (2005) presented a more meaningful definition of neutrality. His approach is to compare the average fitness (1) of the original population with genotype A_1A_1 and that $(1 + 2s)$ of the population fixed with the mutant genotype A_2A_2. Because the difference between the average fitnesses of two populations is $2s$ and the standard error of this difference can be shown to be $\sqrt{2/N}$ when the number of offspring of an individual follows the Poisson distribution with mean 1, the normal deviate (Z) of the difference in mean fitness between the two populations is given by $s\sqrt{2N}$ (Table 4.1). Therefore, to make the average fitness of the new population significantly higher than that of the original population at the 5 percent level, Z $(= s\sqrt{2N})$ must be equal to or greater than 2. That is, s must be equal to or greater than $\sqrt{2/N}$. For example, if $N = 10^6$, s must be greater than $\sqrt{2/N}$ $=0.0014$. This is much larger than the value (5×10^{-7}) given by Kimura's definition.

In other words, in my definition mutations with an s value much larger than $1/2N$ could behave effectively as neutral alleles. This is true even if we use $Z = 1$ as the cut-off point (30 percent significance level). In this case mutant alleles with $s < 1/\sqrt{2N}$ will be regarded as neutral. Thus, if $N = 10^6$, we have $s < 0.0007$. Therefore, even with a less stringent definition, many mutations are expected to behave effectively as neutral alleles. It should be noted that in the real world even the above definition of neutrality is not sufficient because the selection coefficient (s) would never be constant for different generations. In Chapter 2, we have seen that if s fluctuates from generation to generation because of environmental changes, an allele even with $s = 0.001$ would behave effectively as a neutral allele if the variance (V_s) of s is of the order of 0.01 and the effective population size is about 10^4 or greater (Fig. 2.6). It is also known that the level of gene expression varies from cell to cell in unicellular organisms and this expression noise introduces another source of fitness variation (Wang and Zhang 2011). These observations indicate that the proportion of effectively neutral mutations is much greater than previously thought.

However, the ideal definition of neutrality is to consider protein function and determine whether a new mutation affects the function appreciably. Figure 4.1A shows a schematic representation of the fitness effects of a series of mutations. Many

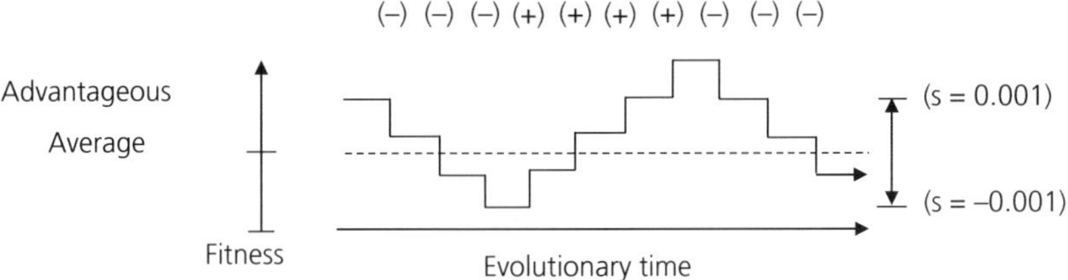

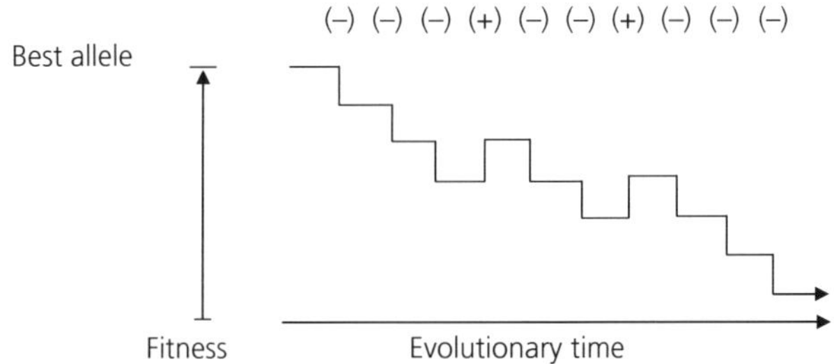

Fig. 4.1. Evolutionary processes of average neutral mutations (A) and slightly deleterious mutations (B). The *s* values for the neutral zone are preliminary. + and – represent positive and negative mutations, respectively. From Nei (2005).

mutations may occur over evolutionary time, but as long as they do not change the protein function significantly, they can be regarded as neutral mutations. In fact, molecular biologists have had a relaxed attitude towards neutral mutations, and when a mutation does not change the gene function appreciably, they call it more or less neutral (Freese 1962; Wilson et al. 1977; Perutz 1983). According to this definition, s or $|s|$ for neutral mutations is likely to be at least as great as 0.001 in mammalian organisms, though it is not easy to relate the functional difference to the fitness difference (see Chapter 6). If we accept this definition, we do not have to worry about minor allelic differences in fitness and can avoid unnecessary controversies concerning the effects of selection.

Nevertheless, various statistical methods developed by Kimura and others are still useful for testing the null hypothesis of neutral mutations. If the neutral hypothesis is not rejected by the methods, the newly defined neutral theory discussed above certainly cannot be rejected. Furthermore, even if the strict neutral theory is rejected, the new neutral theory may not be rejected. This is somewhat uncomfortable, but the same thing is true with Kimura's original theory, because his theory allows for the existence of deleterious mutations or a small proportion of advantageous mutations. At any rate, the biological definition of neutral alleles is more appropriate in the study of evolution than the mathematical definition, and the neutrality of mutations should eventually be studied experimentally.

King and Jukes's View

King and Jukes (1969) took a different route to reach the idea of neutral evolution. They examined extensive amounts of molecular data on protein evolution and polymorphism and proposed that a large proportion of amino acid substitutions in proteins occur by random fixation of neutral or nearly neutral mutations and that mutation is the primary force of evolution. This idea was against the then popular neo-Darwinian view in which a high rate of evolution is achieved only by natural selection (Simpson 1964; Mayr 1965). According to King and Jukes, proteins requiring rigid functional and structural constraints (e.g. histone and cytochrome c) are expected to be subject to stronger purifying selection than proteins requiring weak functional constraints (e.g. fibrinopeptides), and therefore the rate of amino acid substitution would be lower in the former than in the latter. Extending the results obtained by Zuckerkandl and Pauling (1965) and Margoliash (1963), they also emphasized that the functionally important parts of proteins (e.g. the active center of cytochrome c) has a lower substitution rate than the less important parts. Later, Dickerson (1971) confirmed this finding by using an even larger data set. He also noted that the cytochrome c proteins from different mammalian species were fully interchangeable when their function was tested in vitro with respect to the interaction with intact mitochondrial cytochrome oxidase (COX) (Jacobs and Sanadi 1960). For many biologists, these data were more convincing in supporting neutral theory than Kimura's computation of the cost of natural selection.

King and Jukes (1969) used a simple logic to visualize neutral molecular evolution. For example, they showed that the rate of amino acid substitution per year per locus (r) is given by

$$r = v \qquad (4.1)$$

because the fixation probability of a single neutral mutation in a population of effective size N is $1/(2N)$ and the total mutations generated at a locus in a population of N is $2Nv$, where v is the mutation rate per locus per year. Therefore, the rate of amino acid substitution per year is given by $r = 2Nv \times (1/2N) = v$. This equation is often credited to Kimura (1968b), but this is incorrect because what he showed is that the probability of fixation (U) of a mutant gene becomes equal to the initial frequency (p) when the selection coefficient becomes 0. Historically, this issue even goes back to Wright (1938b), who showed that the gene frequency flux (gene substitution) under the irreversible mutation model is equal to the mutation rate (v). However, I do not think Wright had any idea about molecular evolution at that time.

King and Jukes (1969) were also the first to show that the rate of amino acid substitution by advantageous selection is

$$r = 4Nsv \qquad (4.2)$$

where s is the selection coefficient for a semidominant gene in the heterozygous condition.

This equation was derived by noting that the probability of fixation of an advantageous mutation is approximately $2s$, as mentioned in Chapter 2, and the total number of mutations generated is again $2Nv$. This equation is also often credited to Kimura and Ohta (1971) erroneously.

However, the biological interpretation of this equation is not as straightforward as that of Equation (4.1). The reason is that in Equation (4.2) every new mutation is assumed to be more advantageous than pre-existing alleles by the amount of s. This means that the allelic fitness increases continuously in the evolutionary process, but is it biologically reasonable? At this moment, there is no evidence for such an assertion. However, when s is very large, the extent of polymorphism in the population is expected to be very low. Therefore, the equation is approximately applicable. Note that this equation is based on the infinite-site model.

Definition of Neutral Theory

As mentioned above, there are a number of difficulties in the definition of neutral mutations. The mathematical definition alone is not sufficient, because we are dealing with a biological problem here. Therefore, the neutral theory of molecular evolution should be defined in a more flexible way. According to Kimura (1983, p. 34): "*the neutral theory holds that at the molecular level most evolutionary change and most of the variability within species are not caused by Darwinian selection but by random genetic drift of mutant alleles that are selectively neutral or nearly neutral. The essential part of the neutral theory is not so much that molecular mutants are selectively neutral in the strict sense as that their fate is largely determined by random drift. In other words, the selection intensity involved in the process is so weak that mutation pressure and random drift prevail in molecular evolution.*"

Unlike some recent misunderstandings (e.g. Dawkins 1987; Ohta 1992), Kimura's theory was the neutral or nearly neutral theory of evolution from the beginning (Kimura 1968a). At the present time, we of course know some additional factors that cause random evolution, as will be discussed in the following chapters.

It should also be noted that the neutral theory allows for the existence of a large amount of deleterious mutations which will be eliminated by natural selection as well as a small proportion of advantageous mutations. In this sense Kimura's mathematical definition of $|2Ns| \leq 1$ should not be taken seriously. For many purposes Nei's relaxed definition ($|s| < 1/\sqrt{2/N}$) should be sufficient. In mammalian organisms even a more crude definition of $|s| < 0.001$ may be used. This will avoid many trivial controversies over neutral theory. Of course, even this definition should be used with caution, because it would not be appropriate when s varies from generation to generation.

4.3. Molecular Clocks

One of the interesting properties discovered by early molecular evolutionists is the approximate constancy of the rate of amino acid substitution in such proteins as hemoglobins, cytochrome c, and fibrinopeptides. This discovery of "molecular clocks" was immediately criticized by Simpson (1964) and Mayr (1965), who were authorities on morphological evolution. For these traditional evolutionists, it was apparently unthinkable that any character evolves at a constant rate over a long period of time. However, the approximate constancy of evolutionary rate was later observed in many other proteins, though the molecular clock was not always very accurate (Dayhoff 1972; Langley and Fitch 1974).

One solution to this puzzling observation was to assume that most amino acid substitutions are neutral and do not change protein functions appreciably. In fact, Kimura (1968b) and King and Jukes (1969) showed that if neutral mutations occur and are fixed by random genetic drift the rate of amino acid substitution can be constant. Kimura (1969) then took this rate constancy as support for the neutral theory of molecular evolution.

Evolutionary Rate under Purifying Selection

However, there were a few problems with this proposal. First, although the rate of molecular evolution was roughly constant for a particular protein, the rate varied considerably among different proteins. This puzzle was solved when Dickerson (1971) showed that the variation is apparently caused by the differences in functional constraints of protein molecules. For example, histones require a rigid structure for their function and maintain similar structure in both animals and plants. For this reason, there are only a few amino acid differences between animal and plant histones. By contrast, fibrinopeptides have few functional constraints and evolve very fast, because they are cleaved out from fibrinogen in the process of production of fibrin, a protein involved in blood clotting, and have virtually no function. In these cases if we assume that functionally important amino acid sites of a protein remain unchanged in the evolutionary process but functionally unimportant sites change with a neutral rate, the rate of amino acid substitution (r) for the entire protein may be expressed by

$$r = fv \tag{4.3}$$

where f is the proportion of functionally unimportant amino acid sites and v is the mutation rate (Kimura 1983).

In reality, the distinction between functionally important and unimportant sites may be difficult, but the above formula has a symbolic meaning and makes it easy to understand one of the important factors affecting the rate of amino acid substitution.

Evolutionary Rate and Generation Time

The second problem with the molecular clock concept was the fact that the rate of amino acid substitution was apparently constant per year rather than per generation. Because classical genetics had established that the mutation rate was constant per generation in *Drosophila*, humans, and maize, it was a problem how to reconcile these two sets of observations. Ohta (1974) proposed that this dilemma can be resolved by assuming that most mutations are slightly deleterious, that these mutations can be fixed in small populations more easily by genetic drift than in large populations, and that large organisms such as mammals generally have a smaller population size than small organisms such as *Drosophila*. In other words, a larger proportion of mutations may behave just like neutral alleles in large organisms than in small organisms. Since large organisms tend to have a longer generation time (smaller number of generations per unit time) than small organisms, the rate of amino acid substitution per year may be similar for both large and small organisms if the mutation rate per generation is constant. However, this argument is quite unlikely to apply to all groups of eukaryotic and prokaryotic organisms, for which rough molecular clocks apply almost universally (Hedges and Kumar 2009). Note also that if this argument is correct, the genomes of large organisms are expected to deteriorate gradually because of the accumulation of deleterious mutations, but in reality these organisms are more advanced in terms of organismal complexity than small organisms.

Actually, a much simpler solution to this problem was to assume that the rate of non-deleterious mutations is roughly constant per calendar year whereas the rate of deleterious mutations is approximately constant per generation (Nei 1975). In classical genetics the mutation rate was determined almost always by using highly deleterious mutations many of which were homozygous lethal (Muller 1950) and these mutations appear to occur at the time of meiosis (Muller 1959; Magni 1969). It is therefore understandable that classical Mendelian geneticists were led to believe that the mutation rate is constant per generation. However, some bacterial geneticists who studied phage resistance had reached the conclusion that the mutation rate is proportional to chronological time (Novick and Szilard 1950). Because phage-resistance mutations are non-deleterious, this observation suggests that non-deleterious mutations occur roughly at a constant rate per year. For these reasons, Nei (1975) argued that the constancy of amino acid substitution per year can be explained by the neutral theory if most amino acid substitutions are more or less neutral.

However, whether the mutation rate is constant per generation or per year has been controversial for a long time (Laird et al. 1969; Wilson et al. 1977; Wu and Li 1985; Easteal et al. 1995). Kohne (1970) argued that the evolutionary rate of hominoid genes should be lower than that of monkey genes because hominoids have a longer generation time than the latter. The logic behind this argument was that if the generation time is long the number of cell divisions per year in the germline cells is small and therefore the evolutionary rate should be lower if the mutation rate is proportional to the number of cell divisions. This is called the generation time hypothesis, and Li et al. (1987) presented some data supporting this view (see also Tsantes and Steiper 2009). However, recent studies have shown that the average rate of nucleotide substitution for the entire genomic sequence has been virtually the same for the hominoid and monkey lineages (Gibbs et al. 2007). This suggests that the generation time hypothesis is not important for hominoid species. Using the published genomic sequences, Nei et al. (2010) reexamined this problem for vertebrate model organisms. Figure 4.2 shows the numbers of amino acid substitutions per residue (d_A) and synonymous (d_S) and nonsynonymous (d_N) nucleotide substitutions per site for humans and other vertebrate species in relation to divergence time. The d_A, d_S, and d_N values are the averages for 4198 nuclear genes which appear to be orthologous among the 10 species used. The evolutionary times in this figure refer to the estimates obtained from the fossil record (Benton et al. 2009). In this case the numbers of amino acid and nucleotide substitutions increase almost linearly with chronological time.

Functional Constraints of Proteins

However, if we look at individual genes separately, the molecular clock does not necessarily work, and in some cases the rate of amino acid or nucleotide substitutions varies considerably among different evolutionary lineages. One of the well known examples is guinea pig insulin. Mammalian insulins are generally composed of 51 amino acids and highly conserved. Exceptions are those from hystricomorphic rodents such as guinea pigs and chinchillas, and the insulins from these species have been shown to evolve more than 10 times faster than other mammalian insulins (King and Jukes 1969; Opazo et al. 2005). Initially, this high rate of evolution was thought to be due to positive selection (King and Jukes 1969), but later Kimura (1983) proposed that this high rate of evolution is due to relaxation of selection caused by the absence of the zinc ion in the insulin molecules. In fact, several studies have shown that the biological activity

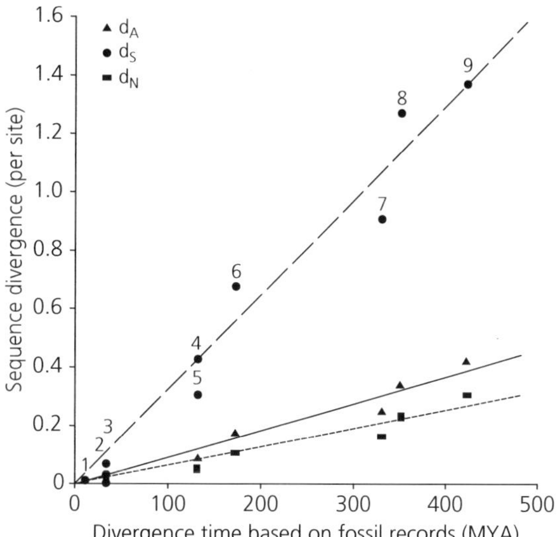

Fig. 4.2. Linear relationships of the number of amino acid substitutions per residue (d_A) and the numbers of synonymous (d_S) and nonsynonymous (d_N) nucleotide substitutions per site (sequence divergence) with divergence times based on the fossil record. Each point represents the average sequence divergence of 4198 nuclear genes with ≥100 codons from 10 vertebrate species (human vs. (1) chimpanzee, (2) orangutan, (3) macaque, (4) mouse, (5) cow, (6) opossum, (7) chicken, (8) western clawed frog, (9) zebrafish). The d_A distance was computed by the Poisson correction method, whereas d_S and d_N were obtained by the modified Nei-Gojobori method (Zhang et al. 1998) with the transition/transversion ratio of 2. The divergence times used are from Benton et al. (2009). MYA: million years ago. From Nei et al. (2010).

(balancing of blood glucose level) of insulins from these groups of species is only 3~30 percent of that of other mammalian species (Horuk et al. 1979; Bajaj et al. 1986).

Many examples of relaxation of selective constraints are now known. The relaxation of selective constraints is usually associated with the degeneration of phenotypic characters, which will be discussed in Chapter 8. Under certain conditions, functional constraints may be enhanced. A well known example is the evolutionary rates of histone H4 protein. In animals and plants, this protein is known to evolve very slowly, but it evolves reasonably fast in protists (Katz et al. 2004). Therefore, it appears that the evolutionary rate of this protein decreased when animal and plant histones evolved.

Variation in Mutation Rate

However, there are other factors that cause the variation of evolutionary rate of proteins. One of them is the change in mutation rates. Because synonymous substitutions are generally believed to be caused by neutral mutation, changes in mutation rates are often studied by examining the rate of synonymous substitution. The average synonymous substitution rate of nuclear genes appears to be nearly the same for animals and plants (Wolfe et al. 1987; Mower et al. 2007). However, the synonymous rate of animal mitochondrial genes is about 10 times higher than that of nuclear genes, whereas plant mitochondrial genes evolve about 10 times slower than nuclear genes (Wolfe et al. 1987). The fast evolutionary rate of animal mitochondrial genes was first thought to be due to Muller's ratchet effect, which would enhance the fixation of slightly deleterious mutations in asexual haploid populations because of the lack of recombination (Lynch 1996). However, this explanation is unsatisfactory because plant mitochondrial genes, which have the same mode of inheritance as that of animal mitochondrial genes, evolve very slowly as mentioned above. It now seems that the fast evolution of animal mitochondrial genes is due to a higher mutation rate partly because of the absence of the DNA repair gene *RecA*, which is present in plant mitochondrial genomes (Lin et al. 2006).

In addition, the evolutionary rate of plant mitochondrial genes is known to vary enormously with gene or evolutionary lineage (Mower et al. 2007). For example, the mitochondrial genes *Atp1* and *Cox1* evolve hundreds of times faster in the genera *Pelargonium*, *Plantago*, and *Silene* of seed plants than those in most other genera. Interestingly, not all genes in the same species evolve at the same rate; some genes in these genera evolve as slowly as the genes in other species. Furthermore, phylogenetic analysis showed that this enhancement of evolutionary rate has occurred only during the last 5 million years in the case of genus *Silene*. Figure 4.3 shows the trees of mitochondrial genes (7 concatenated loci) and chloroplast genes (5 concatenated loci) for 5 species of the genus *Silene*. While chloroplast genes evolved more or less in a linear fashion, the mitochondrial genes in *S. conica* and *S. noctiflora* evolved much faster than the genes in other *Silene* species. The reason why the rate varies so much among different species of plants is unclear, but for some reason the mutation rate apparently has increased in the *S. conica* and *S. noctiflora* lineages. It appears that the mutation rate of plant mitochondrial genes varies according to the genetic background and the environmental condition. Recently, Sloan et al. (2010) showed that the acceleration of nucleotide substitution is closely related to the loss of RNA editing sites in the mitochondrial genome. In nuclear genes, however, this type of extreme variation in mutation rate seems to be rare.

Molecular Clocks and Neutral Theory

Kimura (1969) believed that the molecular clock occurs by accumulation of neutral mutations and therefore the clock can be used for testing the neutral theory. For this reason, a number of authors have attempted to disprove the neutral theory by finding cases where the molecular clock fails (e.g. Ayala 1986; Gillespie 1991). However, since the evolutionary rate of a protein is affected by functional constraints of the protein as well as the mutation rate, the relationship between neutral theory and molecular clocks is complicated. If the mutation rate varies with time as in the case of plant mitochondrial genes, the molecular clock would not hold even if all mutations are neutral. By contrast, if Equation (4.2) holds for a long evolutionary time, the molecular clock would not be rejected even if selection is involved. In practice, however, N and s are likely to change with time and mutation type, so that the rate constancy would rarely hold with advantageous mutations. Note also that Equation (4.2) would hold only when successive mutations are all advantageous and therefore improve gene function continuously. In reality, such mutations are rare, and a more likely scenario would be that once the function of a gene is improved by some mutations, the next step of evolutionary change would be to maintain the function established. In this case most new mutations would be eliminated by purifying selection, and therefore the molecular clock

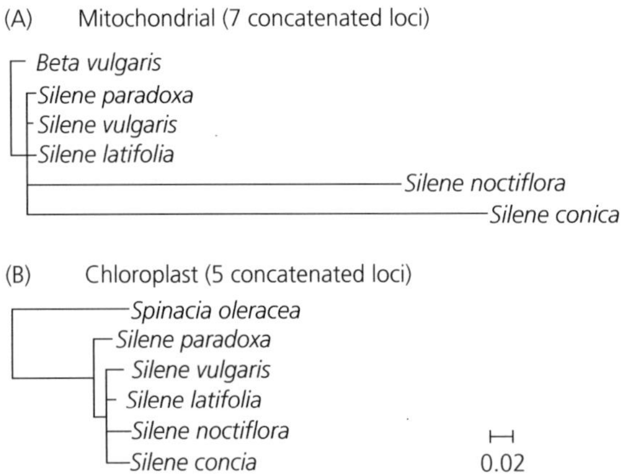

Fig. 4.3. Mitochondrial rate accelerations in the genus *Silene*. Branch lengths correspond to the number of synonymous substitutions per site for mitochondrial DNA (A) and chloroplast DNA (B). Data from Sloan et al. (2010).

not hold. Figure 4.4 shows the distribution of d_N/d_S values for 15 350 orthologous genes between humans and mice. The d_N/d_S value is often used as a measure of the extent of purifying selection. This figure shows that in 99 percent of the genes the d_N/d_S value is less than 1 and therefore most genes are subject to purifying selection.

If we consider the above properties of molecular evolution, Equations (4.1) and (4.3) are likely to hold more often than Equation (4.2), and amino acid substitutions (d_A) and nucleotide substitutions (d_S) would increase approximately at a constant rate. In Fig. 4.2 we have already seen that this is indeed the case approximately.

4.4. Evolution of Protein-Coding Genes

The mammalian genome consists of about 3×10^9 nucleotides, but the number of protein-coding genes in the human genome has been estimated to be about 25 000, and the remaining 95 percent of the genome belongs to noncoding regions (Lander et al. 2001; Waterston et al. 2002). Previously, the noncoding DNA regions were considered to be nonfunctional (Ohno 1972a), but recent studies indicate that many genetic elements that control gene expression reside in the noncoding regions (ENCODE Project Consortium 2012) and therefore they are not necessarily "junk DNA." It is therefore important to consider both protein-coding genes and regulatory elements in noncoding regions in the study of molecular evolution. However, because the function of noncoding regions is still poorly understood, I would like to consider the evolution of protein-coding genes in this section.

General Properties of Evolution of Protein-Coding Genes

One of the salient features of evolution of genes is that new genes are generated by gene duplication or gene transposition, but once the function of a gene is established it tends to maintain the same function for a long time even if the number of genes continues to increase to produce complex organisms. A typical example is the *RecA/RAD51* genes that are required for DNA repair. Both prokaryotes and eukaryotes have only a few copies of the genes and

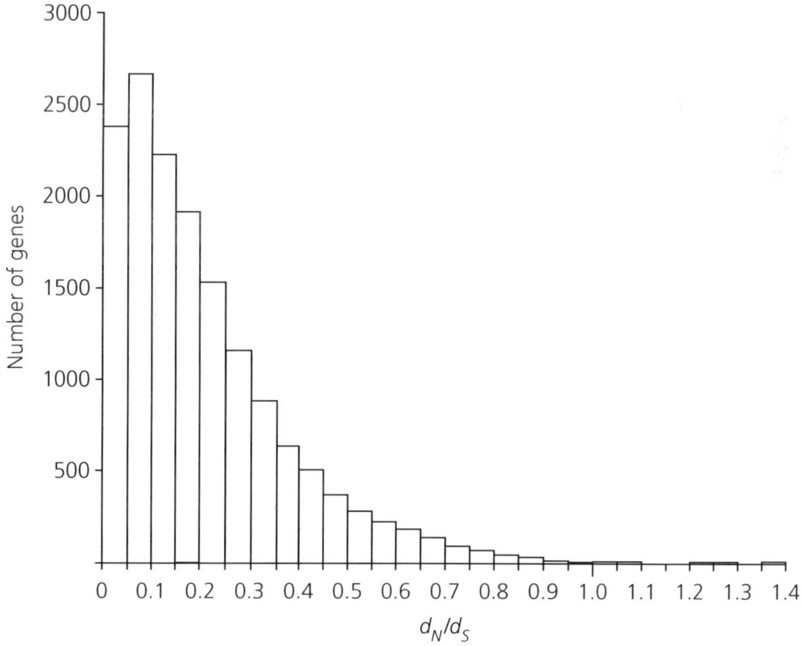

Fig. 4.4. Distribution of the *w* value (d_N/d_S) between human and mouse one-to-one orthologous genes. 15 350 genes with ≥100 codons were used. From Nei et al. (2010).

their gene structures remain largely unchanged in animals and plants (Lin et al. 2006).

This conservative nature of gene evolution is universal except for a few groups of genes. In Fig. 4.4 we have seen that the d_N/d_S ratio is less than 1 for 99 percent of the orthologous genes between humans and mice. The average d_N/d_S is 0.21. Therefore, if synonymous substitutions are approximately neutral, these results indicate that about 80 percent of nonsynonymous mutations are eliminated by purifying selection. Similar results have been obtained with primate species. We can therefore conclude that most mammalian genes are evolving under purifying selection. Because purifying selection is an important feature of the neutral theory, the results in Fig. 4.4 are consistent with the theory.

Fast-Evolving Genes

Although the majority of genes are functionally constrained and many mutant genes are eliminated by purifying selection, there are a few groups of genes that have a rather high d_N/d_S value. The first group of genes belongs to multigene families whose gene products interact with several different ligands. One example is the olfactory receptor (OR) genes in mammalian species (Buck and Axel 1991). The human genomes contain about 400 functional OR genes, whereas the mouse has more than 1000 functional genes (Chapter 5). In mammalian organisms one odorant is perceived by several ORs, and one OR molecule identifies several different odorants (Malnic et al. 1999). For this reason, the functional constraints of OR genes are generally weak, and the genes evolve relatively fast (e.g. Go and Niimura 2008; Nei et al. 2008). (Several authors reported positive selection operating for these genes, but their conclusions are questionable as will be mentioned in Section 4.8.) The genes for pheromone and taste receptors also appear to evolve in the same fashion (Nei et al. 2008).

A second group of fast evolving genes are those involved in unimportant functions. For example, fibrinopeptides do not have important biological functions and evolve in a more or less neutral fashion as mentioned above. More clear evidence for fast evolution of unimportant genes comes from nonfunctional genes or pseudogenes, where the evolutionary rate is expected to be equal to the mutation rate because of no functional constraints (Li et al. 1981; Miyata and Yasunaga 1981). This observation is now regarded as one of the strongest pieces of evidence for neutral theory, as discussed in Section 4.6 below.

In addition, immune system genes such as immunoglobulins and major histocompatibility complex (MHC) genes are known to evolve relatively fast. In this case, however, there is a misconception, and some believe that these genes evolve fast because positive selection accelerates amino acid substitutions. Actually, this is not true. At some codon sites, amino acid substitutions are certainly accelerated, but the proportion of such sites is small and therefore the average substitution rate for the entire genes is lower than the pseudogene rate (Klein and Figueroa 1986; Hughes and Nei 1988). This issue will also be discussed in more detail in Sections 4.5 and 4.6.

4.5. Protein Polymorphism

In the late 1960s and the 1970s, there was another controversy about the maintenance of protein polymorphism, as mentioned earlier. This controversy was initiated by the discovery that natural populations contain a high level of protein polymorphism (Shaw 1965; Harris 1966; Lewontin and Hubby 1966). At this time protein polymorphism was studied by electrophoresis, and the alleles detected by this technique were called allozymes. Because electrophoresis distinguishes between proteins whose electric charges are different from one another, it is believed to detect about 25 percent of amino acid differences (Nei 1975, pp. 25–26). Therefore, it was not a perfect method, but it could detect protein polymorphism inexpensively at many loci in different species. The controversy was a new version of the previous controversy concerning the classical and balance theories of maintenance of genetic variation mentioned in Chapter 2. The "classical" theory asserted that most genetic variation within species is maintained by the mutation-selection balance, whereas the "balance" theory proposed that genetic variation is maintained primarily by overdominant selection or some other types of balancing selection. Initially, the high degree of protein polymorphism observed appeared to support the balance theory, because this theory proposed that most genetic loci are heterozygous. However, there was a nagging problem with this theory. That is, the genetic polymorphism maintained by natural selection generates a substantial amount of genetic load or genetic deaths. As mentioned in Chapter 2, Kimura and Crow (1964) had computed the amount of genetic load generated by overdominant selection and shown that the load will be unbearably high for mammalian organisms if a large number of polymorphic loci are to be maintained by overdominant selection. For this reason, Lewontin and Hubby (1966) could not decide between the two hypotheses when they discovered extensive electrophoretic variation.

Sved et al. (1967), King (1967), and Milkman (1967) then proposed that this genetic load can be reduced substantially if competitive selection occurs so that individuals with a number of heterozygous loci greater than a certain number are selected. As mentioned earlier in Fig. 2.6, this type of truncation selection is unlikely to occur in nature. In fact, Nei (1971, 1975) showed that competitive selection does not occur in the form of truncation selection. As long as natural selection occurs independently for different loci, the genetic load is expected to be high. Meantime, Robertson (1967), Crow (1968), and Kimura (1968a) suggested that most allozyme polymorphisms are probably neutral and that the wild-type alleles in the "classical" hypothesis are actually composed of many isoalleles or neutral alleles. However, the "balance" camp did not accept this argument, because they believed that almost all genetic polymorphisms were maintained by balancing selection though the selection coefficient may be small (Dobzhansky 1970; Clarke 1971).

One important progress made during this period was that the neutral theory generated many theoretical predictions about the allele frequency distribution within populations and the relationships between genetic variation within and between species, etc., so that one could use these theories to test the applicability of the neutral theory to actual data. In other words, we could use the neutral theory as a null hypothesis for studying molecular evolution. This type of statistical study of evolution was almost never done before the neutral theory was proposed. The results of these studies are summarized by Lewontin (1974), Nei (1975, 1987), Wills (1981), and Kimura (1983). Although the interpretations of the results by these and other authors were not necessarily the same, it was clear by the early 1980s that the extent and pattern of protein polymorphism

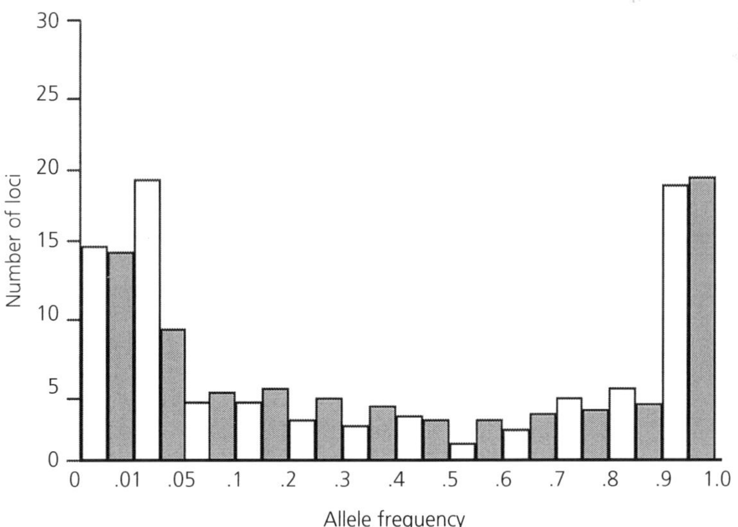

Fig. 4.5. Observed and expected distributions of allele frequencies in *Drosophila heteroneura*. The observed distributions (25 loci, $\hat{H} = 0.162$, $n = 605$) are represented by open columns, and the expected distributions by shaded columns. n: number of individuals studied. Here the infinite-allele model of neutral mutation is used. From Chakraborty et al. (1980).

within species roughly agree with what would be expected from the neutral theory (e.g. Yamazaki and Maruyama 1972; Nei et al. 1976; Skibinski and Ward 1981; Nei and Graur 1984). One such example is given in Fig. 4.5, in which the expected distribution of allele frequencies under the neutral theory and the observed frequency distribution are presented. The two distributions agree with each other reasonably well.

Of course, this does not mean that all amino acid substitutions are neutral or nearly neutral. There must be some amino acid substitutions that are adaptive and change protein function. In fact, this was one of the important subjects of molecular evolution, and many examples of such substitutions were later discovered. There were also many deleterious mutations that were polymorphic but later eliminated from the population, as expected from the "classical" theory of polymorphism. Many of these deleterious mutations appeared to reduce the fitness of mutant heterozygotes slightly, but some had lethal effects in the homozygous condition.

4.6. Neutral Evolution at the DNA Level

Synonymous and Nonsynonymous Nucleotide Substitutions

In the study of evolution, DNA sequences are more informative than protein sequences, because a large part of DNA sequences are not translated into protein sequences and there is degeneracy of the genetic code. The genetic variation in the noncoding regions of DNA such as the intergenic regions, introns, flanking regions and synonymous sites can only be studied by examining DNA sequences. Because of degeneracy of the genetic code, a certain proportion of nucleotide substitutions in protein-coding genes are expected to be silent and result in no amino acid substitution. King and Jukes (1969) predicted that these silent or synonymous nucleotide substitutions should be more or less neutral and therefore the rate of synonymous nucleotide substitution should be higher than the rate of amino acid substitution if the neutral theory is correct. One of the first persons to study this problem empirically was Kimura (1977), who compared the rate of amino acid substitution (r_A) with that of nucleotide substitution at the third codon position (r_3) of histone 4 mRNA sequences from two species of sea urchins. The reason why he used the third codon position was that a majority of synonymous substitutions occur at this position and there were no statistical methods for estimating the numbers of synonymous and nonsynonymous substitutions separately at that time. The highly conserved protein histone 4 showed an extremely low value of r_A, which was estimated to be 0.006×10^{-9} per site per year. Yet, the rate of nucleotide substitution at the third codon position was $r_3 = 4 \times 10^{-9}$. The latter rate is nearly the same as that of synonymous substitution for other nuclear genes studied later.

These results supported King and Jukes's idea that synonymous substitutions are more or less neutral.

Table 4.2. Rates of nucleotide substitution per site per year (r) of mouse (ψα 3), human (ψα 1), and rabbit (ψ β2) globin pseudogenes in comparison with those of the first (r_1), second (r_2), and third (r_3) codon positions of their counterpart functional genes.

	Functional Genes		Pseudogene (ψ)	
	r_1	r_2	r_3	b
Mouse α3	0.69	0.69	3.32	5.0
Human α1	0.74	0.67	2.51	5.1
Rabbit β2	0.71	0.51	2.09	3.6
Average	0.71	0.62	2.64	4.6

Note. All rates should be multiplied by 10^{-9}. From Li et al. (1981).

Pseudogenes as a Paradigm of Neutral Evolution

In 1981 even stronger support of the neutral theory came from studies of the evolutionary rate of pseudogenes (Li et al. 1981; Miyata and Yasunaga 1981). Pseudogenes are nonfunctional genes because they contain nonsense or frameshift mutations, and presumably no positive or negative selection operates in these genes. For this reason, the rate of nucleotide substitution is expected to be high and more or less equal to the rate of neutral mutations. By contrast, if neo-Darwinism is right, one would expect that virtually no nucleotide substitutions occur because they are functionless and there is no way for positive selection to operate. When Li et al. (1981) studied the rate of nucleotide substitution for three globin pseudogenes from the human, mouse, and rabbit, the average rate was about 5×10^{-9} per site per year and was much higher than the rates for the first, second, and third codon positions of the functional genes (Table 4.2). An independent study by Miyata and Yasunaga (1981) about a mouse globin pseudogene also showed a high rate of substitution. These observations clearly supported the neutral theory rather than neo-Darwinian evolution.

In recent years, a large number of pseudogenes have been discovered in various organisms. For example, the human genome contains about 17 000 pseudogenes in contrast to about 23 000 functional genes (Podlaha and Zhang 2010). Similarly, the zebrafish genome has about 16 000 pseudogenes in contrast to 24 000 functional genes. Many of these pseudogenes are nonfunctional and evolve faster than functional genes (Ota and Nei 1994). However, some pseudogenes are known to be transcribed, and therefore they are suspected to have some biochemical function, possibly regulatory functions of gene expression. These pseudogenes evolve as slowly as functional genes. Therefore, the boundary between functional genes and pseudogenes is not always clear cut. A detailed account of this issue has been presented by Podlaha and Zhang (2010).

Slightly Deleterious or Nearly Neutral Mutations

In the study of molecular evolution, many different selection theories such as overdominant selection, frequency dependent selection, and varying selection intensity due to environmental factors have been proposed particularly with respect to the maintenance of genetic polymorphism (Lewontin 1974; Nei 1975, 1987; Wills 1981; Gillespie 1991). Most of them are no longer seriously considered as a general explanation, but Ohta's (1973, 1974) theory of slightly deleterious mutation has recently received considerable attention. In early studies of protein polymorphism detected by electrophoresis, Lewontin (1974, p. 208) and Ohta (1974) noticed that the average gene diversity or heterozygosity (H) for protein loci was about 6–18 percent for both human and *Drosophila* populations, and therefore heterozygosity appeared to have no relationship with species population size. If this is true, it is certainly inconsistent with the neutral theory, because in this theory the average heterozygosity should increase with population size if the mutation rate remains the same. For this reason, these authors criticized the neutral theory.

Ohta's original proposal of the slightly deleterious mutation theory was to explain this apparent constancy of average heterozygosity for species with different population sizes. She argued that if a population contains the wild-type alleles and many slightly deleterious mutations at a locus the average heterozygosity in small populations would be relatively high because slightly deleterious alleles would behave as though they were neutral. In large populations, however, the effect of selection is stronger, and many deleterious mutations would be eliminated. Therefore, average heterozygosity could be more or less the same for different population sizes (Ohta 1974). However, the observation by

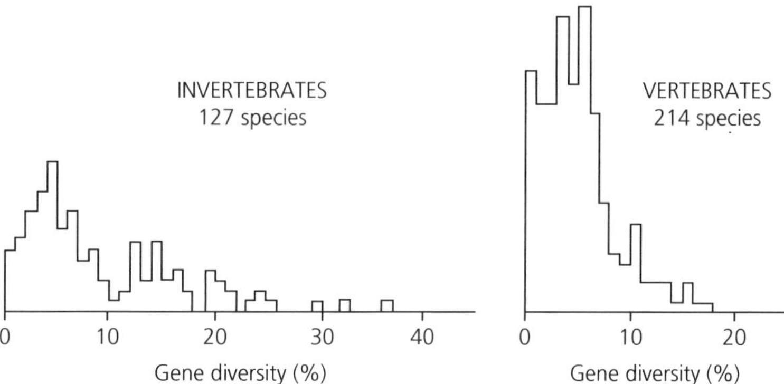

Fig. 4.6. Distributions of average gene heterozygosity for species of invertebrates and vertebrates. Only species in which 20 or more loci were examined are included. The ordinate represents the number of species. From Nei and Graur (1984).

Lewontin and Ohta was based on data from a small number of species, and when many different species were examined, average heterozygosity was generally lower in vertebrates with small population sizes than in invertebrate species with large population sizes (Nei 1975). Later Nei and Graur (1984) studied this problem using data from 341 species and reached the conclusion that average heterozygosity generally increases with increasing species size particularly when bottleneck effects are taken into account (Fig. 4.6). Therefore, Ohta's original explanation is no longer applicable.

Nevertheless, investigators studying the mechanism of maintenance of DNA polymorphism have often concluded that there is a substantial amount of polymorphism of slightly deleterious mutations whose frequencies have been enhanced by genetic drift (Sunyaev et al. 2001; Hughes et al. 2003; Hughes and Friedman 2008). These types of observations have often been taken as support for Ohta's hypothesis. Actually, however, these observations do not necessarily refute the neutral theory, because in the pre-molecular era it was already known that most outbreeding populations contain a large number of deleterious alleles in heterozygous condition (Muller 1950; Simmons and Crow 1977). Note that the neutral theory never claims that all alleles are neutral but that the majority of mutations fixed in the population are neutral or nearly neutral, as mentioned above.

Another problem with Ohta's theory is that if deleterious mutations accumulate continuously in a gene, the gene will gradually deteriorate and eventually lose its function (Fig. 4.1B). If this event occurs in many important genes, the population or species would become extinct (Ohta 1973; Kondrashov 1995). In some genes such as rRNA or tRNA genes the effects of initial mutations occurring in the stem regions may be detrimental because the mutations would impair the proper base pairing, but the effects may be rectified by subsequent compensatory mutations (Hartl and Taubes 1998). Ohta (1973) included these mutations in the category of slightly deleterious mutations. However, a small number of mismatches of nucleotides do not seem to affect the function of rRNAs or tRNAs, and they are effectively neutral when long-term evolution is considered (Fig. 4.1A). Therefore, they should be called neutral mutations (Nei 2005). Note that evolution cannot occur by deleterious mutations alone when long-term evolution is considered. It requires some advantageous or neutral mutations.

In recent years Ohta (1992, 2002) modified her theory calling it the nearly neutral theory. In this theory she now calls mutations with the $|Ns| \le 4$ nearly neutral. However, this theory is essentially the same as the original neutral theory conceived by many early molecular biologists (Fig. 4.1A). At that time no one believed that there are strictly neutral alleles with $Ns = 0$. Actually, mutations with even larger $|Ns|$ values can be called neutral if N is sufficiently large, as mentioned earlier. We have seen in the presence of random fluctuation of selection intensity even mutations with $|N\bar{s}| \approx 200$ could behave as neutral alleles.

4.7. Advantageous Mutations

Evolution of New Protein Function

As mentioned in Sections 4.3 and 4.4, a large proportion of amino acid substitutions appear to be more or less neutral except for certain groups of genes mentioned later in this section, and only a relatively small number of amino acid substitutions affect protein functions. There are now many such examples, particularly about proteins controlling physiological characters (Table 4.3). One of the earliest studies supporting this view was Perutz et al.'s (1981) work on the evolutionary change of hemoglobin in crocodiles. Crocodilian hemoglobin lost its original function (the binding of organic phosphate, chloride, and CO_2) and gained a new function (bicarbonate ion binding). This functional change represents an adaptive response to the blood acidity that occurs during the prolonged stay of crocodiles under water and can be explained by five amino acid substitutions. This is a small portion of the total number of amino acid substitutions (123) between crocodiles and humans. In general, the tertiary and quaternary structures of hemoglobins have remained virtually unchanged during vertebrate evolution, and most amino acid substitutions in hemoglobins do not appear to be related to any significant functional change (Perutz 1983).

Another interesting case of hemoglobin adaptation is that of the goose that flies over the Himalayan Mountains. The blood of the greylag goose (*Anser anser*), which lives in the plains, has the normal oxygen affinity. By contrast, the blood of the bar-headed goose (*Anser indicus*), which migrates across the Himalayas at an altitude of 9000 m, has an unusually high oxygen affinity (Petschow et al. 1977). There are four amino acid differences between the hemoglobins of the two species, but the oxygen affinity difference is caused by only one amino acid substitution (Jessen et al. 1991). Similarly, the functional change of a stomach lysozyme of ruminants can also be explained by a small proportion of amino acid changes (Jolles et al. 1984).

The "red" and "green" color vision genes in humans are contiguously located on the X chromosome and are believed to have been generated by gene duplication that occurred just before humans and Old World monkeys diverged. The proteins (opsins) encoded by these two genes are known to have 15 amino acid differences (Nathans et al. 1986). However, only two or three amino acid differences are responsible for the functional difference of the two proteins, and other amino acid differences are virtually irrelevant (Yokoyama and Yokoyama 1990). Yokoyama and Radlwimmer (2001) have shown that most evolutionary changes of red-green color vision in vertebrates can be explained by amino acid changes at five critical sites of the protein.

It should be noted that although the functional changes of all proteins listed in Table 4.3 are caused by a few amino acid changes, there are genes whose adaptive evolution is affected by many nonsynonymous nucleotide substitutions. One example is the evolution of the gene for eosinophil cationic protein (ECP) (Fig. 4.7). This gene was generated by gene duplication from the gene for eosinophil-derived neurotoxin (EDN) in the ancestral lineage of hominoids and Old World (OW)

Table 4.3. Examples of functional changes of proteins caused by a few amino acid (A.A.) and other changes. The characters affected here are primarily physiological.

Protein/gene	Organism	A. A. changes	Character involved	Reference
A and B alleles	Human	2	Blood group ABO	Yamamoto and Hakomori (1990)
β+-Lactamare (PBP)	Bacteria	1–4	Antibiotic resistance	Hedge and Spratt (1985)
EDN	Human	2	Antiviral activity	Zhang and Rosenberg (2002)
Hemoglobin	Alligator	5	Underwater living	Perutz (1983)
Hemoglobin	Bar-headed Goose	1	High altitude	Jessen et al. (1991)
Hemoglobin	Llama	1–2	High altitude	Piccinini et al. (1990)
Heterochrony	*C. elegans*	1	Cell differentiation	Ambros and Horvitz (1984)
Lysozyme	Ruminants	2	Stomach acidity	Jolles et al. (1984)
Opsins	Human	3	Red/green color vision	Yokoyama and Yokoyama (1990)
Opsins	Vertebrates	5	Color vision variation	Yokoyama and Radlwimmer (2001)
Period (per) gene	*Drosophila*	1	Courtship song rhythm	Yu et al. (1987)
psb A gene	Plants	1	Herbicide resistance	Hirschberg and McIntosh (1983)
TFL1/FT	*Arabidopsis*	1	Flower activation	Hanzawa et al. (2005)
TαFT/FT	Wheat	Retropt	Vernalization	Yan et al. (2006)
TνFT/FT	Barley	Intron-1	Vernalization	Yan et al. (2006)

Retropt: retrotransposon insertion. Intron-1: Loss of introns 1.

```
           1                                                50
Human      .......... ........M. ......AR.. ...R....A. ...SLN.PR.  |
Chimp      .......... ........M. ......AR.. ...R....A. ...SLN.PR.  |
Gorilla    .......... ........M. ......AR.. ...R....A. ...SLN.PR.  | ECP
Orang      .......... ........S. .G....A..R ...R....A. ..VSLN.P..  |
Macaque    .......... ........M. ......AR.. ...K....A. ....VN.PR.  |
Node b     .......... ........M. ......AR.. ...R....A. ....LN.PR.  |
Node a     MVPKLFTSQI CLLLLLGLLG VEGSLHVKPP QFTWAQWFEI QHINMTPQQC
Node c     .......... .......... .......... .......... ......S...
Human      .......... .........A .......... .........T ......S...  |
Chimp      .......... .........A .......... .........T ......S...  |
Gorilla    .......... .........A .......... .........T ......S...  | EDN
Orang      .......... S........A .D........ .........T ......S...  |
Macaque    .......... ........M. ......A..G .......... ......SG..  |
Tamarin    .......... .V...F...S ..V..Q...Q ..S.....S. ...QT..LH.  |

           51                                              100
Human      .I...A.... RW........ .R........ ....QS.R.. H..T.....R  |
Chimp      .I........ RW........ .R........ ....QS.R.. H..T.....Q  |
Gorilla    .I........ RW........ .R........ ....QS.R.L H..T.....R  | ECP
Orang      .T........ .....D.... .R........ .......... ...T.H...R  |
Macaque    .I........ .......... .R....YTA. ..R.ER.R.. ...T.H...R  |
Node b     .I........ .......... .R........ .......R.. ...T.H...R  |
Node a     TNAMRVINNY QRRCKNQNTF LLTTFANVVN VCGNPNITCP RNRSLNNCHH
Node c     ....Q..... .......... .......... .......... S.........
Human      ....Q..... .......... .......... ......M... S.KTRK....  |
Chimp      ....Q..... .......... .......... ......M... S.KTRK...Q  |
Gorilla    .......... .......... .......... ......M... S.KTRK....  | EDN
Orang      N...Q....F .......... .R........ .......... S...R.....  |
Macaque    ....Q..... .......... ......D..H .....SMP.. S.T.......  |
Tamarin    .S...A..R. .P........ .H........ ....T..... ..A.......  |

           101                                                     157
Human      .RFR...L.. D.I.A..... ...DR.GRR. .......... ..S.R..... .......  |
Chimp      .RFR...L.. D.I.A..... G..DR.GRR. .......... ..S.R..... .......  |
Gorilla    .RFR...L.. D.I.A..... ...DR.GRR. .......... Q.S.R..... .......  | ECP
Orang      .RF....L.. ....A..... K..DRTERR. .......... ..S.R..... .......  |
Macaque    .RYR...L.. D.I.A...T. ...DR.GRR. .....ES... ..S.R..... .......  |
Node b     .RFR...L.. D.I.A..... ...DR.GRR. .......... ..S.R..... .......  |
Node a     SGVQVPLIHC NLTGPQISNC RYAQTPANMF YVVACDNRDP RDPPQYPVVP VHLDTTI
Node c     .......... ...S...... .......... .I........ .......... ....RI.
Human      ..S....... ...S...... .......... .I.......Q .......... ....RI.  |
Chimp      ..S....... ...S...... .......... .I.......Q .......... ....RI.  |
Gorilla    ..S....... ...S...... .......... .I.......Q .......... ....RI.  | EDN
Orang      .......... ...S...... .......... .I........ .......... ....RI.  |
Macaque    .......... ...SRR.... ..T..T..KY .I...N.S.. .......... ....RI.  |
Tamarin    .......TY. .......... V.SS.Q.... .......... .......... .......  |
```

Fig. 4.7. Amino acid sequences of present-day and ancestral ECP and EDN proteins. Amino acids are presented by single-letter codes, and dots show the same amino acids as those of the sequence at node *a*. The arginine (R) residues of ECP and EDN are shown in bold type when they are not identical with those of the ancestral protein at node *a*.

monkeys (Fig. 4.8A). Only one copy of the EDN/ECP gene exists in New World (NW) monkeys, and this gene is called EDN here. In humans, EDN and ECP proteins are found in the large specific granules of eosinophil leukocytes. The EDN protein is known to reduce the infectivity of certain RNA viruses including HIV. By contrast, human ECP has a cell-membrane destructive function that generates the toxicity to bacteria and parasites (Zhang and Rosenberg 2002).

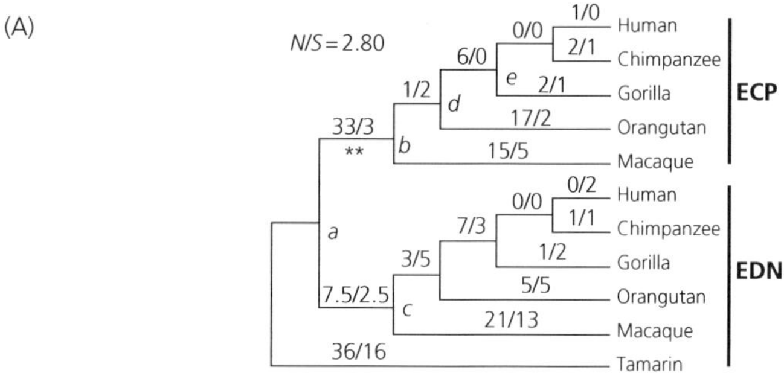

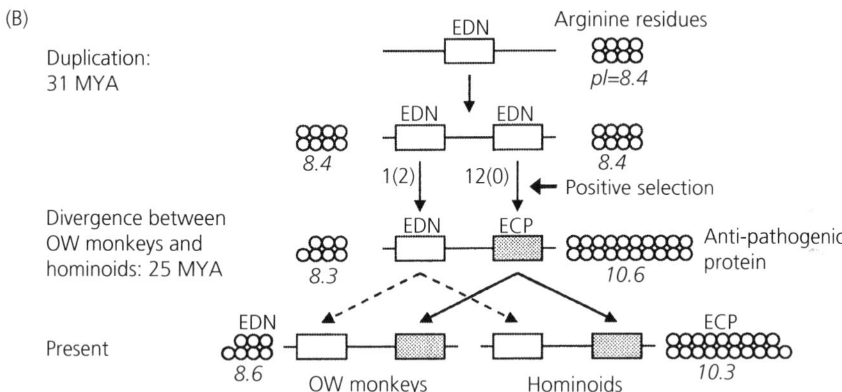

Fig. 4.8. (A) Evolutionary tree of primate ECP and EDN genes. The root of this tree is located on the branch linking node *a* and the NW monkey tamarin. The gene duplication occurred at node *a*. The numbers of nonsynonymous (*n*) and synonymous (*s*) substitutions per sequence per branch are presented as *n*/*s* above each branch. The *N*/*S* (= 347/124 = 2.8) ratio for the entire sequences is given above the tree. (B) Evolution of the novel anti-pathogen toxicity and arginine changes in ECP. Each circle represents one arginine residue in ECP or EDN sequences. For the present-day proteins, only the numbers of arginines for human sequences are presented. *pI*: isoelectric point. MYA: million years ago. From Zhang et al. (1998).

ECP and EDN proteins are both composed of about 160 amino acid residues, but ECP protein sequences are quite different from EDN sequences. ECP has many more arginine residues than EDN, and they are located in many different codon positions (Fig. 4.7). Zhang et al. (1998) conducted a statistical analysis of ECP and EDN sequences from five different species of hominoids and OW monkeys and inferred the evolutionary changes of the amino acid sequences. The results are presented in Fig. 4.8. Diagram A in this figure shows the phylogenetic tree of the genes studied and the number of synonymous (s) and nonsynonymous (n) nucleotide substitutions for each branch. The amino acid sequences for the ancestral species (a, b, c) were inferred by parsimony and other methods. The ratio of n to s for each branch was then compared with the expected ratio (N/S = 2.8), which was obtained under the assumption that all nucleotide changes have occurred by chance alone. In this test, the n/s ratio was significantly higher than N/S only in one branch (between nodes a and b). This result suggests that nonsynonymous substitution was accelerated only in this branch. Examination of amino acid sequences suggested that this high n/s ratio occurred because many amino acids changed from non-arginine to arginine by the aid of natural selection.

Figure 4.8B shows an evolutionary scenario of the amino acid sequences and the function of the ECP genes in OW monkeys and hominoids. As mentioned above, the ECP gene was derived from an ancestral gene of the NW monkey EDN gene by gene duplication. At this time, the primary function of EDN was to digest RNAs as a RNase, but it also had a low degree of antiviral activity. After gene duplication, however, the ECP gene was generated by accumulating many arginine residues and therefore ECP became a highly cationic antipathogenic protein. This evolutionary change occurred before divergence between OW monkeys and hominoids largely by new mutation and positive selection. Positive selection probably occurred when ECP proteins were exposed to bacteria and new mutations were fixed to avoid constantly changing bacterial pathogens.

The above example indicates that the functional change of a protein can be caused by many amino acid substitutions at different residues. In general, it is difficult to predict the locations of positively selected amino acid sites. To know these sites, we have to do experimental studies.

Immune System Genes

As mentioned in the previous subsection, positive Darwinian selection operating for a gene may be detected by comparing the number of synonymous substitutions per synonymous site (d_S) and the number of nonsynonymous substitutions per nonsynonymous site (d_N). One of the first applications of this approach was done by Hughes and Nei (1988, 1989), who compared d_N and d_S for the peptide-binding site (PBS) (or antigen recognition site; ARS) composed of about 57 codons and the non-PBS of major histocompatibility complex (MHC) genes from humans and mice. MHC molecules are for distinguishing between self and nonself peptides and play a role of the initial step of adaptive immunity. Their results clearly showed $d_N > d_S$ for the PBS but $d_N < d_S$ for the non-PBS. These results suggested that in the PBS positive selection operates whereas in the non-PBS purifying selection prevails. Interestingly, vertebrate MHC loci are exceptionally polymorphic often with about 25 alleles per locus in a population, and the cause of this high degree of polymorphism had been debated for more than two decades before 1988. One hypothesis for explaining this polymorphism was heterozygote advantage or overdominant selection (Doherty and Zinkernagel 1975), but there was no evidence supporting this. Knowing that d_N will be greater than d_S under overdominant selection (Maruyama and Nei 1981), Hughes and Nei (1988) proposed that the high degree of MHC polymorphism is probably caused by overdominant selection.

Why is then overdominant selection operating at MHC loci? The reason seems to be as follows. One allele at an MHC locus is known to identify a set of foreign antigens for elimination, whereas another allele identifies a different set of antigens. Therefore, the heterozygote for these two alleles is better protected from different parasites (viruses, bacteria, fungi, etc.) than the homozygote for each of these two alleles. Obviously, the proportion of heterozygotes at an MHC locus is high when the number of alleles is large, and therefore a high degree of polymorphism is generated. In practice, the MHC is composed of a large number of linked genes (Hughes and Yeager 1998), and therefore this immune system is effective for protecting host individuals from various parasites. Later Takahata and Nei (1990) showed mathematically that the overdominance hypothesis can explain various properties of MHC polymorphism such as the large number of alleles at a locus and long-term maintenance of allelic polymorphism. Since then, hundreds of different studies have been conducted about the relative values of d_N and d_S for MHC genes from different species, and most of the studies have shown essentially the same results (Hughes and Yeager 1998; Klein et al. 2007). Some demographic data suggesting heterozygote advantage at MHC loci have also been published (Hedrick 2002).

The relative values of d_N and d_S have been studied for many other immune systems genes including those for immunoglobulins (Tanaka and Nei 1989), T-cell receptors (Su and Nei 2001), and natural killer cell receptors (Hughes 2002). These studies also identified positive selection at the ligand recognition site, but the genes involved are not as polymorphic as MHC genes, and it appears that positive selection is not just for generating genetic polymorphism but for accelerating gene turnover in the population (Tanaka and Nei 1989). It is possible that the accelerated rate of nonsynonymous substitution is caused by protection of the host organism from the attack of ever-changing parasites such as viruses, bacteria, and fungi.

A higher value of d_N than d_S has also been observed in many disease-resistant genes in plants (e.g.

Michelmore and Meyers 1998; Xiao et al. 2004). These genes are essentially immune system genes and defend the host organism from parasites. Another group of genes that often show the $d_N > d_S$ relationship are antigenic genes in the influenza virus (Ina and Gojobori 1994; Fitch et al. 1997; Suzuki and Gojobori 1999), HIV-1 (Hughes 1999a), plasmodia (Hughes 1999a), and other parasites. These genes, especially RNA virus genes, usually show a high rate of mutation and help the parasites to avoid the surveillance systems of host organisms. Here the high rate of nonsynonymous substitution compared with that of synonymous substitution is apparently caused by the "arms race" between hosts and parasites.

The existence of a high degree of polymorphism with a large number of alleles is consistent with Dobzhansky's balance theory of polymorphism mentioned in Chapter 2. However, the number of such loci in the genome is generally small. Furthermore, because the number of alleles at these loci is generally very large, their contribution to genetic load is small (see Equation 2.22 in Chapter 2). Therefore, this observation is not the same as Dobzhansky's conception that almost all loci are polymorphic due to balancing selection. Rather molecular data indicate that most genetic loci are under purifying selection. In this sense molecular data support the classical theory of genetic variation advocated by Herman Muller and his supporters. However, this group of scientists did not anticipate the existence of a large amount of neutral variation, which is invisible at the phenotypic level, and therefore they were also partially incorrect.

Trans-Species Polymorphism

If overdominant selection or balancing selection operates for a long time for a pair of alleles at a locus, one would expect that the polymorphic status may be maintained through one or more speciation events. In this case the same pair of alleles or allelic lineages is observed in two or more species (Takahata and Nei 1990). In fact, Figueroa et al. (1988) showed that two polymorphic allelic lineages marked with insertions/deletions at a MHC class I locus are shared by mice and rats, which diverged at least 10 MYA. Shared polymorphic alleles have also been reported for the class I and class II MHC loci in humans and chimpanzees, which diverged about 6 MYA (Lawlor et al. 1988; Mayer et al. 1988). This indicates that a pair of polymorphic alleles was established before the formation of two species and this pair of alleles may persist in the two descendant populations (Fig. 4.9). This type of polymorphism is called trans-species polymorphism (Klein and Takahata 2002)

Trans-species polymorphism is commonly observed at MHC loci. Particularly, the MHC class II DRB1 locus contains more than 300 registered alleles in the worldwide human population and about 10 percent of the allelic lineages are trans-specific when humans and chimpanzees are considered (Klein et al. 2007). Trans-species polymorphism has also been observed in the cichlid fish species living in Lake Victoria in East Africa. According to geological evidence, this lake completely dried out about 15 400 years ago and then refilled with water about 800 years later. It is believed that the cichlid fishes experienced a rapid speciation in Lake Victoria during the last 14 000 years and there are more than 200 species or forms in the lake at present. How this rapid speciation has occurred is unclear, but it is known that different species often share trans-species polymorphism at the DNA level. Klein et al. therefore speculated that the ancestral population that migrated into the lake was a mixture of different genetic stocks and that rapid speciation has occurred when the initial genetic variation was assorted into different groups of fishes adapted to different environments in the lake. Note that this type of study could not be done before the molecular technique was introduced in population genetics for identifying homologous genes.

In a famous story, Fisher et al. (1939) visited London Zoo and tested the polymorphism of the *T* and *t* alleles at the locus of the phenylcarbamide (PTC) bitter taste receptor in chimpanzees and found that the same type of polymorphism as that of humans exists. They then concluded that the PTC polymorphism was shared by humans and chimpanzees and therefore the polymorphism must have been maintained by overdominant selection. In recent years, however, Wooding et al. (2006) showed that the *T* and *t* alleles in humans and chimpanzees are different and originated independently in the recent past,

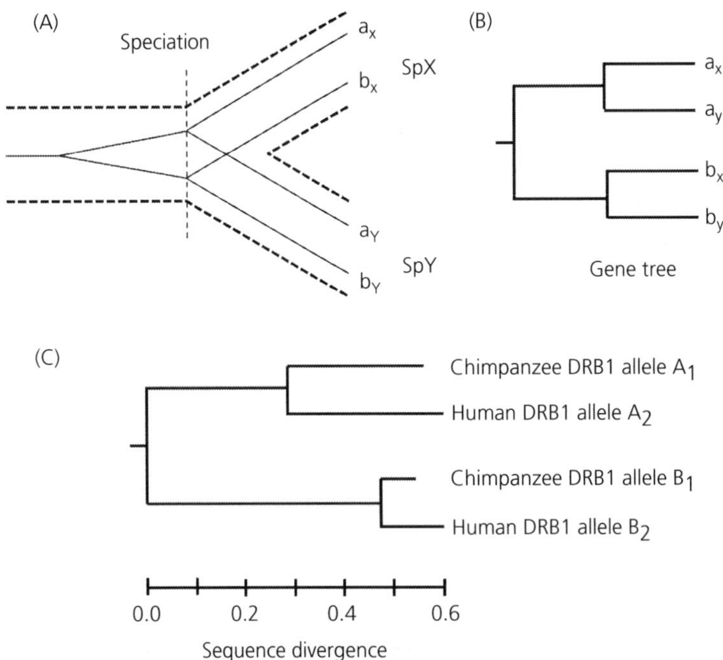

Fig. 4.9. Allelic divergence (A) and the expected allelic tree (B) in the case of trans-species polymorphism. Allele a_X derived from species X (SpX) is more closely related to a_Y derived from species Y (SpY) than to b_X from SpX. Similarly, allele b_X from SpX is more closely related to b_Y from SpY than to a_X from SpX. (C) Phylogenetic tree of the human and chimpanzee alleles at the MHC class II DRB1 locus. MHC data from Klein and Takahata (2002).

and that the *t* allele in chimpanzees contains an interrupted reading frame. This indicates that the PTC polymorphism may not be maintained by overdominant selection.

Some trans-species polymorphisms apparently have been maintained for tens of millions of years. Su and Nei (1999) observed that there are three alleles at a heavy-chain variable region gene 1 locus (V_H1) in rabbits and the nucleotide differences among three alleles are as high as 18 percent. Using information about the divergence between rabbits and related species (hares), they estimated that the polymorphism at the rabbit V_H1 locus has persisted for about 50 MY. This is an awfully long time, but this conclusion was supported by another study of rabbit V_H1 genes (Esteves et al. 2004). Miura et al. (2010a) discovered that two polymorphic alleles (*PSMB8N* and *PSMB8d*) at the immunoproteosome subunit beta type 8 (PSMB8) locus in the teleost fish medaka are highly divergent and have a protein sequence difference of about 20 percent. Showing that this polymorphism is observed in many medaka species in East Asia, they estimated the time of divergence between the two alleles to be 30–60 MY.

The above two examples as well as the trans-species polymorphisms in MHC loci are all related to the immune system in vertebrates, and it is easily understandable why some polymorphisms are maintained in the population for a long time. Although the long-term polymorphism in other genetic systems has not been studied well, it is possible that some of the polymorphisms are also long-lasting. For example, some disease-resistance genes in plants can be trans-specific. The multiple alleles concerned with self-incompatibility in some plant species and sex determination in honeybees have also been maintained for a long evolutionary time (Wright 1939; Yokoyama and Nei 1979; Gempe and Beye 2010). The molecular mechanism of generating self-incompatibility (SI) is complex, but in petunias the SI alleles expressed in the style have been shown to be ribonuclease genes and the pollen determinants are S-locus F-box genes (Kubo et al. 2010).

4.8. Recent Statistical Studies for Detecting Positive Selection

We have seen over previous sections in this chapter that the controversies over the neutral theory of molecular evolution have occurred partly because some authors defined neutrality too strictly and partly because there are some exceptional genes whose evolution pattern deviates from the neutral fashion. It should be kept in mind that no genes would ever evolve in a strictly neutral manner for a long time. As mentioned above, the neutral theory merely asserts that at the molecular level most evolutionary changes of protein or DNA sequences are not caused by Darwinian selection but by random genetic drift of selectively neutral or nearly neutral alleles. If we accept this definition, most amino acid or nucleotide substitutions occur in a more or less neutral fashion, though there are some conspicuous exceptions such as MHC polymorphisms and disease-resistance genes in plants. In Chapter 3 we have seen that the distribution of mutant allele frequencies in human populations follows the neutral pattern quite well when a large number of mutant alleles (several thousands of nucleotide changes) for the protein-coding regions and noncoding regions are considered.

In recent years, however, a substantial number of papers claiming detection of positive selection at the protein level have been published. These papers are based on statistical analyses of genomic data under various assumptions, which are not necessarily satisfied in the real world. It is therefore necessary to scrutinize the validity of the assumptions and the statistical methods used. It is also important to examine the biological meanings of their findings. This has already been done to some extent by Hermisson (2009), Hughes (2008), Nozawa et al. (2009a), and Nei et al. (2010). Here I present a summary of Nei et al.'s review article.

Bayesian Methods for Identifying Positively-Selected Codon Sites

The finding that a limited number of codon sites are subject to positive selection in MHC genes has led a number of theoreticians to suspect that many protein-coding genes might be subject to a similar form of positive selection. They then developed statistical methods to infer the positively-selected codon sites by comparing a set of nucleotide sequences. In these methods a special codon substitution model called the codeml model (e.g. Goldman and Yang 1994; Muse and Gaut 1994) is used, and the w ($= d_N/d_S$) value is assumed to vary from codon site to codon site according to a specific mathematical model (e.g. uniform and beta distributions). Comparing several DNA sequences, one can then estimate the w value for each codon site using Bayesian statistical methods. If this w value for a given codon site is significantly higher than 1, the site is inferred to be under positive selection (e.g. Kosakovsky Pond et al. 2005; Yang 2007).

During the last 10 years, a large number of biologists have used these methods and reported detection of positive selection in many different genes from various organisms including humans, chimpanzees, and macaques (see Nei et al. 2010). For example, analyzing a large number of genes from 10 vertebrates species, Uddin et al. (2008) found signatures of human ancestry-specific adaptive evolution in 1240 genes during their descent from the last common ancestor with rodents and suggested that adaptive evolution of these genes was important for generating human-specific morphological and physiological characters. Interestingly, the set of genes identified included 273 olfaction-related genes, although Gimelbrant et al. (2004) had shown that OR genes evolve more or less in the neutral fashion.

Recent theoretical and empirical studies have shown that these Bayesian methods are quite unreliable and generate a high proportion of false positives (Suzuki and Nei 2002; Hughes and Friedman 2008; Yokoyama et al. 2008; Nozawa et al. 2009a, 2009b). There are several reasons for this. First, the likelihood ratio test (LRT) used in these methods is unreliable because unrealistic mathematical models are used (see Appendix F). The number of nucleotide substitutions at a codon site is also often too small for LRT to be used. In these cases, codon sites may be falsely identified as positively-selected sites because of a high w value generated by chance (Suzuki and Nei 2004; Nozawa et al. 2009a, 2009b). In fact, when positive selection is implicated by this test, the estimate of w often becomes ∞, which is biologically absurd and impossible. In general, the estimate of auxiliary parameters such as the proportions of neutral codons should agree with the true values under computer simulations, but they are usually very different from the latter (Nozawa et al. 2009a).

When a LRT is conducted, it is necessary to compare a null hypothesis model with no selection (M_O) with a selection model (M_S) (Yang 2007; Yang and dos Reis 2011). For example, for a null hypothesis model w is assumed to be 1 for a certain proportion of codon sites but $w < 1$ for other sites. However, this is quite unrealistic, because in real data w should vary from site to site even for the sites without positive selection. The selection model (M_S) used is also unrealistic. In the codeml model, the w value at a given codon site is assumed to remain the same for all nucleotide substitutions. In practice, this assumption does not hold because w would change as the number of nucleotide substitutions increases (see Appendix F).

Nozawa et al. (2009b, 2009a) showed that even when a computer simulation is conducted under a given set of M_O and M_S the results are not always realistic. For example, it is known that the type I error (P) should have a uniform distribution for the test to be used for a general purpose. However, the P values for Yang's Bayesian method shows a U-shaped distribution with a peak near $P = 0$ and $P = 1$ and for certain sets of M_O and M_S the P value near 0 becomes excessively high. This finding is disturbing because the M_O and M_S models used in this method are not realistic, and therefore the LRT may lead to an incorrect P value. Note also that since we really do not know M_O and M_S, it is almost impossible to conduct a proper test of statistical power.

More importantly, these methods are dependent on the assumption that positive selection occurs at a codon site where the number of nonsynonymous substitutions significantly exceeds the number of synonymous substitutions. In practice, this is not the correct assumption needed. Adaptive evolution often occurs by a single nonsynonymous (or amino acid) substitution without repetition. For example, red and green color vision in vertebrates are caused by single amino acid substitutions at positions 277 and 285 of a vision pigment protein. These amino acids remain the same for all vertebrates, though synonymous substitution may occur repeatedly. In this case the w value is expected to be small (see Appendix F).

For the above reasons, it is difficult to predict positively-selected codon sites by using the Bayesian statistical methods. Actually, it is now possible to infer the nucleotide sequences of ancestral organisms by using parsimony or Bayesian methods and reconstruct the ancestral proteins experimentally. One can then study the protein functions of ancestral and extant species and their evolutionary changes (Jermann et al. 1995; Zhang 2006; Yokoyama et al. 2008). Yokoyama (2008) used this type of experiment to study the evolution of visual pigments (color vision genes). When they compared their experimental results with the adaptive sites predicted by Bayesian methods, the agreement between experimental results and statistical predictions was very poor (Fig. 4.10). Similar poor agreements have been obtained for other datasets for visual pigments (Nozawa et al. 2009a) and olfactory receptors (Zhuang et al. 2010).

MK Test and its Extensions

In recent years another class of statistical tests has been used extensively for detecting positive selection. It is McDonald and Kreitman's (1991) test (the "MK test") and its modifications. In recent versions of the MK test, more than a dozen polymorphic genomic sequences (including many protein-coding genes) are sampled from each of two or more closely related species, and the ratio of the number of nonsynonymous polymorphic sites (P_N) to that of synonymous polymorphic sites (P_S) within species is compared with the ratio of the number of nonsynonymous nucleotide substitutions (D_N) to that of synonymous substitutions (D_S) between species. If D_N/D_S is significantly greater than P_N/P_S, positive selection is invoked. In this approach, both P_N and P_S are assumed to represent neutral nucleotide polymorphisms. The argument for this assumption is that all deleterious mutations are quickly eliminated from the population and all advantageous mutations are quickly fixed so that P_N as well as P_S represent mostly neutral mutations. In other words, only strongly advantageous, strongly deleterious, and neutral mutations are assumed to be present. One

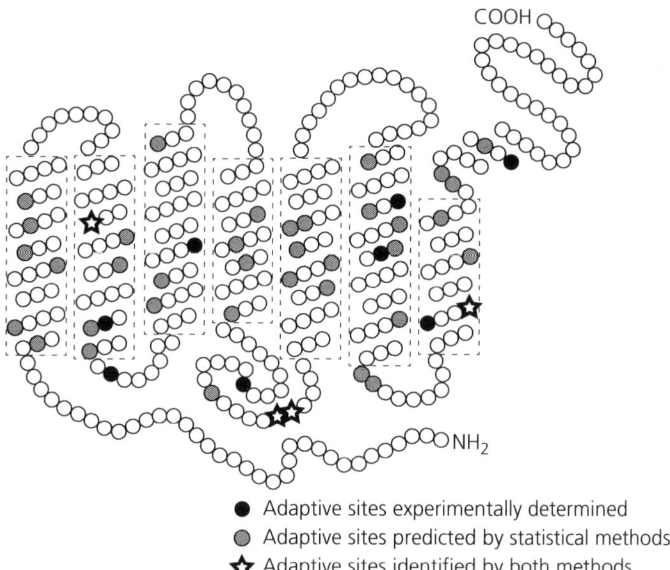

Fig. 4.10. Structure of the bovine rhodopsin protein on which experimentally determined adaptive sites and statistically predicted adaptive sites in vertebrates are shown. The dashed line boxes indicate transmembrane regions. Nei et al. (2010).

can then measure the fraction of neutral polymorphisms by $f = P_N/P_S$ and the fraction of deleterious mutations by $1 - f$ (Sella et al. 2009).

Under this assumption, P_N/P_S is considered to represent the ratio of true numbers of nonsynonymous to synonymous neutral mutations, and the proportion of adaptive nonsynonymous substitutions between species is estimated by $\alpha = [(D_N/D_S) - (P_N/P_S)]/(D_N/D_S)$. In this case α is supposed to be between 0 and 1, but in practice it takes a value between $-\infty$ and 1, because of sampling errors. Therefore, a weighted average ($\bar{\alpha}$) of α for many loci is computed by $\bar{\alpha} = [(\bar{D}_N/\bar{D}_S)-(\bar{P}_N/\bar{P}_S)]/(\bar{D}_N/\bar{D}_S)$, where $\bar{D}_N$, $\bar{D}_S$, $\bar{P}_N$ and $\bar{P}_S$ are the average of D_N, D_S, P_N, and P_S for all loci, respectively, and this $\bar{\alpha}$ is used as a measure of the proportion of adaptive nonsynonymous substitutions between species.

In recent years many investigators estimated the $\bar{\alpha}$ value for a large number of genes in several different groups of organisms. In humans, *Arabidopsis*, and yeast, the $\bar{\alpha}$ value was not significantly different from 0, so that D_N and D_S were thought to have increased by neutral mutations (see Nei et al. 2010). In nuclear genes of *Drosophila* species, however, many authors obtained estimates of $\bar{\alpha} = 0.25\sim0.95$ (e.g. Fay et al. 2002; Smith and Eyre-Walker 2002; Begun et al. 2007; Sawyer et al. 2007; Shapiro et al. 2007). Begun et al. (2007) obtained an estimate of $\bar{\alpha} = 0.54$, examining 10 065 genes from 7 strains of *D. simulans* and a single strain of *D. melanogaster*. One of the most extreme $\bar{\alpha}$ values was obtained by Sawyer et al. (2007) for 91 genes from *D. melanogaster* and *D. simulans*. This study suggested that about 95 percent of amino acid differences between *D. melanogaster* and *D. simulans* are caused by positive selection ($\bar{\alpha} = 0.95$).

These results are very different from those derived from the studies of molecular evolution for the last 40 years. Is there any problem with the MK test? First, we note that the MK test depends on several simplifying assumptions. For example, P_N is assumed to represent only neutral nonsynonymous polymorphism. This assumption is clearly wrong, because every population contains some mildly deleterious nonsynonymous mutations. For this reason, some researchers eliminated low frequency polymorphisms (e.g. Fay et al. 2002; Smith and Eyre-Walker 2002). However, elimination of genes (or nucleotide sites) with low frequency alleles is not justified, because most of the neutral mutations are also in low frequency and some slightly deleterious mutations can be of moderate frequency (Wright 1938a; Wright 1969, p. 385). In other words the selection coefficient *s* against a deleterious mutation is continuous, and it is very difficult to determine the threshold of removal of nucleotide sites with low frequency alleles. The assumption that no advantageous mutations are included in P_N is also incorrect, because the *s* value for advantageous mutations is continuous and the frequency of mutations with small positive *s* is expected to be higher than those with large *s*. Many mutations with small positive and small negative *s* must be included in P_N as well as in D_N. This makes the interpretation of $\bar{\alpha}$ very difficult.

Second, there are alternative explanations for the positive α value in MK tests. It is already known that slightly deleterious mutations can be fixed in the population when population size fluctuates over evolutionary time. In this case α may become greater than 0 (McDonald and Kreitman 1991; Hughes 2008). Another factor that makes α positive is the fluctuation of *s* for nonsynonymous mutations. In the real world the *s* value of a mutation would never be constant but vary from generation to generation for the reasons mentioned earlier in Chapter 2. In this case variation of *s* would act as another factor that causes a random fluctuation of allele frequencies (Wright 1948a), and a mutant gene may evolve as though it is neutral if the mean of *s* is effectively 0. However, the extent of P_N would be reduced substantially because of random fluctuation of *s*, whereas P_S would not be affected (Nei and Yokoyama 1976). Thus, a positive α value does not necessarily mean a signature of positive selection. If D_N/D_S (or d_N/d_S) remains more or less constant over evolutionary time when a large number of loci are considered (Fig. 4.2), α will be a measure of deficiency of nonsynonymous polymorphisms. This indicates that the MK test is not really for detecting positive selection (Nei et al. 2010).

Another problem is the assumption that all excess nonsynonymous substitutions in D_N/D_S relative to P_N/P_S are adaptive and caused by positive selection. As mentioned earlier, only about 5 percent of amino acid substitutions seem to affect protein function. If this estimate applies to many other proteins, even an excess of 95 percent amino acid substitutions observed by Sawyer et al. (2007) would not be important as a selective force (Nei et al. 2010). In fact, Sawyer et al.'s estimate of *Ns* was about 4, suggesting that amino acid substitution has been effectively neutral (Chapter 2).

Extended Haplotype Homozygosity and F_{ST} Tests

There are many other population genetics tests of positive selection. One group of tests is Tajima's (1989) *D* statistic and its modifications. In these methods the consistency of the intrapopulational pattern of nucleotide frequency distribution with the neutral expectation is tested by using various statistics. A typical example is Tajima's *D* statistic, which suggests balancing selection when $D > 0$ and purifying selection or directional positive selection when $D < 0$. Hudson et al. (1987) proposed another method that examines the consistency of the nucleotide frequency distribution within and between species. The null hypothesis of these methods depends on the assumption that the population is in the mutation-drift balance. In practice, this assumption is almost never satisfied, and therefore it is generally difficult to obtain definitive conclusions from this type of statistical tests. Furthermore, even if positive selection is hinted at, it is difficult to identify the amino acid change that is responsible.

Another group of methods is the analysis of extended haplotype homozygosity (EHH) data. An increasing number of investigators are now using these methods to detect a signature of positive selection by examining the pattern of EHH with single nucleotide polymorphism (SNP) data. The principle of these methods is that if a particular nucleotide mutation at a SNP site is strongly selected for some reason the other SNP sites closely linked with this mutation would also increase in frequency because of the hitchhiking effect and they may generate a high degree of homozygosity for the haplotypes carrying the mutant nucleotide for an extended chromosomal region (Sabeti et al. 2002). By contrast, the haplotypes associated with the original nucleotide are expected to show no enhanced homozygosity because of the recombination that may have occurred many times between the nucleotide site and other sites in the past. Therefore, the homozygosity of SNP sites for the haplotypes associated with the mutant nucleotide is expected to be high for an extended chromosomal region compared with the haplotypes associated with the ancestral nucleotide. One may therefore be able to detect a signature of selection by comparing the extent of haplotype homozygosity for the mutant nucleotide (EHH_M) with that for the ancestral nucleotide (EHH_A). The ratio of EHH_M to EHH_A is called the relative EHH (rEHH) (Sabeti et al. 2002). This method apparently worked well with the G6PD and CD40 ligand (TNFSF5) genes, which have been presumably under positive selection.

Many investigators have used this type of statistical method to identify SNP sites or genomic regions that may be under positive selection (see Nei et al. 2010). For example, Voight et al. (2006) suggested that 250 genomic regions are under positive selection in human populations. Similarly, Sabeti et al. (2007) identified about 300 selected regions in the human genome.

There are some problems with these methods. First, it is difficult to set up a neutral null hypothesis against which the likelihood of the selection hypothesis can be tested. For this reason, no statistical test of neutral evolution is conducted, and the genomic regions that show the top 1 percent or 5 percent of rEHH values are chosen as the regions under positive selection. Therefore, if millions of SNP sites are tested, the top 1 percent may include hundreds of genomic regions. Because rEHH (or any other statistic) is affected by random events such as mutation, recombination, and genetic drift as well as the size and quality of SNP data, the statistic used would be subject to a substantial amount of random errors. Therefore, a high rEHH value may not necessarily be due to selection.

Second, these estimates have been obtained under the assumption that a particular SNP site or a particular set of SNP sites are positively selected and nearby SNP sites are dragged by hitchhiking to become homozygous because of the low recombination rate. In practice, this is a mere assumption, and no investigators have identified any driving nucleotide site except for a few sites for which natural selection has been suspected before the study. Wang et al. (2006) showed that ~35 percent of selected SNPs are not within 100 kb of known genes. Have these SNPs really been affected by hitchhiking even though there is no known functional element? It is therefore important to have some empirical evidence of selection for each putatively selected genomic region. Until this evidence is presented, the results of these studies remain mere speculations.

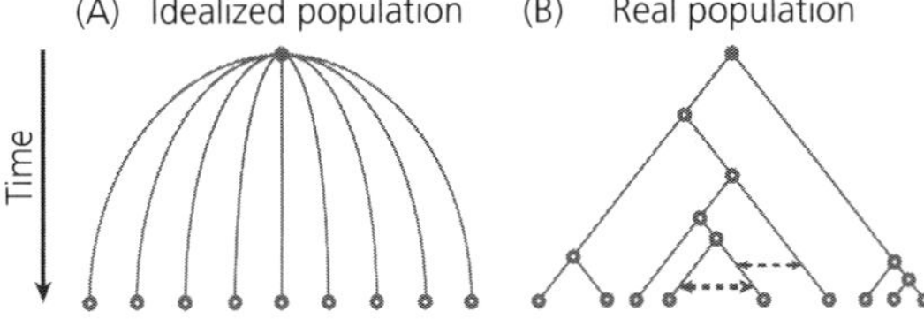

Fig. 4.11. Simple examples of population structures. (A) An idealized model of population structure, in which subpopulations evolve independently after population splitting. (B) A realistic model of population structure, in which some subpopulations are more closely related than others. Partial gene exchange also occurs between some subpopulations (broken lines with arrows).

There is another statistical method for predicting selected genomic regions. It is the F_{ST} statistic method, which can be used when a population is divided into subpopulations. The F_{ST} is computed for each locus or SNP site by $F_{ST} = V_x / [\bar{x}(1-\bar{x})]$, where V_x is the variance of allele frequency x among subpopulations and $\bar{x}$ is the mean of x for all subpopulations. When all subpopulations are derived from a parental population at the same time and gene migration occurs among them at a rate of m per generation, following the island model (Fig. 4.11A), the expectation of equilibrium F_{ST} for neutral alleles can be written as $1/(1 + 4Nm)$ approximately (Wright 1931). Here N is the effective population size of each subpopulation. If all loci are neutral and evolve independently, the variance of F_{ST} among different loci is given by $k\bar{F}_{ST}^2/(n-1)$, where n is the number of subpopulations, $\bar{F}_{ST}$ is the average F_{ST} for the loci examined, and $k = 2$. Lewontin and Krakauer (1973) proposed that this equation can be used for testing the neutral theory. However, this approach was criticized by Nei and Maruyama (1975) and Robertson (1975a, 1975b), who indicated that k can be much greater than 2 when some subpopulations are more closely related than others and the mutation rate varies with locus. In reality, the population differentiation is always more complicated than the idealized model assumes (Fig. 4.11B), so that the Lewontin-Krakauer test was soon abandoned by the authors themselves (Lewontin and Krakauer 1975).

In recent years, however, a modified form of the Lewontin-Krakauer test was proposed. In this modified form F_{ST} is computed for a large number of loci (SNP sites) for each organism (genome-wide analysis), and the loci showing the highest or lowest 1 percent or 5 percent of F_{ST} values are assumed to be under positive selection (e.g. Akey et al. 2002; Barreiro et al. 2008; Myles et al. 2008). Again, however, this outlier method is not justified, because this is not a test of the null hypothesis of neutral evolution (see Nei et al. 2010). Furthermore, the original criticism by Nei, Maruyama, and Robertson still applies.

In the genome-wide analysis the false-positive rate of identifying positive selection may be revealed by comparing the putatively selected genomic regions obtained by different statistical methods in the same species. In humans, Nielsen et al. (2007) compared 713 such genes included in the selected genomic regions identified by Voight et al. (2006) with 90 genes identified by Wang et al. (2006) and found that only 7 genes were shared between them. Akey (2009) also compared the positively selected genomic regions identified by 9 genome-wide studies by Akey et al. (2002), Carlson et al. (2005), Kelley et al. (2006) and others. The total number of regions identified by these studies was 5110, but only 722 regions (14.1 percent) were shared by at least two studies, 271 regions (5.3 percent) by at least three studies, and 129 regions (2.5 percent) by at least four studies. These results indicate how unreliable these statistical methods are and that a large proportion of putatively selected sites are apparently false positives. This conclusion again raises questions about the results of genome-wide analysis.

Statistical Studies and Biochemical Verification

Because genome sequences and SNP data are now available for many different organisms, there is a new trend of scanning all genes or all SNP sites available to find positively selected genomic regions. In general, this genome-wide analysis is merely a collection of single-locus analyses for a large number of loci. Although genome-wide analysis gives a large amount of information about natural selection if the analysis of each locus is done properly, it may also give an erroneous conclusion as in the case of the outlier methods. When natural selection is studied only for a single locus, the study is usually done carefully by examining the molecular nature of allelic differences and allele frequency changes. Therefore, a solid conclusion is obtainable at least for the locus. In a genome-wide analysis every locus is analyzed in the same way by using a particular computer program, and therefore we tend to miss special features of individual loci.

This does not mean that all statistical studies are useless. If a statistical analysis is conducted in combination with biochemical studies, one may obtain deep insights into the evolutionary process, as in the case of MHC genes. As emphasized above, the eventual goal of molecular evolutionary biology is to understand the mechanism of evolution in biochemical terms. At the present time, many authors seem to be satisfied with just finding signatures of positive selection. In my view these signatures of selection are not sufficient. We have to know the biological mechanism of natural selection, which operates among different individuals. Furthermore, once we identify a particular mutation that is responsible for selection, we should examine the nature of the mutation at the molecular level. It is important to know the molecular basis of mutational change, because mutation is the ultimate source of all phenotypic innovations and natural selection is merely for shifting allele frequencies.

In this chapter we are primarily concerned with the evolutionary changes of genes due to mutation and selection at single loci. In reality, any phenotypic change is controlled by a large number of genes, and therefore the identification of a single mutation that causes a particular phenotypic change does not necessarily mean that the mechanism of evolution has been solved. For example, I previously mentioned (in Section 4.7) that the difference in blood oxygen affinity between the greylag goose and the bar-headed goose is caused by a single amino acid substitution in hemoglobin molecules. However, the single amino acid substitution in hemoglobins may not be sufficient to explain the ability of the bar-headed goose to fly over the Himalayas. This ability must be controlled by the genes that make the bar-headed goose fly high in the sky, generate resistance to cold temperature, navigate the mountain crossing, control the expression pattern of hemoglobin genes, etc., in addition to the amino acid substitution. It is therefore necessary to know the changes to other genes as well as the functional change in hemoglobins.

This indicates that to understand the evolutionary change of any phenotypic character we must study the evolutionary changes of multiple genes simultaneously and the developmental process of the character involved. The study of this problem is much harder than the identification of DNA changes that are related to natural selection. Despite this difficulty, however, it would be important to identify major mutations that affect protein function significantly, particularly for the loci that are expressed at a later stage of development (mostly the genes concerned with physiological characters).

Frequency Distributions of Mutant Nucleotides

Under certain conditions, however, statistical studies alone may give a relatively solid conclusion. One of them is the analysis of frequency distributions of mutant nucleotides (SNPs) for the protein-coding and noncoding regions of DNA discussed in Chapter 2, Section 2.4. In this case the frequency distribution for the noncoding region agreed well with the expected distribution for neutral mutations, as might be expected (Fig 2.9B). However, the neutral distribution also agreed with the observed distribution for the coding region as well (Fig 2.9A). Furthermore, the observed distribution was considerably different from the expected distribution for a small degree of positive selection ($2Ns = 4$) or negative selection ($2Ns = -4$) (see Fig. 2.5). Therefore, this observation suggests that nucleotide substitution in the protein-coding region is caused mostly by neutral mutations.

This conclusion agrees with that reached from the pattern of nucleotide substitutions in the protein-coding region (Fig. 4.2). Actually, a similar conclusion was obtained from a study of the frequency distribution of nucleotide mutations obtained from the genomic sequences of the African American population (Miura et al. 2013). Furthermore, Caicedo et al. (2007) conducted a similar study about the SNP data from the wild rice *Oryza rufipogon* and showed that the observed distribution agreed very well with the expected distribution for neutral mutations. These results refute some recent statistical studies that suggested the prevalence of positive selection at the nucleotide level (e.g. Begun et al. 2007; Shapiro et al. 2007).

Nevertheless, the study of mutant allele frequencies at the nucleotide level cannot identify a small proportion of nucleotide substitutions that would have caused innovative changes in phenotypic characters. These mutations must be studied by using molecular techniques.

4.9. Summary

Comparison of orthologous amino acid sequences from different species has made it possible to study the extent of genetic variation within and between species by the same genetic scale for the first time in the history of the study of evolution. Unexpectedly from the neo-Darwinism view, this study has shown that the extent of sequence divergence increases as the time of divergence between species increases. This is also approximately true for synonymous and nonsynonymous nucleotide substitutions in the coding region of genes or nucleotide substitutions for noncoding regions of DNA. These observations indicate that molecular evolution occurs primarily by random fixation of neutral or nearly neutral mutations and that mutation is the primary force of evolution. The definition of neutral mutations currently used is not on a logical basis, and if we consider the effect of random fluctuation of allele frequencies properly, the definition can be relaxed considerably, and most mutations with selection coefficients of $Ns < 200$ can be shown to be effectively neutral.

The rate of amino acid substitution varies enormously among different proteins, and the substitution rate is lower in proteins that have strong functional constraints than in proteins that do not have many constraints. Statistically, the extent of functional constraint can be measured by $w = d_N/d_S$. In general, most proteins evolve conservatively, and w is much lower than 1. However, although many mutations are eliminated by natural selection, the remaining portion of amino acid substitutions appears to occur in a more or less neutral fashion. Furthermore, a small proportion of amino acid substitutions may change protein function drastically and may contribute to the generation of innovative phenotypic characters. The w value also may change drastically when the mutation rate or the extent of functional constraints changes. In general, pseudogenes in small gene families such as globin gene

Darwinian selection. Most of them are involved with immune systems, antigenicity in microorganisms, and disease resistance in plants. Particularly well studied are MHC and immunoglobulin genes in vertebrates. These genes are composed of the antigen-binding regions, and positive selection occurs primarily in these antigen-binding regions, where nonsynonymous nucleotide substitutions are favored to avoid the invasion of new types of foreign antigens (arms race).

Some immune system genes such as MHC or immunoglobulin genes are often subject to balancing selection and show a high degree of polymorphism. This is caused by the occurrence of new pathogenic mutations and the host protection by new immune system mutations. In this case allelic turnover occurs, but polymorphic alleles may stay in the population over millions of years generating trans-species polymorphism. There are also some pairs of alleles that apparently have been maintained as polymorphic for about 50 million years in vertebrates.

However, in a majority of genetic loci, the pattern of protein polymorphism within species is consistent with the expectation from the neutral theory. In recent years a large number of statistical analyses reporting the importance of natural selection have been published, but these statistical analyses are often based on faulty mathematical or biological assumptions. To identify the cause of natural selection, it is important to know the molecular differences between mutant and original alleles.

CHAPTER 5

Gene Duplication, Multigene Families, and Repetitive DNA Sequences

5.1. New Genes Generated by Gene Duplication

Historically, the word mutation has been used to refer to any kind of genetic changes including gene duplication, chromosomal changes, genome duplication and gene transposition as well as nucleotide changes. In the Chapter 4 we discussed primarily single-locus mutations and their roles in genomic and phenotypic evolution. In this chapter I now want to consider various forms of genetic elements and their changes in relation to phenotypic evolution.

In the first half of the twentieth century, the existence of duplicate genes and duplicate genomes (polyploids) was well known. Establishment of new species by genome duplication was also known (see Taylor and Raes 2004). This clearly indicated that mutations play important roles in evolution. Yet, evolutionists did not pay much attention to the importance of gene duplication (see Dobzhansky 1937; Huxley 1942; Stebbins 1950), apparently because at that time they were busy proving that natural selection is the most important factor in evolution (Ford 1964; Kettlewell 1973). Furthermore, because there was no reliable technique to estimate the number of genes in an organism, it was difficult to study the roles of gene or genome duplication in evolution. Only after Watson and Crick (1953b) discovered that DNA is the genetic material was it possible to show that complex organisms such as vertebrates generally have a higher amount of DNA content and a larger number of genes than simple organisms such as bacteria and fungi. However, it was soon discovered that the relationship between DNA content and organismal complexity is not very simple because a genome may contain a large proportion of noncoding DNA. For example, some species of salamanders contain more than two times the DNA than the human DNA content. In the following I will discuss recent developments in the study of evolution by gene duplication with some historical background.

The importance of gene duplication in generating new genes was suggested by Sturtevant (1925), who studied the genetic origin of the *Bar* eye mutant in *D. melanogaster* and conjectured that it was generated by gene duplication caused by unequal crossover. This conjecture was proven to be correct when Bridges (1935) and Muller (1936) examined the salivary chromosome and showed that the *Bar* mutant was caused by the duplicate bands in the *Bar* locus region. This and other similar findings later led to the proposal that duplicate genes are an important source for creating new genes (Lewis 1951; Stephens 1951). However, molecular evidence for this idea was lacking until Ingram (1961, 1963) showed that myoglobin and hemoglobin α, β, and γ chains in humans are products of a series of gene duplications that occurred a long time ago (Fig. 5.1). Because the occurrence of gene duplication is a mutational event and has nothing to do with natural selection, this example clearly shows the importance of mutation in evolution. Surely, natural selection should have played some role in the fixation and maintenance of duplicate genes, but this is not the cause of innovation in evolution. Furthermore, the role of natural selection operating with these duplicate genes is primarily purifying selection to save advantageous mutation as in the case of gene mutations. In the 1960s many more different multigene families

Mutation-Driven Evolution. First Edition. Masatoshi Nei.

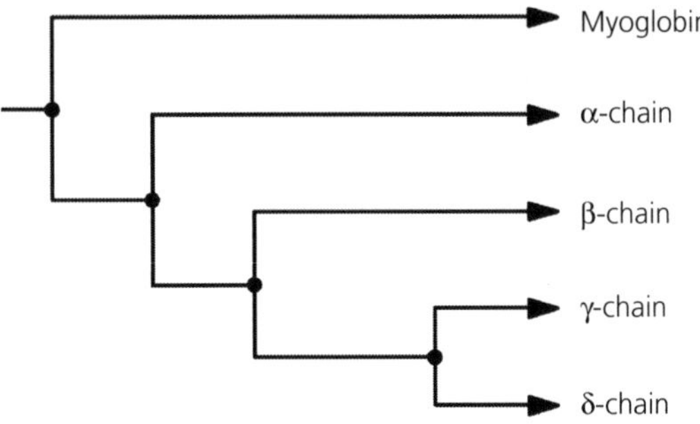

Fig. 5.1. Evolution of hemoglobin chains. The point in time of a gene duplication is indicated by a solid circle. From Ingram (1961). Reproduced with permission from Nature.

were discovered (Dayhoff 1969), and this discovery set forth the study of gene duplication and evolution of multigene families. However, the real magnitude of the importance of gene duplication was not recognized until genomic sequence data became available from different model organisms.

Increase in the Number of Genes by Gene Duplication

There are three major mechanisms for producing duplicate genes: (1) genome duplication; (2) tandem gene or segmental duplication; and (3) gene transposition. Genome duplication does not necessarily double the number of functional genes because some genes are quickly silenced or eliminated from the genome (Kellis et al. 2004; Adams and Wendel 2005; Scannell et al. 2007). Yet, this is the most effective mechanism for increasing the number of genes in the genome. By contrast, the increase of gene number by segmental duplication is generally small at any one time, but it may produce thousands of genes if it occurs repeatedly, as in the case of olfactory receptor genes in vertebrates (Glusman et al. 2001; Young et al. 2002; Nei et al. 2008). Transposition of genes from one genomic location to another generally occurs by the aid of transposons and may produce hundreds of thousands of copies.

Considering the rate of increase of DNA content from bacteria to mammals and the rate of amino acid substitution, Nei (1969b) predicted mathematically that the genomes of current vertebrates contain a large number of duplicate genes that are important for the formation of complex organisms.

Table 5.1. Number of genes in multigene families of single-cell and multicellular organisms

Protein family	Yeast	Worm	Fruitfly	Human
Immunoglobulin domain	0	64	140	765
Zinc finger	48	151	357	706
Protein kinase	121	437	319	575
GPCR	0	358	97	569
P-loop motif	97	183	198	433
Reverse transcriptase	6	50	10	350
mm Domain	54	96	157	300
G protein β WD-40 repeats	91	102	162	277
Ankyrin repeats	19	107	105	276
Homeobox domain	9	109	148	267

From International Human Genome Sequencing Consortium (2001).

This study provided a theoretical basis for Ohno's (1967, 1970) idea that gene duplication is an important mechanism of evolution of complex organisms. Although Nei's prediction was very rough as it was based on the poor molecular data available at that time, we now know that a majority of vertebrate genes exist as gene families. Recent genome sequencing in vertebrates has shown that this is indeed the case (Table 5.1). Probably because polyploidization occurs more often in plants than in animals, plants tend to have more multigene families than animals. The numbers of member genes in a gene family varies enormously in both plants and animals. In some gene families the number is as high as thousands.

Nei (1969b) also predicted that vertebrate genomes would contain a large number of pseudogenes (nonfunctional genes) because duplicate copies of genes can be nonfunctionalized. It is interesting to note that the mammalian genome is now known to have about 20 000 pseudogenes (Torrents et al. 2003; Podlaha and Zhang 2010). This is nearly as large as the number of functional genes.

Ohno (1970) has presented a treatise on evolution by gene duplication. He emphasized his view of evolution by gene duplication in a broader context and tried to explain major events of vertebrate evolution by gene duplication. One of his main arguments was that genome duplication has an advantage over tandem gene duplication in the formation of new genes because in genome duplication both protein-coding and regulatory regions of genes are duplicated at the same time whereas tandem duplication may disrupt the coordination of regulatory elements and protein-coding regions of genes. On the basis of genome sizes, he then proposed that the vertebrate genome experienced about two rounds of genome duplications before the evolution of the X and Y or the Z and W chromosomes in reptiles. Later, several authors called this the 2-round (or 2R) hypothesis of genome duplication (e.g. Kasahara et al. 1996; Hughes 1999b). However, this hypothesis has been controversial (Makalowski 2001).

Recent studies have shown that a large portion of duplicate genes are lost from the genome after genome duplication and most polyploid genomes are quickly diploidized (Kellis et al. 2004; Adams and Wendel 2005). It is also known that a large number of tandem or segmental duplications have occurred during the past several hundred million years and the duplicate genes have often been transferred to different chromosomes or different chromosomal segments as in the case of olfactory receptor genes (Zhang and Firestein 2002). Therefore, even if the 2R hypothesis is correct, it would be very difficult to prove it now because the history of genome duplication has been largely erased. Note also that unlike Ohno's original argument tandem duplication has no disadvantage in creating new genes compared with genome duplication because tandem duplication usually includes the regulatory region at the same time. However, whatever the mechanism is, we should note that gene duplication is a form of mutation that generates various evolutionary innovations. In plants, there is no dispute about the importance of genome duplication in evolution.

Genome Size and Number of Genes

Table 5.2 shows the DNA contents and the numbers of protein-coding genes in various groups of organisms. The genome size and the number of genes are very small in viruses, because the survival and

Table 5.2. Genome sizes and the number of protein-coding genes in various groups of organisms.

Organisms	Genome size/Mb	# genes	Organisms	Genome size/Mb	# genes
Viruses			Fungi		
T4 phage	(170 kb)	280	Neurospora	40	10 000
T5 phage	(122 kb)	168	Saccharomyces	12	6300
λ phage	(49 kb)	73	Oyster mushroom	34	12 000
ϕX174	(5.4 kb)	11			
Flu A type	(14 kb)	11	Animals		
HIV	(9.2 kb)	9	*C. elegans*	100	19 000
			Fruitfly	160	14 000
Eubacteria			Amphioxus	520	22 000
Escherichia coli	5.5	4300	Ciona	120	16 000
Pseudomonas	5.1	4600	Zebrafish	1400	24 000
Cyanobacteria	6.4	5400	Fugu	390	26 000
Rickettsia (par)	1.1	830	*Xenopus*	1700	28 000
Buchnera (par)	0.62	560	Chicken	1000	22 000
Mycoplasma (par)	0.58	476	Opossum	3500	19 000
			Platypus	440	21 000
Archaebacteria			Mouse	2500	24 000
Thermotoga	1.9	1800	Dog	2400	19 000
Methanosarcina	4.8	3600	Macaque	2900	20 000
Haloterrigena	3.9	3700	Chimp	3100	28 000
			Human	2900	25 000
Protists					
Diatom	2.5	11 000	Plants		
Tetrahymena	100	27 000	Moss	500	39 000
Slime mold	34	13 000	Arabidopsis	120	25 000
Protozoa	23	5000	poplar	550	46 000
Plasmodium	24	5400	Rice	420	41 000
Choanoflagellate	42	9200	Corn	2800	32 000

Mb: megabase. kb: kilobase. par: parasitic. Data were compiled from various sources.

reproduction of viruses are dependent on the biochemical machinery of the host organism, whether the host is a bacterium or a eukaryote. Bacteria have a much larger number of genes, and they are generally capable of living themselves. However, some parasitic bacteria such as *Mycoplasma* and some symbiotic bacteria such as *Buchnera* have a small number of genes. Apparently, these bacteria do not need many genes that are normally required for metabolic pathways, because metabolites are provided by the host organism. For example, the *Buchnera* bacteria, which are symbiotic with aphids, originated from a relative of *E. coli* some 200 million years ago, and apparently they lost many genes for metabolism. They now have only about 560 genes compared with 4300 genes of the free-living bacterium *Escherichia coli.*

However, even the genomic sizes of free-living bacteria are much smaller than those of eukaryotes, which have complex metabolic and reproductive systems. Unicellular protists and fungi generally have smaller genome sizes and smaller numbers of genes than animals and plants, though some protists (e.g. *Tetrahymena*) have a large number of genes. In animals, simple organisms such as *Caenorhabditis elegans* and fruitflies generally have smaller amounts of DNA than more complex organisms such as vertebrates. However, the number of genes is not necessarily smaller in simple animals than in complex animals, though the current estimates of gene numbers are quite crude. For example, the number of genes of *Amphioxus* is as large as that of humans. In plants, genome size varies considerably with species, and so does the number of genes. Some plants such as poplar and rice have two times more genes than vertebrates.

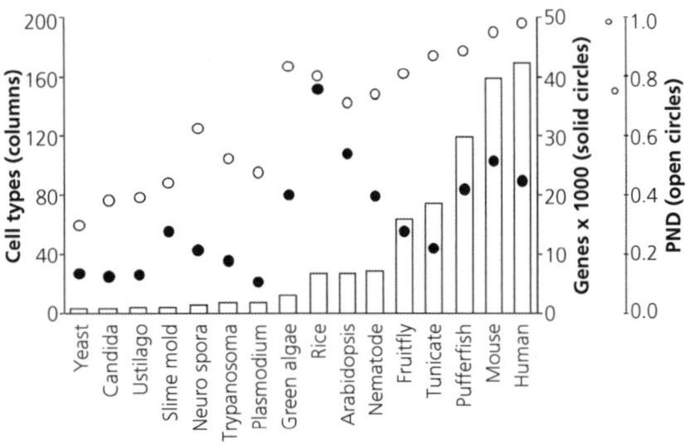

Fig. 5.2. Eukaryotic complexity (column) is not well correlated with the total number of genes (closed circle) ($r = 0.44$), but it is correlated with the proportion of noncoding DNA (open circles) ($r = 0.78$). Data from Vogel and Chothia (2006) and Taft et al. (2007).

Gene Numbers and Phenotypic Complexity

Table 5.2 and Fig. 5.2 clearly show that the DNA content or the number of genes in the genome is not necessarily high in complex organisms. What is then correlated to organismal complexity? To answer this question, we must have some quantity that measures the extent of organismal complexity. In practice, it is not easy to find such a measure, but at present the complexity of an organism is often measured by the number of cell types in the organism. If we use this measure, the relationship between the number of genes and the complexity becomes as given in Fig. 5.2. There is some correlation between the two quantities, but the correlation is still poor. Interestingly, Taft et al. (2007) reported that the proportion of noncoding DNA in the genome is roughly correlated to the complexity as measured by the number of cell types (Fig. 5.2). If these results are reliable, organismal complexity appears to be controlled by the noncoding regions of DNA, where many different regulatory elements of gene expression reside. However, the real reason for this relationship should be studied in more detail.

Vogel and Chothia (2006) used a different approach to study this problem. They considered the size of each gene superfamily (a group of related gene families) and studied the relationship between this family size and the number of cell types for 38 unicellular and multicellular eukaryotes. They found that of the 1219 superfamilies examined, there were 194 whose sizes are strongly correlated with the number of cell types. Half of these superfamilies were involved in extracellular processes or regulation. Half of all superfamilies examined had no significant correlation with complexity. These results suggest that organismal complexity is generated not by all gene families but by some specific ones.

It is therefore important to know the gene families that really contribute to the emergence of complex organisms. Actually, there are many examples

Table 5.3. Numbers of functional genes and pseudogenes in the multigene families for chemosensory receptors and immunoglobulins.

	Olfactory	Pheromone	Taste	Immunoglobulin		
	OR	V1R	T2R	IgVH	IGVλ	IGVκ
Humans	388 (414)	5 (115)	25 (11)	46 (84)	33 (38)	34 (38)
Mouse	1063 (328)	187 (121)	35 (6)	89 (64)	3 (0)	80 (78)
Dog	822 (278)	8 (33)	16 (5)	81 (66)	43 (61)	16 (9)
Cow	1152 (977)	40 (45)	19 (15)	12 (5)	23 (9)	9 (13)
Opossum	1198 (294)	98 (30)	29 (5)	24 (7)	45 (27)	76 (48)
Platypus	348 (370)	270 (579)	5 (1)	43 (21)	14 (7)	9 (9)
Chicken	300 (133)	0 (0)	3 (0)	1 (48)	1 (24)	0 (0)
Xenopus	1024 (614)	21 (2)	52 (12)	38 (42)	8 (4)	45 (10)
Zebrafish	155 (21)	2 (0)	4 (0)	38 (9)	0 (0)	8 (5)
Lamprey	40 (27)	3 (?)	0 (0)	0 (0)	0 (0)	0 (0)
Amphioxus	34 (9)	0 (0)	0 (0)	0 (0)	0 (0)	0 (0)

IgVH, IGVλ, and IGVκ represent the numbers of immunoglobulin variable region of H, λ, and κ genes, respectively (see Fig. 5.7). Functional genes are given together with pseudogenes (parentheses). From Das et al. (2008), Nei et al. (2008), Grus and Zhang (2009), and others.

of gene families the size of which have increased with evolutionary time. This is particularly so in the gene families concerned with phenotypic characters that are expressed in later stages of development. A conspicuous example is the gene family encoding olfactory receptors (ORs) (Table 5.3). In primitive chordates (*Amphioxus*) and jawless fishes, the number of OR genes is quite small. In teleost fishes (e.g. zebrafish), however, the number is in the order of 100. In land animals, the number of OR genes is even larger, and it ranges from 300 to 2000. Frogs, opossums, cows and mice have more than 1000 functional genes. These results suggest that the increase of OR genes was an important factor for the transition of animals from aquatic life to terrestrial life. Similarly, pheromone and taste receptor genes are considerably larger in mammals than in fishes. It is also known that the sizes of many homeobox gene families have increased from simple animals to complex animals (Nam and Nei 2005). These results suggest that the increase in the number of genes has been important for generating complex organisms.

Immunoglobulins or antibodies are known to exist only in jawed vertebrates. Therefore, amphioxus and lamprey do not have any immunoglobulins. However, the number of variable region genes (*H*, λ, κ chain genes) (see Fig. 5.7) are nearly the same for all jawed vertebrates, though some variable region genes are absent in certain groups of species (Table 5.3). The chicken genome contains only one functional gene in the heavy (*H*) and λ variable gene regions, but there are many pseudogenes. Actually, these pseudogenes are not functionless, because they are used for diversification of the functional gene by somatic gene conversion. Immunoglobulin genes are quite complicated, but these genes apparently have played important roles in vertebrate evolution.

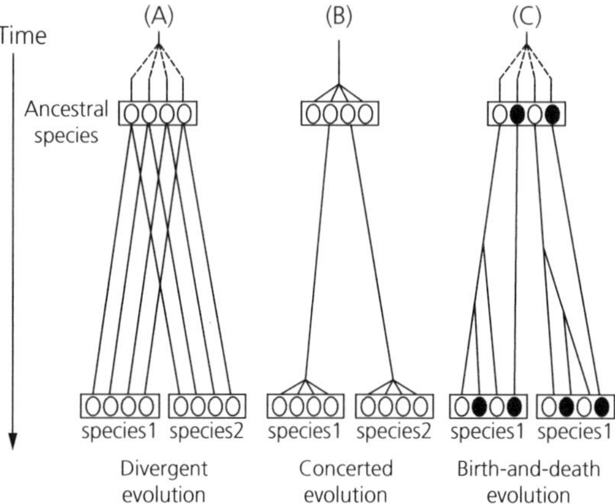

Fig. 5.3. Three different models of evolution of multigene families. Open circles stand for functional genes and closed circles for pseudogenes. Nei and Rooney (2005).

5.2. Evolution of Multigene Families

In the 1960s and 1970s it was customary to study gene evolution by examining each gene or a few related genes separately as in the case of cytochrome c and globin genes. However, we now know that most genetic systems or phenotypic characters are controlled by many multigene families (Nei and Rooney 2005; Vogel and Chothia 2006). Here a genetic system means any functional unit of biological organization such as the adaptive immune system in vertebrates, flower development in plants, meiosis, and mitosis. Therefore, it is important to understand the evolution of multigene families and their interaction.

The evolution of multigene families has been a subject of controversy for many years. The paradigm of evolution of multigene families before 1970 was that of hemoglobin α, β, γ, and δ chains and myoglobin (Fig. 5.1). The genes encoding these polypeptides or proteins are phylogenetically related and have diverged gradually as the duplicate gene acquired new gene functions. This mode of evolution is called divergent evolution (Fig. 5.3A). Around 1970, however, a number of researchers showed that ribosomal RNAs (rRNAs) in *Xenopus* are encoded by a large number of tandemly repeated genes and that the nucleotide sequences of the intergenic regions of the genes are more similar within a species than between two related species (e.g. Brown et al. 1972). These observations were difficult to explain by the model of divergent evolution, and a new model called concerted evolution was proposed (Fig. 5.3B). In this model all the members of a gene family are assumed to evolve in a concerted manner rather than independently, and a mutation occurring in a repeat spreads through the entire member genes by repeated occurrence of unequal crossover or gene conversion. This model is capable of explaining previously puzzling observations about the evolution of rRNA genes (see Section 5.3).

This apparent success led many authors to believe that most multigene families evolve following the model of concerted evolution, and a number of authors investigated the evolutionary modes of various multigene families (Hood et al. 1975; Zimmer et al. 1980; Ohta 1983). Later, however, the applicability of concerted evolution to some gene families was questioned as more DNA sequence data became available (Gojobori and Nei 1984; Hughes and Nei 1990), and another model called birth-and-death evolution (Nei and Hughes 1992) was proposed. In this model new genes are created by gene duplication, and some duplicated genes are maintained in the genome for a long time, whereas others are deleted or become nonfunctional through deleterious mutations (Fig. 5.3C). This model was originally proposed to explain the modes of evolution of multigene families concerned with immune systems genes such as immunoglobulin and major histocompatibility complex (MHC) genes (Hughes and Nei 1993; Ota and Nei 1994; Nei et al. 1997) and disease-resistance genes (Zhang et al. 2000). In recent years, however, most multigene families have been shown to evolve following the model of birth-and-death evolution (Nei and Rooney 2005). Yet the

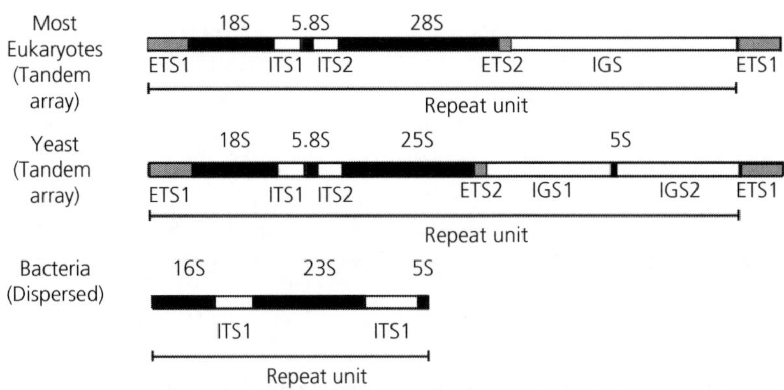

Fig. 5.4. Molecular structures of rRNA gene repeats in different organisms. From Nei and Rooney (2005).

controversy over the evolution of multigene families continues, partly because there are so many different types of gene families and partly because the general mechanism of gene conversion is still unclear (Klein et al. 2007). In the following two sections I would like to present the main features of concerted and birth-and-death evolution.

5.3. Concerted Evolution

Unequal Crossover, Gene Conversion, and Purifying Selection

One of the best-known examples of concerted evolution comes from the study of rRNA genes. The rRNA gene family of the African toads *Xenopus laevis* and *X. mulleri* consists of about 450 repeat genes or members. Each member gene consists of the 18S, 5.8S, and 28S RNA genes, external transcribed spacers (ETS1 and ETS2), internal transcribed spacers (ITS1 and ITS2), and an intergenic spacer (IGS) (Fig. 5.4). Using DNA or RNA hybridization techniques, Brown et al. (1972) showed that the nucleotide sequences of IGS are very similar among member genes of the same species but differ by about 10 percent between *X. laevis* and *X. mulleri*. This observation could not be explained by the then popular model of divergent evolution. According to this model, the differences in nucleotide sequence between different repeats of the same species are expected to be as large as those between repeats of different species. The explanation becomes more difficult to accept if we note that the nucleotide sequences of the 18S and 28S coding regions are virtually identical between *X. laevis* and *X. mulleri*. Actually, the 18S and 28S coding regions are very similar even among distantly related organisms such as animals and plants.

This puzzling observation can be explained by the model of concerted evolution originally proposed by Brown et al. (1972). According to this model, unequal crossover occurs randomly among members of a gene family, and repeated occurrence of unequal crossover has an effect to homogenize the member genes. In this case the number of member genes may increase or decrease by chance, but a certain range of the number of genes is maintained because of the functional requirement. In the absence of mutation this process will eventually make all member genes identical (Fig. 5.3B). In reality, of course, mutation always occurs, so that a gene family is expected to have some variant genes. It should now be clear that when a species diverges into two species and the gene clusters in each descendant species evolve independently, the clusters within species tend to have similar gene copies because of unequal crossover, whereas the genes belonging to the clusters of the two species gradually diverge by mutation. This is exactly what we observe in the IGS regions of rRNA genes in *Xenopus*. Later Smith (1976) conducted computer simulations to show that concerted evolution indeed can explain the observation about the evolutionary change of IGS region. As was previously mentioned, the 18S and 28S coding regions are virtually identical between *X. laevis* and *X. mulleri* as well as between different copies of the same species. This identity has apparently been maintained by strong purifying selection that operates in the coding regions. Thus, we can explain the entire observation about the rRNA gene family in *Xenopus* in terms of unequal crossover, mutation, and purifying selection.

In addition to these factors, gene conversion (Jeffreys 1979; Slightom et al. 1980) was also proposed to explain the homogenization of member genes of multigene families. Here the gene conversion hypothesis is different from the gene conversion event studied in fungal species (e.g. *Neurospora*). In this hypothesis the DNA sequence of one of several tandem copies of a gene was assumed to be converted by another copy so that the sequence of the converted gene becomes identical with the donor copy. For example, the amino acid sequence of the two γ-globin genes in humans are identical with each other except for one residue, where one copy has glycine (G) and the other has alanine (A). However, the DNA sequences of the flanking regions of the two genes were different. For these reasons, Slightom et al. (1980) proposed that the two globin genes are homogenized by an occasional gene conversion event, which was clearly different from the gene conversion event studied in fungi, but the authors did not provide any biological mechanism.

Yet, this hypothesized gene conversion was simpler to explain the homogenization of multiple copies than the original idea of unequal crossovers.

This also simplified the mathematical treatment of concerted evolution (Birky and Skavaril 1976; Nagylaki and Petes 1982; Ohta 1982). For this reason, the gene conversion hypothesis became popular even in the case of rRNA genes. However, recent molecular studies have shown that homogenization of rRNA gene copies is caused primarily by unequal crossover rather than by gene conversion (Eickbush and Eickbush 2007). Furthermore, in most mathematical formulations the effects of purifying selection operating in the coding regions of 18S and 28S genes are neglected. Therefore, caution should be exercised in the application of the mathematical formulas to real data. The gene conversion theory has also become popular among researchers of MHC polymorphism, as will be discussed in Section 5.4.

Note that the relative contributions of unequal crossover (or gene conversion) and purifying selection to the homogenization of the rRNA genes have rarely been discussed. For this reason, the homogeneity of the rRNA-coding regions (18S and 28S) was often attributed to unequal crossover rather than to purifying selection. Actually, even the IGS regions appear to be subject to purifying selection in *Xenopus* because this region contains elements of promoters and enhancers (Robinett et al. 1997; Caudy and Pikaard 2002). It is therefore necessary to keep in mind that concerted evolution applies primarily to the IGS region, and even in this region a substantial proportion of mutations may be eliminated by purifying selection. It should also be noted that the RNA gene cluster usually contains many pseudogenes caused by deleterious mutations in the coding or regulatory regions (Brownell et al. 1983).

The 5S rRNA genes form separate gene clusters in most eukaryotic organisms and are located in different genomic regions. This gene family includes 9000–24 000 member genes in *Xenopus* (Brown and Sugimoto 1974) and about 500 members in humans (Gonzalez and Sylvester 2001). These 5S rRNA genes are also known to undergo concerted evolution (Brown and Sugimoto 1974). Furthermore, the gene families of small nuclear RNAs (snRNA) involved in intron splicing and other important cellular functions apparently undergo concerted evolution. The study of concerted evolution of U2 snRNA genes in primates has been conducted by Weiner and his group (Pavelitz et al. 1995; Liao et al. 1998), and these authors have shown that the coding regions of U2 snRNA gene members are very similar to one another but the intergenic regions are heterogeneous within each species. These results again demonstrate the importance of purifying selection in the coding regions.

The above survey of the evolutionary changes of various forms of RNA genes indicates that these genes have been maintained to serve essentially the same function for a long time and this has been achieved by either concerted evolution or purifying selection or both. This is understandable because all the RNA genes considered here play a pivotal role for the maintenance of cellular or physiological functions and they should not change so easily, to avoid any disruption of their functions. It should be noted that for this purpose unequal crossover and gene conversion play the same role and the distinction between the two mechanisms is not of biological importance.

Tandemly Arrayed Histone Genes

Not long after the evolution of rRNA gene families was explained by the model of concerted evolution, many researchers began to assume that this model applies to various other multigene families. The general view at that time was that a gene family that produces a large amount of gene products is subject to concerted evolution to homogenize the genes. One such family was the histone gene family of sea urchins (Kedes 1979; Hentschel and Birnstiel 1981; Holt and Childs 1984). This is a large multigene family with several hundred members that are divided into four different classes on the basis of developmental and tissue-specific expression patterns: (a) "early histone genes" that are active during late oogenesis through the blastula stage of embryogenesis, (b) "cleavage stage histone genes" that encode the first histones expressed after fertilization, (c) "late histone genes" that are active from the late blastula stage onwards, and (d) "sperm histone genes" that are expressed only during spermatogenesis (Maxson et al. 1983; Mandl et al. 1997).

The chromosomal arrangement of histone genes varies with class of genes and species (Fig. 5.5). In most sea urchin species the early histone genes are

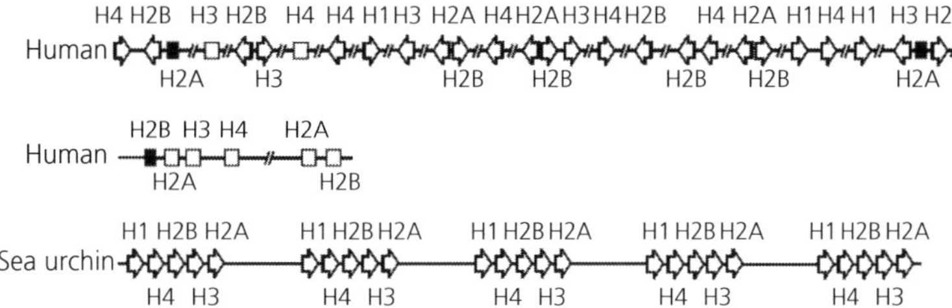

Fig. 5.5. Genomic structures of histone genes in humans and sea urchins The black square box in human histone gene repeats represents a pseudogene. Expressed histone genes are shown in white, and an arrow indicates the transcriptional direction when known. From Nei and Rooney (2005).

present in about 300–500 repeat units. In the sea urchin *Lytechinus pictus*, they are arranged in tandem arrays that consists of virtually identical repeating units of the 5 histone genes (H1, H2A, H2B, H3, and H4), each of which is separated from the other by noncoding intergenic sequence (IGS) regions. In this species, only the early genes are present in tandem array, whereas the other three classes of genes appear to be dispersed throughout the genome and present in significantly fewer copy numbers. The IGS regions of the early gene tandem arrays in *L. pictus* show a considerable amount of variation, whereas the protein-coding regions are highly conserved. This was taken as evidence for concerted evolution in the early study of histone genes. Not long afterwards, researchers studying the sequence divergence of late histone genes of *L. pictus* claimed that these genes also undergo concerted evolution, as did another group of researchers studying the late genes of the sea urchin *Strongylocentrotus purpuratus* (Maxson et al. 1983).

By the end of the 1980s, most researchers in the field had concluded that virtually all multigene families evolve in a concerted fashion. Therefore, the studies of histone genes from sea urchins and other species were perceived as confirmation of the general view. Many authors have claimed that even the genetic variability of MHC loci is caused by gene conversion, and this was thought to be a source of enhancing genetic variability within loci rather than a homogenizing factor (Mellor et al. 1983; Ohta 1983; Weiss et al. 1983). These views remained popular until DNA sequence data became available in the late 1980s and a new model of evolution called birth-and-death evolution was shown to be more appropriate (see Section 5.4).

It is interesting to note that concerted evolution primarily depends on unequal crossover or gene conversion, which is a mutational process, and positive selection was rarely considered. The selection needed here was purifying selection. Yet, this model was often considered to belong to the realm of neo-Darwinism (e.g. Arnheim 1983). In the following I will consider the birth-and-death model of evolution, which is also largely dependent on a mutational process but positive selection is not excluded.

5.4. Birth-and-Death Evolution

MHC Genes

The birth-and-death model of evolution of multigene families was first proposed to explain the unusual pattern of evolution of MHC genes in mammals (Nei and Hughes 1992; Nei et al. 1997). The function of MHC genes is to bind foreign peptides and present them to T lymphocytes, thereby triggering an immune response. MHC genes can be divided into class I and class II genes on the basis of molecular structure and function of the polypeptide encoded. Class I genes can be further divided into classical and nonclassical genes. The classical class

I (Ia) genes are highly polymorphic, and the number of alleles per locus within a species sometimes exceeds 100. This high degree of polymorphism is important for protecting the host from the attack of various types of parasites (viruses, bacteria, fungi, and others), which are always changing with time (Chapter 4). By contrast, the nonclassical Class I (Ib) genes are less polymorphic and their functions may be quite different from those of Ia genes.

In the 1970s and 1980s when most investigators believed that multigene families were generally subject to concerted evolution, the MHC gene family was no exception. Therefore, some authors attempted to explain the polymorphism of MHC genes by means of unequal crossover or gene conversion (Lopez de Castro et al. 1982; Ohta 1983; Weiss et al. 1983). In particular, Ohta (1983) and Weiss et al. (1983) proposed that the high degree of polymorphism at Ia loci could be explained by gene conversion. This view was based on the idea that if some parts of a sequence at a monomorphic locus are converted by a nucleotide sequence from another monomorphic locus, polymorphism may be generated at the first locus. Enhancement of polymorphism would occur even if both loci are polymorphic to some extent as long as the nucleotide sequences between the two loci are sufficiently different and gene conversion occurs in both ways between the two loci. One problem with this idea is that it does not explain why and how gene conversion starts to occur between two previously monomorphic loci. The coexistence of Ia and Ib MHC genes in the same DNA region is also difficult to explain by the gene conversion hypothesis. If gene conversion occurs continuously between the two loci, the extent of polymorphism should be essentially the same for the two loci. Furthermore, if phylogenetic analysis is conducted for the alleles from different loci, there would be no monophyletic clades formed for each locus. In reality, this is not the case. Kriener et al. (2000a) critically examined data that seemingly supported the idea of concerted evolution and concluded that the evidence is weak. They argued that some of the data showing the identical gene segments between paralogous pairs of genes can be explained by co-ancestry of the segments or even clustered mutations (Kriener et al. 2000b).

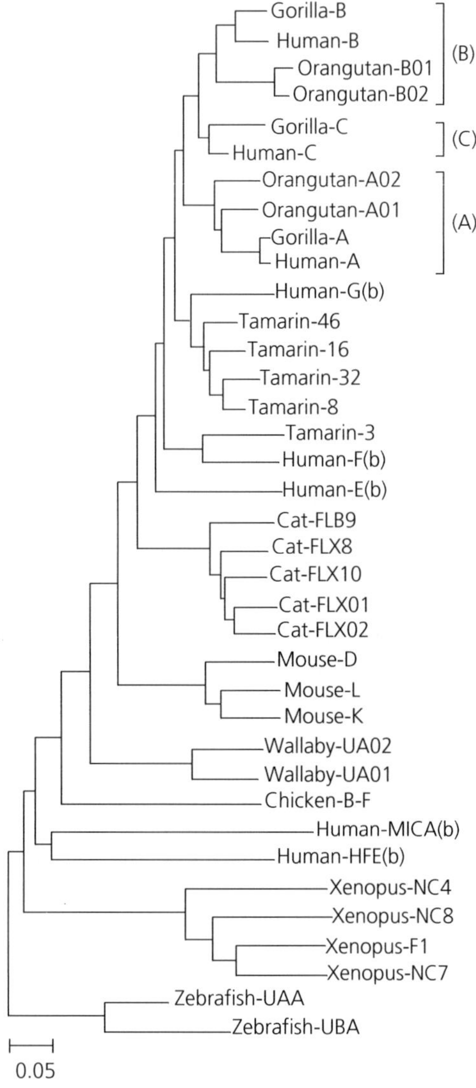

Fig. 5.6. Phylogenetic tree of class I MHC genes from different vertebrate species. Class I A, B, and C loci are shared by hominoids and Old World monkeys, but they are absent in other species, where other classical class I loci are polymorphic. Symbol (b) indicates a nonclassical locus. Modified from Nei et al. (1997).

The idea of gene conversion was weakened considerably when Hughes and Nei (1988, 1989b) showed that MHC polymorphism is primarily caused by overdominant selection that operates at the peptide-binding region of MHC molecules. This finding made it unnecessary to invoke gene conversion as an explanation for MHC polymorphism. It also provided a theoretical basis for the concept of trans-species polymorphism (Chapter 4) previously discovered by Figueroa et al. (1988), Lawlor et al. (1988), and McConnell et al. (1988). In a phylogenetic analysis of MHC class I and class II genes, Hughes and Nei (1989, 1990) and Nei et al. (1997) showed that the evolutionary pattern of these genes is very different from what would be expected under concerted evolution. The phylogenetic tree of class I genes from a number of vertebrate species is presented in Fig. 5.6. It indicates that different orders or families of mammals generally have different genes or genetic loci. For example, the classical loci A, B, and C are shared only by hominoid species (e.g. human, gorilla, and orangutan) and Old World monkeys, and New World monkeys (e.g. tamarin) and nonprimate mammals do not have the genes. Similarly, cats and mice have different Ia loci. In other words, different families or orders of mammals do not share truly orthologous genes. This evolutionary pattern indicates that some genes were generated by gene duplication and some duplicate genes were lost after the divergence of mammalian orders. Actually, the genomic regions of human and mouse MHC genes contain a large number of pseudogenes, exactly as would be expected under the birth-and-death model (Fig. 5.3C). This conclusion makes sense biologically because the genetic variability at MHC loci is generated to defend the host from many new types of parasites. Gene conversion is not useful for this purpose, though it may occur.

Figure 5.6 also shows that a few nonclassical loci as indicated by (b) diverged from Ia loci a long time ago and now have different functions. For example, the human nonclassical gene HFE(b) was separated from the classical genes A, B, and C by gene duplication and now has acquired a new function. It has the ability to form complexes with the receptor for iron-binding transferrins and thus regulate the uptake of dietary iron by cells of the intestine (Feder et al. 1998). A mutation of this gene is known to cause the genetic disease *hemochromatosis*. This type of acquisition of new functions by duplicate genes is also an important feature of birth-and-death evolution. Phylogenetic analysis of class II region genes showed essentially the same evolutionary pattern as that of class I genes (Gu and Nei 1999). However, the rate of gene birth and gene death is much lower in the class II gene family than in the class I gene family.

(A) Immunoglobulin

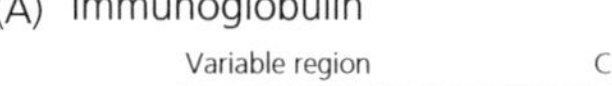

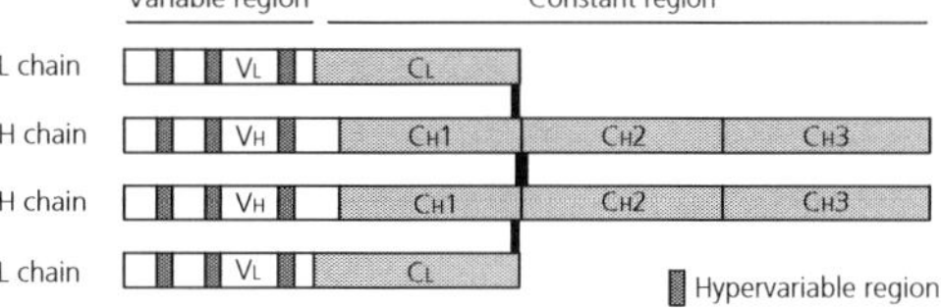

(B) Cartilaginous fish (sharks)

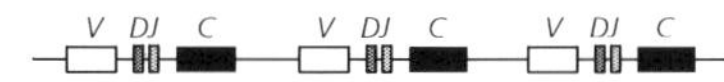

(C) Mammals

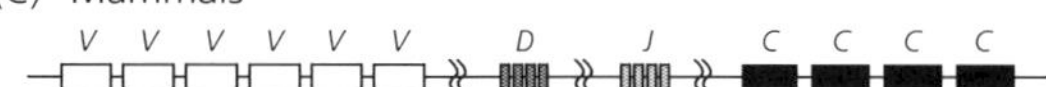

Fig. 5.7. (A) Basic structure of an immunoglobulin (Ig) molecule. Genomic structures of Ig genes in (B) cartilaginous fish and (C) mammals. An Ig is composed of two heavy (H) chain polypeptides and two light (L) chain polypeptides. Each chain consists of one or three constant (C) regions (C_H or C_L) and a variable region (V). The antigen binding of an immunoglobulin is achieved primarily by the variable region. In mammals the L chains can be classified into the λ and the κ chains. In the formation of a mammalian H chain, one of each of the V, D, and J domains in the genome is combined with C regions. In the L chain the J domain is missing.

Immunoglobulins and other Immune Systems Genes

An *immunoglobulin*, also known as an *antibody*, is a large Y-shaped protein used by the immune system to identify and neutralize foreign parasites such as bacteria and viruses. Human immunoglobulins are composed of heavy (H) chains and light chains, and the light chains can be further divided into λ and κ chains (Fig. 5.7). Each of the three groups of chains consists of constant (C) and variable (V) regions, and the polypeptides of variable regions in each category are primarily responsible for binding foreign antigens. They are encoded by a genomic cluster, called a variable region gene family. There are about 70–130 genes in each of the H, λ, and κ variable region gene families in humans (Table 5.3). All of these multigene families are subject to birth-and-death evolution (Ota and Nei 1994), and about 50 percent of the genes are pseudogenes (Matsuda et al. 1998). Because immunoglobulins are composed of constant regions and variable regions of different polypeptides, together with additional D (diversity) and J (joining) components (Fig. 5.7), the number of different types of immunoglobulins produced is very large. For this reason, immunoglobulins are capable of protecting the host from the attack of millions of different foreign antigens (Tonegawa 1983). It is interesting to note that this system of host protection is different from that of the MHC system, in which a large number of alleles from a limited number of loci are used for protecting the host from different foreign antigens.

Another interesting point to be noted is the difference in the genomic structure of immunoglobulin genes between jawless cartilaginous fishes (e.g. sharks) and other jawed vertebrates. In cartilaginous fishes, many units of the gene cluster V-D-J-C are present in the genome, and each unit produces the immunoglobulin heavy chain. In mammals, however, the V, D, J, and C region gene segments are present as tandem clusters in the genome, and in the formation of a mature H-chain polypeptide one of each of the V, D, and J clusters are combined with three of the constant region genes. It is often stated that the mammalian system is superior to the cartilaginous system because the former can generate a large number of immunoglobulin molecules from a limited amount of genomic region. In practice, this argument is controversial since the cartilaginous system works well in this group of organisms. However, it is interesting to note that the two different systems were generated from the same ancestral (simpler) system probably accidentally by mutational processes.

At this stage, it should be mentioned that camels and llamas, which belong to the family *Camelidae*, have an unusual type of immunoglobulin. These immunoglobulins do not contain the light chain (neither λ nor κ chain) and consist of only the heavy chain (Hamers-Casterman et al. 1993). About 50 percent of the immunoglobulins used in these organisms are of this type, and the other 50 percent are the normal immunoglobulins containing the light chain. However, this new type of immunoglobulin has a normal function, and there is no reason to believe that this unusual immunoglobulin molecule is necessary for camels and llamas in addition to the ordinary ones. The immunoglobulin

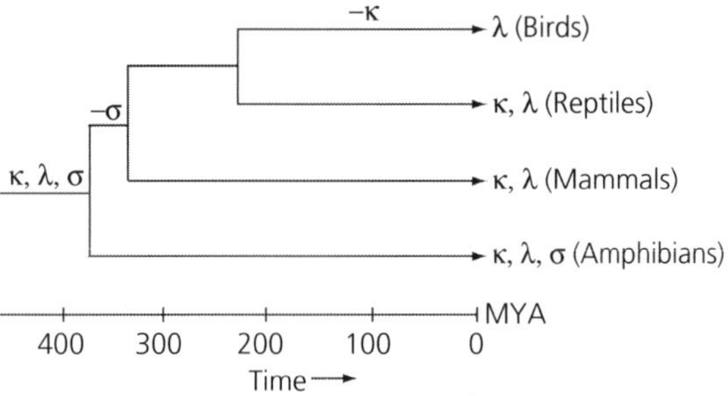

Fig. 5.8. Evolutionary changes in the three different types (κ, λ, σ) of immunoglobulin light chain genes in tetrapods. MYA: million years ago. Modified from Das et al. (2009).

molecule without the L chain apparently occurred accidentally by mutation(s) and has spread through the camel and llama genomes by selection or genetic drift. This example also indicates the importance of mutation in evolution.

Another interesting observation about light chains is that the presence of both the λ and the κ chains is not a requirement for the function of immunoglobulins. In fact, bird species and a group of microbats do not have κ chain genes at all. It should also be noted that *Xenopus* has an additional light chain called the σ chain, but the gene for this chain seems to have been lost in the process of evolution of reptiles, birds, and mammals (Fig. 5.8). This observation suggests that the light chains of immunoglobulins are relatively unimportant and different chains are interchangeable.

There are many other gene families that play important roles in the adaptive immune system in vertebrates. One of them is that for T-cell receptors (Klein and Horejsi 1997). The molecular structure and the genomic organization of T-cell receptors are similar to those of immunoglobulins, and the variable region gene families for different classes of T-cell receptor genes are also known to be subject to birth-and-death evolution (Su and Nei 2001). Most of these gene families include many pseudogenes.

In addition to the above multigene families, the gene families concerned with innate immunity (immune systems without life-long memory) have also been shown to undergo birth-and-death evolution (Hao and Nei 2005; Nikolaidis et al. 2005). For example, the natural killer (NK) cell receptors of humans are composed of immunoglobulin-like domains (KIR), but the rodent receptors are of the lectin type called lectin-like killer cell receptors (KLR or Ly49), and the molecular structures of these two groups of receptors are very different (Klein and Horejsi 1997). It is unclear how these different types of NK cell receptors evolved in two different orders of mammals. Both KIR and KLR gene families are known to be subject to birth-and-death evolution. KIR genes are also subject to domain shuffling as well as to nucleotide changes (Trowsdale et al. 2001; Rajalingam et al. 2004). Furthermore, the number of member genes of these families has expanded very rapidly by gene duplication during the past 20–30 million years (Khakoo et al. 2000; Hao and Nei 2005). Yet about half of these genes are apparently nonfunctional (Kelley et al. 2005).

Olfactory and other Chemosensory Receptor Genes

The number of member genes (family size) of the MHC and immunoglobulin variable gene families are quite large, but the largest gene family in mammals is that of olfactory receptor (OR) genes. Olfactory (odor molecules) receptors are G protein coupled receptors that contain 7 α-helical transmembrane regions. OR genes are expressed in sensory neurons of olfactory epithelia in nasal cavities. The human and mouse genomes contain about 800 and 1400 genes, respectively. However, 50 percent of human genes and about 30 percent of mouse OR genes are pseudogenes (Table 5.3). These genes are scattered over many different locations of almost all chromosomes, and they are generally located as a tandem array in each genomic region. It is relatively easy to identify the orthologous gene pairs between humans and mice by phylogenetic analysis. This analysis suggests that gene conversion or unequal crossover has not occurred frequently and that the number of OR genes apparently has increased by tandem gene duplication and chromosomal rearrangement (Niimura and Nei 2005).

It is interesting to note that olfactory receptors are for detecting different odor substances (odorants) among millions of odorants that are floating in the air or water. It is known that one odorant receptor is capable of perceiving a few different odor molecules and one odorant can be detected by several different receptors. Therefore, a few hundred to about 1000 OR genes are capable of detecting millions of different odors. This genetic system of coping with odorant diversity is quite different from that of the immune systems for protecting the host from millions of different parasites.

In reality, the ability to smell is controlled not only by the number of OR genes but also by the brain function for odor recognition, and the human brain possibly has a higher power of distinguishing between subtle differences in odor molecules than the mouse brain (Shepherd 2004). At present, however, the mechanism of odor recognition in the brain is virtually unknown, and therefore we do not consider this factor any further here.

Table 5.3 shows the number of functional OR genes and OR pseudogenes in various species of vertebrates. The extent of variation of the number among different species is enormous, and this variation must have been caused largely by the environmental conditions and the lifestyle to which each species is adapted. In general, terrestrial animals have more OR genes than marine animals, apparently because terrestrial animals utilize both aquatic and airborne odorants for their successful life. This suggests that the increase of OR genes in terrestrial animals was an important factor for the transition of animals from aquatic to terrestrial life. It is also interesting to see that the platypus genome has a relatively small number of functional OR genes and a large fraction of pseudogenes (52 percent) than other mammals. This is probably caused by their special lifestyle. Platypuses are semiaquatic animals and have a special sense in their bills, which combines electroreception and mechanoreception (Pettigrew 1999). Platypuses can find prey with their eyes, ears, and nostrils closed. Therefore, they don't need many airborne OR genes (Nei et al. 2008).

However, it is not always easy to explain the number of OR genes by the environmental requirements or lifestyle of the organism. For example, dogs are known to be very sensitive to odor perception, but their number of OR genes is smaller than that of cows and opossums. Here again the number of OR genes alone cannot explain the capability of olfaction. To understand the biological basis of olfaction, we must consider the brain function as well.

Pheromones are water-soluble chemicals that are emitted and sensed by individuals of the same species to elicit reproductive behaviors or other changes to physiological characters. In terrestrial vertebrates pheromones are perceived by the vomeronasal organ (VNO), which is located at the base of the nasal cavity and is separated from the main olfactory epithelium. One supergene family that controls the VNO receptors is called the V1R (vomeronasal receptor 1) gene family (Dulac and Axel 1995). Pheromone receptors are G protein coupled receptors similar to olfactory receptors, but there is little sequence similarity between the two receptor proteins (Nei et al. 2008).

The mouse genome has about 310 V1R genes, but the number of functional genes is 187. The rat genome has 102 functional genes and about 50 pseudogenes (Grus et al. 2005). Similarly, opossum and cow have a substantial number of functional genes. By contrast, the human has only 5 functional genes and 115 pseudogenes. The number of functional genes in dogs is also quite small. The difference between the mouse and human genomes apparently occurred by massive pseudogenization or deletion of V1R genes in the human lineage. In fact, some primate species including humans do not have functional VNOs and therefore are thought to have no perception of vomeronasal pheromones. This pseudogenization or deletion of V1R genes in human lineage apparently occurred because humans use visual and auditory senses for sexual and physiological behavior. This example explains why the number of copies of a multigene family can vary so much among different orders of mammals.

Vertebrate species are known to have many other types of chemosensory receptor genes such as those for taste perception. These genes are also known to vary among different species considerably and to be subject to birth-and-death evolution (Nei et al. 2008).

Birth-and-Death Evolution with Strong Purifying Selection

In the 1980s when genetic variation of multigene families was studied by restriction enzyme analysis, many gene families that are required to produce a large quantity of gene products were assumed to be subject to concerted evolution. One example is the histone gene family, as mentioned earlier. In this gene family, even the authors who studied nucleotide sequences believed that histone gene families were subject to concerted evolution (Matsuo and Yamazaki 1989). This view apparently arose from their preconception about the prevalence of gene conversion.

By the 1990s, however, a substantial number of sequences of histone genes had been obtained from various species of animals, plants, fungi, and protists. Rooney et al. (2002) and Piontkivska et al. (2002) conducted an extensive statistical analysis of these data to examine whether the histone gene families are subject to concerted evolution or birth-and-death evolution. They reasoned that if concerted evolution is the main factor, both the number of synonymous differences per synonymous site (p_S) and the number of nonsynonymous differences per nonsynonymous site (p_N) must be virtually 0 for any pair of genes because gene conversion affects both synonymous and nonsynonymous sites in the same way. By contrast, if protein similarity is caused by purifying selection and every member gene evolves independently, p_S is expected to be greater than p_N because in this case synonymous substitutions accumulate continuously whereas nonsynonymous substitutions are eliminated by purifying selection.

When this approach was applied to histone H3 and H4 genes from diverse groups of eukaryotic species, p_S was clearly higher than p_N in almost all cases (Piontkivska et al. 2002; Rooney et al. 2002; Kim and Yamazaki 2004). Similar results were also obtained from an extensive study of the histone H1 gene family (Eirin-Lopez et al. 2004). These results therefore clearly show that the histone gene families are subject to strong purifying selection but all member genes evolve according to a birth-and-death process.

Ubiquitin is also a highly conserved protein and plays a major role in both cellular processes and protein degradation in eukaryotes. Because of the high degree of protein sequence similarity among paralogous proteins, the gene family for this protein was also thought to be subject to concerted evolution (Sharp and Li 1987; Nenoi et al. 1998). This view changed when Nei et al. (2000) conducted an extensive statistical analysis of sequence data using p_S

and p_N values mentioned above. The results of this study clearly showed that it is purifying selection rather than concerted evolution that homogenizes protein sequences. In most species, nonsynonymous nucleotide differences among the member genes were 0, whereas the synonymous differences were virtually saturated.

5.5. Multigene Families and Evolution of New Genetic Systems

So far we have considered the evolution of each gene family without considering the interaction with other gene families. In practice, most genetic systems or phenotypic characters are controlled by the interaction of many multigene families. Here a genetic system means any functional unit of biological organization such as olfaction (odor recognition) and adaptive immunity in vertebrates, flower development in plants, meiosis, and mitosis. Evolution of these genetic systems is obviously very complicated, and our understanding of the evolutionary mechanism is very poor. At the present time, what we can do is to study the evolution of each component multigene family and speculate the possible course of evolution of interaction between different gene families. In the following sections, we will briefly consider a few examples of multigene families that are involved in the evolution of a specific genetic system.

Adaptive Immune System

One of the best-studied genetic systems in this respect is the evolution of the adaptive immune system in jawed vertebrates. In the adaptive immune system (AIS), lifelong immunity is maintained for certain groups of parasites (viruses, bacteria, fungi, and others) once the host is attacked by them. The jawless vertebrates and other nonvertebrate animals do not have this system, though most animals have so-called innate immune systems. How did the adaptive immune system evolve in jawed vertebrates? This is still an unsolved problem and is

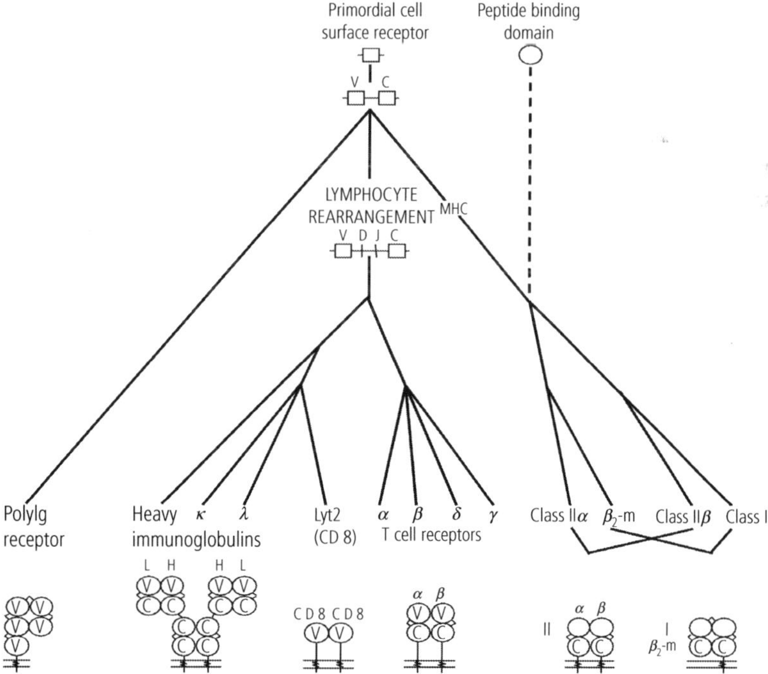

Fig. 5.9. Evolutionary relationships of immune system genes. Class I, MHC class I; class II, MHC class II; V, variable domain; C: constant domain. The peptide-binding domains of MHC molecules are structurally different from Ig variable domains. Modified from Hood et al. (1985). Reproduced with permission from Elsevier.

currently under investigation (van den Berg et al. 2004; Klein and Nikolaidis 2005). However, it is well known that this system works through the interaction of many different multigene families such as the MHC, immunoglobulin, and T-cell receptor gene families. Most of these multigene families are evolutionarily related (Fig. 5.9) and are apparently products of repeated processes of birth-and-death evolution. In other words, continuous operation of birth-and-death evolution appears to have generated a supergene family and this supergene family is the structural basis of the genetic system.

Of course, there are many other genes involved in the function of the AIS. The thymus, where the V, D, J, and C region genes are joined, also exists only in jawed vertebrates. Therefore, we have to explain how this organ evolved. Because the AIS evolved only in jawed vertebrates, some authors (e.g. Kasahara et al. 2004) proposed a big bang theory of evolution that occurred suddenly when jawed vertebrates evolved. However, examining the evolutionary histories of the component gene families, Klein and Nikolaidis (2005) rejected this hypothesis. They then proposed that the adaptive immune system evolved by recruiting genetic elements that have evolved primarily to serve other functions and by refining existing molecular cascades. This resulted in the appearance of new organs and new types of cells. In this view, the AIS has evolved gradually following Jacob's (1977) theory of evolutionary tinkering. However, it is important to remember that repeated processes of birth-and-death evolution also played a pivotal role in the evolution of the AIS. This type of evolutionary change is often called gene co-option or gene recruitment. A more detailed discussion will be presented in Chapter 6.

Homeobox Genes Involved in Animal and Plant Development

Homeobox genes are members of an important supergene family that control animal and plant development. They encode transcription factors that interact with *cis*-regulatory elements of protein-coding genes. They can be divided into two different groups: Typical and Atypical groups. Typical homeobox genes contain a homeobox of 60 codons, whereas Atypical group genes have homeobox genes with a few more or a few less codons (Burglin 1997). The Typical group includes several dozen gene families. A well-known example is the HOX gene family that plays a key role in determining the body pattern in animals. This family has also undergone birth-and-death evolution (Amores et al. 2004). However, because the genes in this family are linearly arranged as clusters in chromosomal regions and the arrangement is collinear with the development of body segments (e.g. head, thorax, and abdomen) controlled by the genes, there must be intricate interaction among the HOX genes. In bilaterian animals the HOX gene cluster is composed of 13 cognate group genes, and it is possible to infer the evolutionary history of the 13 cognate genes though the inference is very crude (e.g. Zhang and Nei 1996; Gehring et al. 2009). According to these studies, the HOX gene clusters evolved by sequential duplication and functional differentiation. Apparently because the HOX genes interact with one another in a very specific way, essentially the same gene cluster is maintained in almost all bilaterian animals, as mentioned in Chapter 3. Another important typical homeobox gene is *Pax6*, which is the master control gene of eye development (Gehring 1998).

The Atypical group includes about seven gene families, five of which are called the TALE group. The TALE group genes are characterized by three extra codons between the helix 1 and helix 2 regions. These gene families were all concerned with some aspects of development in eukaryotes and were derived from a common ancestor that existed before the separation of animals, plants, and fungi. In other words, these diverse gene families were generated by successive gene duplication and differentiation over a long period of time. Interestingly, occasional loss of paralogous genes also appears to contribute to the differentiation of phenotypic characters (Nam and Nei 2005).

However, for a particular morphological character to be formed, a large number of other genes are involved. At present, we know very little about the number of genes involved in the development of a particular character and how they interact except for simple characters that are expressed in a late stage of development (Carroll et al. 2005).

Multigene Families and Flower Development in Plants

Flowers of angiosperms (flowering plants) are composed of sepals, petals, stamens, pistils, etc., and differ from poorly developed flowerlike organs in gymnosperms. One group of genes that play important roles in flower development are transcription factors called MADS-box genes. There are several classes of MADS-box genes that are essential for flower development (Weigel and Meyerowitz 1994; Ma and dePamphilis 2000; Theissen 2001). In a phylogenetic analysis of MADS-box genes, Nam et al. (2003) suggested that MADS-box genes controlling flower development (floral MADS-box genes) originated from a common ancestral gene about 650 MYA. Tanabe et al. (2005) identified floral MADS-box like genes in three species of green algae, which are believed to have originated about 700 MYA. If we note that the oldest fossil records of angiosperms and gymnosperms are about 150 and 300 million years old, respectively, it appears that the ancestral genes of floral MADS-box existed long before the flowering system evolved. Tanabe et al. speculate that this group of genes originally controlled the development of haploid and diploid stages of green algae. MADS-box genes are ancient genes and are known to exist in plants, animals, and fungi. In animals they control muscle development. In a process of evolution of gymnosperms and angiosperms, however, a different groups of MADS-box genes seems to have evolved to form flowers (Nam et al. 2003). MADS-box genes also seem to have evolved through a birth-and-death process (Nam et al. 2004).

5.6. Genomic Drift and Copy Number Variation

In the process of birth-and-death evolution the number of gene copies of a gene family is expected to vary from time to time. This temporal change would occur partly because of random duplication and inactivation of genes and partly because of changes in environmental conditions. In the previous section we have seen that the number of gene copies per genome often varies extensively among different species.

One of the most well-studied multigene families with respect to this problem is the olfactory receptor genes in mammals. Figure 5.10 shows the number of functional OR genes in eight different mammalian species and estimates of the numbers

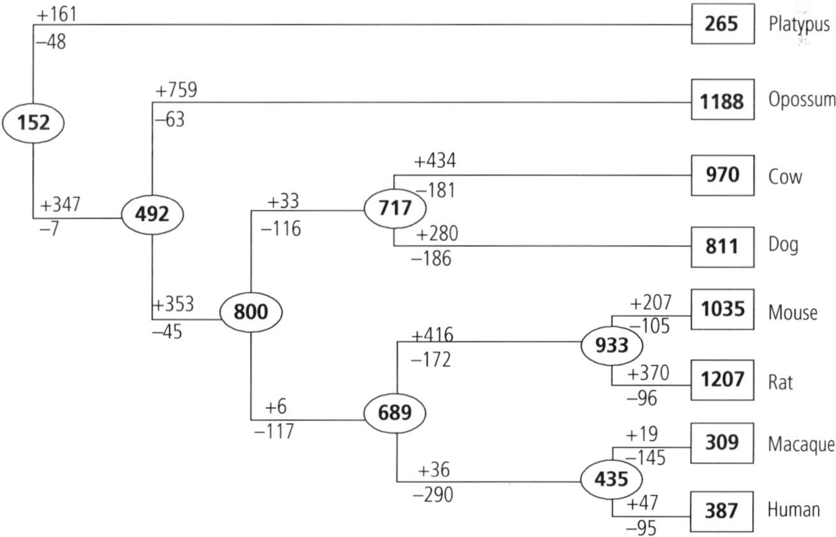

Fig. 5.10. Evolutionary changes in the number of olfactory receptor genes in mammalian evolution. The number of gene gains and gene losses for each evolutionary branch are given with + and − signs, respectively. From Nei et al. (2008).

of gene gains and gene losses in each evolutionary lineage from the common ancestor. This figure shows that the number of OR genes varies extensively among different mammalian species and that the numbers of gains or losses in the evolutionary process are also enormous. For example, the evolutionary lineage of mice gained 207 genes and lost 105 genes from the common ancestor of mice and rats. Similarly, the rat lineage gained 370 genes and lost 96 genes during the same evolutionary period. It should also be noted that opossums gained 759 genes but lost only 63 genes after divergence from placental mammals and that they now have as many as 1188 OR genes. By contrast, platypuses have only 265 genes. Generally speaking, a mammalian species with a small number of functional OR genes tends to have a large number of pseudogenes (Table 5.3).These extensive changes of OR genes in the evolutionary process is undoubtedly related to changes in the environment in which the organism lives. For most mammalian species, detection of millions of different odorants is crucial for their survival. Yet, animals living in different environments require different numbers of ORs (Table 5.3). For example, olfaction seems to be less important for primate species that are endowed with trichromatic vision than for other dichromatic mammalian species, because trichromatic vision is very powerful for perceiving environmental signals. This could be the reason why humans or macaques have a smaller number of OR genes than rodents. However, this view is controversial, and some experimental data do not support it (Matsui et al. 2010). Platypuses also show a smaller number of functional OR genes than the number of pseudogenes (Table 5.3). The real reason for this is unclear, but it may have to do with their semiaquatic lifestyle, as mentioned earlier.

Actually, if we consider the evolutionary changes of OR genes in many mammalian species, its relationship with environmental factors is not always clear, and there seem to be random elements that determine the number of OR genes. These random elements are of course caused by random duplication and random inactivation of genes. In other words, the number of OR genes may fluctuate around the most appropriate number of the genes for a given species, and this fluctuation appears to be quite high if we consider the existence of a large number of pseudogenes in many species. This type of random change of gene copy number is called the genomic drift (Nei 2007). Genomic drift has been observed with many other gene families (Nozawa et al. 2007; Hollox et al. 2008), and they now appear to be an important evolutionary factor at the genomic level. They are also known to cause various genetic diseases (Gonzalez et al. 2005; Hollox et al. 2008).

When genomic drift occurs frequently, we would expect that different individuals have different gene copy numbers with respect to a multigene family. That this is indeed the case has been shown by Nozawa et al. (2007) with respect to chemosensory receptor genes as well as with all annotated gene copies (Sebat et al. 2004; Redon et al. 2006). Figure 5.11A shows the distributions of relative copy number of functional OR genes and pseudogenes in the human population. It is interesting to note that both the distributions approximately follow the normal distribution suggesting that the gain and loss of OR genes occur more or less randomly and copy number variation (CNV) evolves in a neutral fashion.

Nevertheless, this does not mean that the number of copies of OR genes is unimportant for the ability of human olfaction. On the contrary, the individuals with a larger number of OR genes may have a higher level of sensitivity to different odorants than those with a smaller number. Actually, it has been shown that polymorphism in OR genes contributes to variability of odorant perception in humans (Keller et al. 2007). Nevertheless, olfaction is only one component of fitness for humans, and its contribution to total fitness is probably minor in the presence of many other factors. For this reason, the number of OR genes may not be directly related to fitness in human populations. In human populations there are people with no sense of smell (anosmics), but these people do not seem to have any fertility problems.

However, the copy number change due to genomic drift may occasionally play important roles in phenotypic evolution (Nei 2007). That is, when a new environmental niche is open for a species and this niche requires a large number of gene copies, a group of individuals with large numbers

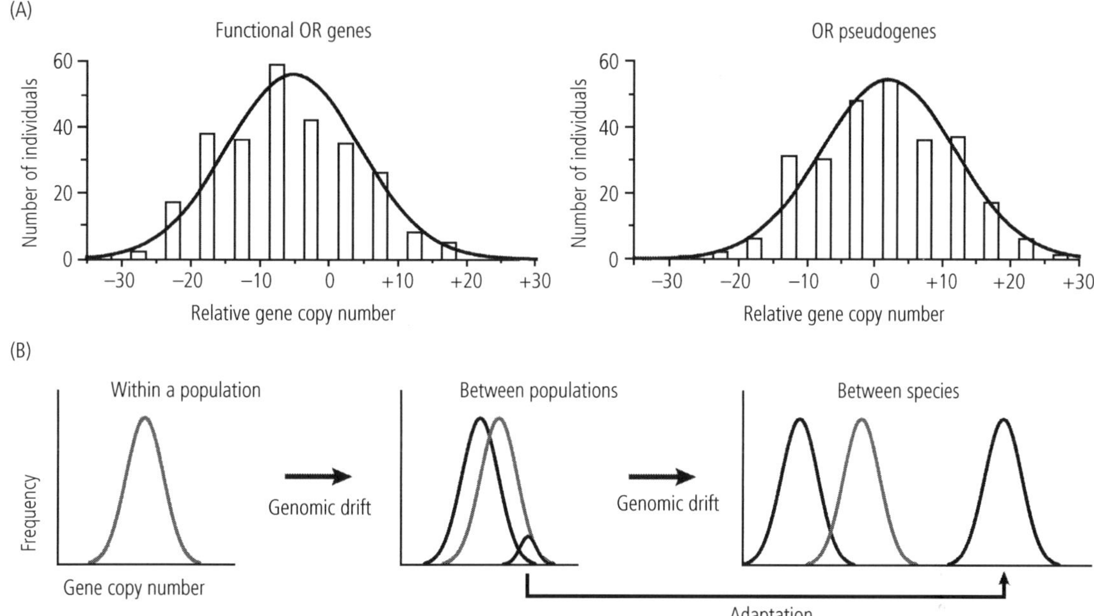

Fig. 5.11. Copy number variation and genomic drift. (A) Distributions of relative copy number of OR genes in humans. The relative copy number represents the difference in copy number between a sampled individual and the reference individual. The curve represents the normal distribution. (B) Genomic drift is a random process of copy number changes that occur by duplication, deletion and inactivation of genes. In this case, a distribution of gene copy number will follow the normal distribution if the number of gene copies is sufficiently large. Consequently, a natural population has a substantial amount of copy number variation by genomic drift as long as the copy number is within a range determined by functional requirements. In the case of chemosensory receptor genes, this copy number range is generally large. Therefore, when a population is separated into two geographic populations, these populations can have different distributions of copy number. When these populations evolve into different species, the copy number difference may be even larger owing to genomic drift. A new species may also be generated when a group of individuals who have a large number of genes (the small peak in the middle diagram) moves to a new niche where having a larger number of genes is more advantageous. From Nozawa et al. (2007).

of genes generated by genomic drift may move to this niche and eventually establish a new species. Figure 5.11B shows a simple model of copy number evolution with respect to chemosensory receptor genes. Here copy number is assumed to change mostly at random as long as the number is within the upper and lower boundaries determined by physiological requirements. When a population is separated into two populations, these populations may have different distributions of copy number largely due to genomic drift. This type of differentiation of populations may proceed even to generate different species. By contrast, a new species may be generated when a group of individuals having a large number of gene copies move to a new niche, where a large number of genes is favorable.

5.7. Noncoding DNA and Transposable Genetic Elements

Eukaryotic genomes are known to contain a high proportion of DNA that is not directly involved in protein production (about 95 percent of the human genome). This noncoding DNA includes intergenic spacers, introns, regulatory regions of genes, transposable genetic elements, tandemly repeated DNA, and others. Recent studies have shown that a substantial portion of noncoding DNA has some roles in the regulation of gene expression (Lynch 2007; Mattick 2011; ENCODE Project Consortium 2012). In general, the evolutionary dynamics of noncoding DNA is controlled largely by mutational events, and the role of natural selection appears to be minimal. In the following I present a brief description of the evolution of noncoding DNA which is related to the subject of this book.

Exons and Introns

The protein-coding genes of eukaryotes are divided into exons that encode amino acid sequences and introns that are excised out before the translation of mRNAs into polypeptides. The number of introns varies considerably with the gene, and some genes contain a few dozen introns. When introns were discovered, Gilbert (1978) suggested that introns are remnants of noncoding regions of ancient genes in early life, which were later combined to produce larger units of genes, and that the introns were useful for reshuffling exons to generate functionally more efficient genes. This view is often called the intron-early theory. However, sequencing of prokaryotic genes showed that they generally lack introns and that the number of introns tends to increase as the complexity of the organism increases (Palmer and Logsdon 1991). These observations suggested that introns were inserted into genes in later stages of the evolution of life. This view is called the intron-late theory.

Of course, splicing out of introns in the process of producing mature mRNAs is a complicated biochemical process and requires several enzymes. Why and how then did the insertion of introns evolve? This is still a mystery, particularly if we consider the fact that insertion and deletion of introns in eukaryotic genes have occurred frequently (Lynch 2007). Here we note that insertion and deletion of introns are mutational events and no clear selective advantage or disadvantage appears to be associated with them, particularly at the individual level at which natural selection operates.

Although the intron-early theory is generally disfavored now, there are eukaryotic gene families in which different types and numbers of exons are associated with functional differentiation. One good example is the killer cell immunoglobulin-like receptor (KIR) genes in primates. Natural killer (NK) cells are essential for the early immune response against tumor and virus-infected cells. The cytotoxic activity of NK cells is regulated through the interaction of the receptors on the NK cell surface with MHC class I molecules. The human genome contains a large number of KIR gene loci and polymorphic alleles, and the KIR genes can be classified into at least six different types of domain structures (Fig. 5.12). Type 3DL has three extracellular domains and a long intracellular domain where two immune-receptor tyrosine inhibitory motifs (ITIMs) are present, whereas 3DS has three extracellular domains, but the number of ITIMs varies (Fig. 5.12). These different domain structures apparently have been generated by exon shuffling, and they now have different immunological functions (Rajalingam et al. 2004). In rodents different NK receptor molecules called Ly49s or KLRs play the same functional roles as that of KIRs and have evolved in a similar fashion (Hao and Nei 2004).

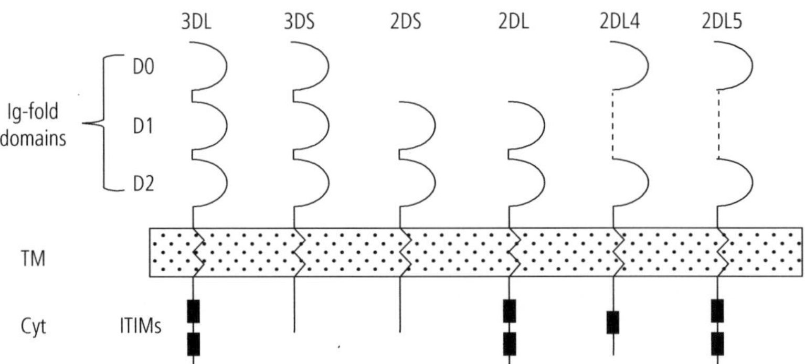

Fig. 5.12. Domain organization of KIRs. Most KIRs can be classified into six subgroups according to the domain organization. One recently identified cattle KIR molecule (2DS1) has an unusual domain organization (D0 and D1) (Storset et al. 2003), which is not shown here. TM, transmembrane region; Cyt, cytoplasmic region; ITIM, the immunoreceptor tyrosine inhibitory motif.

A more extreme case of evolution by exon shuffling has been reported in the F-box gene superfamily in plants (Xu et al. 2009). F-box proteins are substrate-recognition components of the SCF ubiquitin ligases. The plant F-box gene superfamily consists of about 700 genes and can be classified into 42 different families, and each of these gene families has a distinct exon-intron organization and distinct substrate specificity. Comparison of F-box genes from different families of *Arabidopsis*, poplar, and rice has shown that much of the variation in exon-intron organization has been generated by exon shuffling.

Transposable Genetic Elements

Transposable genetic elements are DNA sequences that can be transposed from one genomic location to another. The number of transposable elements is very large, and they make up about 50 percent of the mammalian genome. They can be classified into transposons and retrotransposons. Transposons are DNA sequences that move from one genomic location to another by the enzyme transposase carried by the sequences. When this transposition occurs, the transposon may or may not carry one or more extra genes (exogeneous genes) that encode other proteins. One example of such transposons is the *P* element in *Drosophila*, which may transfer exogenous genes. For this reason, the *P* element has been used as a means of transferring genes from one individual to another in *Drosophila* experiments. The *P* elements are transferable to different *Drosophila* species by horizontal gene transfer (Clark et al. 1994).

By contrast, retrotransposons are DNA sequences that are first transcribed into RNAs and then reverse-transcribed into cDNA before they are inserted into another genomic location. They are classified into LTR retrotransposons and non-LTR retrotransposons, but the majority are non-LTR retrotransposons in the human genome. Non-LTR retrotransposons can further be divided into two groups: LINEs (long interspersed elements) and SINEs (short interspersed elements) (Singer 1982). Functional LINEs are about 6500 bp long, whereas SINEs are usually 100 to 500 bp long. These retrotransposons are abundant in the eukaryotic genome, and they comprise about 42 percent of the human genome (Brouha et al. 2003). The majority of LINEs in the human genome belong to the L1 family, which contains about 500 000 member copies (17 percent of the genome) (Cordaux and Batzer 2009). By contrast, the largest family of SINEs in humans is the Alu family, which makes up about 11 percent of the genome. However, more than 99 percent of L1 elements are inactive in the sense that they do not have the reverse transcriptase. In human populations Brouha et al. (2003) found that only 90 L1 sequences contain the reverse transcriptase (*RT*) gene and are therefore capable of self-replication.

The SINE sequences do not have any RT, and their replication depends on the use of the RT of their partner LINE (Weiner 2000; Kajikawa and Okada 2002). Yet, there are a large number of SINEs in the eukaryotic genome, and they decay gradually by mutation during evolution. In humans the majority of SINEs belong to the Alu family, which was derived from 7SL RNA, part of the signal recognition particle (Ullu and Tschudi 1984). This family expanded rapidly in the human lineage, and there are about 300 000 Alu sequences in the human genome. In nonprimate organisms, however, SINEs are apparently derived from several different tRNAs (Daniels and Deininger 1985; Sakamoto and Okada 1985). Therefore, different organisms may have different families of SINEs, but the evolutionary mechanism appears to be similar.

Transposons and retrotransposons are known to cause some new mutations for protein-coding genes and generate chromosomal changes (Cordaux and Batzer 2009). However, despite their abundance in the genome, the evolutionary changes of these elements are largely controlled by mutational inputs and genetic drift. Because transposons and retrotransposons are abundant in eukaryote genomes, they can be used as genetic markers to identify protein-coding genes in the genome. Retrotransposons are also useful in constructing phylogenetic trees. In fact, Nikaido et al. (1999) showed that whales originated from a relative of the extant hippopotamus by using retrotransposon markers. Furthermore, Sasaki et al. (2008) reported that two SINE elements acquired the roles of enhancer of the genes involved in the brain formation of mammals. Therefore, transposable genetic elements may play some important roles in evolution (Lynch 2007, pp 189–191).

Tandem Repetitive Sequences

In addition to the above transposons and retrotransposons, there are several kinds of repeated DNAs that are abundant in the genome and appear to have little effect on phenotypic characters, though they occasionally cause certain kinds of complex genetic diseases. Most well known are *microsatellite DNA* and *minisatellite DNA*, and they are called *variable numbers of tandem repeats* (*VNTRs*).

Microsatellite DNAs are tandem repeats of 1–5 base pairs. They are also called *short tandem repeat* (*STR*) DNAs. The human genome contains a few million microsatellite loci (Takezaki and Nei 2009). For example, one of the common microsatellite loci is the dinucleotide repeat such as CACACACACACACA. However, such a locus is usually polymorphic and may have alleles with different repeat numbers such as 6, 7, 8, or 9 repeats with CA dinucleotides. Most microsatellite DNA loci are subject to no obvious selection and highly polymorphic. For this reason, microsatellite DNA loci are often used for constructing population trees (Takezaki et al. 2010).

Minisatellite DNAs are similar to microsatellite DNAs but consist of 10–60 bp repeats (Wyman and White 1980). The human genome contains about 1000 minisatellite loci, and at each locus there is polymorphism with respect to repeat numbers. However, minisatellite repeats may not be as regular as microsatellite repeats, and different repeats may contain slightly different nucleotide sequences. Different alleles at this locus represent different lengths of nucleotides, and the number of alleles can be a hundred or more, though only 15–25 can be distinguished in practice. The heterozygosity for a VNTR locus can be more than 0.9. Because of this extraordinarily high degree of polymorphism, VNTR loci are often used for identification of individuals or paternity tests in forensic science (Jeffreys 2005).

Eukaryotic genomes contain several other groups of repetitive DNA sequences. One group constitutes *heterochromatins*, which are concentrated in centromeric and telomeric regions of chromosomes. Heterochromatin is a tightly packed form of DNA and controls gene expression through regulation of the transcription initiation. Heterochromatin is composed of largely inactive repetitive DNA sequences of various lengths.

The above survey of noncoding DNA elements indicates that most of the variations in these repetitive DNAs are generated by mutational processes and they are maintained primarily by the balance between mutation and genetic drift.

5.8. Summary

Gene duplication is one of the most important mutational mechanisms that introduce new genetic variation in the evolutionary process. Duplicate genes generated by genome duplication or segmental duplication may diverge in function by nucleotide substitution, deletions/insertions, and recombination, generating innovative genetic and phenotypic characters. The number of genes in the genome has increased substantially from unicellular organisms to multicellular organisms. In vertebrates or flowering plants, however, the organismal complexity as measured by the number of cell types is not always correlated to the number of genes.

The number of gene families and their family size are generally greater in complex organisms than in simple organisms. However, the total number of genes in the genome of an organism is not necessarily correlated with the number of cell types, an index of organismal complexity. Rather, the proportion of noncoding DNA in the genome appears to show a higher correlation with organismal complexity. This observation suggests the importance of noncoding DNA in gene regulation, but at the present time the real reason for this observation is unclear. Despite this uncertainty, about 20 percent of gene families (or superfamilies) show a positive correlation between family size and organismal complexity. There are several gene families which are much larger in complex organisms than in simple organisms. One of the largest gene families in land animals is that of olfactory receptor genes, and the number of gene copies in this family expanded enormously when animals evolved from aquatic to terrestrial life. There is also a large amount of variation in the number of the copies among different terrestrial organisms, some of which are apparently caused by species-specific adaption.

Several decades ago, most multigene families were thought to evolve according to concerted evolution that homogenizes the nucleotide sequences of member genes. Recent genome sequence data do not support this view except for a few gene families. Rather they support the model of birth-and-death evolution, in which new genes are assumed to be generated by gene duplication and some duplicate genes are maintained in the genome for a long time whereas others are deleted or become nonfunctional. This model is capable of explaining the evolution of genes with new functions. It can also explain the existence of many paralogous genes with nearly equal function.

Eukaryotes contain several important genetic systems that are crucial for their survival. Examples are the adaptive immune system in jawed vertebrates, flowering system in plants, meiosis, and mitosis. They are usually controlled by many multigene families that are historically related and interact with one another. This suggests that the evolution of genetic systems has occurred by the interaction of expanding multigene families. However, these interactions seem to have been generated by birth-and-death evolution of each gene family. The adaptive immune system in jawed vertebrates has evolved recruiting different various multigene families that evolved earlier to adapt for various purposes. A similar conclusion may be derived for the evolution of the flowering system in angiosperms. These findings suggest that new genetic systems have evolved gradually in a long process of repeated gene duplication.

Recent studies of genomic evolution have shown that genetic loci are not fixed entities but are subject to duplication, deletion, and transposition quite often. This variation is caused by genomic drift. For this reason, the number of genes per genome or per gene family varies considerably in the evolutionary process. It also varies among different individuals in the same species. One of the most extreme cases is the variation of gene copy number of olfactory receptor genes. In this gene family, the number of gene gains and losses in an evolutionary lineage can be of the order of tens or hundreds. The variation of the number of OR genes among different individuals of a human population is quite large.

Previously the noncoding regions of DNA in the genome were thought to be largely nonfunctional. However, recent studies indicate that the noncoding DNA often encodes small RNAs that control the level of gene expression at the transcriptional or post-translational level. Some examples are small interfering RNAs (siRNAs), microRNAs (miRNAs), and Piwi RNAs (piRNAs), which control gene expression level at the post-translational level. Noncoding DNAs also contain a large amount of transposons and retrotransposons. The primate retrotransposon Alu sequences are derived from 7SL RNAs, but nonprimate retrotransposons are primarily derived from tRNAs. Because transposons and retrotransposons are abundant in eukaryotic genomes, there are many different levels of regulation of gene expression. Retrotransposons have been shown to be useful as genetic markers for constructing phylogenetic trees.

Eukaryotic genomes are also known to contain a large number of tandem repetitive sequences. Most well known are microsatellite DNA and minisatellite DNA, which are often called variable numbers of tandem repeats (VNTRs). The variation of these sequences appears to be essentially neutral and their fate is determined largely by genetic drift. Some of this group of DNAs have become useful for forensic science (Committee on DNA forensic science: update – National Research Council 1996).

In this chapter we considered various forms of gene duplication, including genome duplication, segmental gene duplications, genomic drift, intron gain/loss, gene transposition, horizontal gene transfer, formation of microsatellite and minisatellite DNAs, etc. At the present time, some authors study these DNA changes without calling them as mutations. This has occurred because these mutations are conceptually different from those considered by Mendelian geneticists in the early 20th century. At this time, the molecular basis of mutation and the relationship of mutation and phenotypic change were not known, and when they discovered chromosomal changes, they simply treated them as separate genetic changes.

However, the first type of phenotypic mutations identified by de Vries (1901–1903) were later shown to be polyploids and other chromosomal changes as will be discussed in Chapter 7. Furthermore, these chromosomal changes are known to cause gene silencing and gene losses and therefore they are regarded as integrated forms of mutations. Similarly gene duplications and gene transposition belong to mutation rather than natural selection when we consider the evolutionary process. This consideration makes it clear that mutations play important roles in evolution.

CHAPTER 6

Evolution of Phenotypic Characters

In the last two chapters we discussed the evolutionary change of protein-coding genes and the roles of gene duplication in genomic evolution without considering the mechanisms of gene expression. In multicellular organisms all cells have the same set of genes, yet some genes are expressed only in certain tissues at a certain stage of development whereas others are expressed in other tissues. Some genes are expressed only after some other genes are expressed. These differential gene expressions are controlled by a set of regulatory genes that operate cooperatively. The mechanism of this gene regulation is quite complicated, and the detailed aspects are not well understood. However, the evolution of this complex system of gene regulation is an important factor for generating complex organisms. In this chapter we consider only the general principles of gene regulation and their implications in evolution and present a few examples of phenotypic evolution.

6.1. Changing Concepts of the Gene and Gene Expression

Definition of a Gene

When Mendel (1866) discovered the laws of inheritance of discrete characters, he used the word element to denote the unit of genetic substances that are inherited from parents to offspring. This element was later called the gene (Johannsen 1909), which is now widely used. Initially, this gene was an abstract concept, but Thomas Morgan and his associates (Morgan et al. 1915) presented evidence that the gene is a real physical entity that is located on a given position on a chromosome. However, the real chemical material that composes a gene was not known until Watson and Crick (1953a) showed that the gene is a piece of a DNA molecule. This discovery immediately provided the physical basis of Beadle and Tatum's (1941) one-gene-one-enzyme hypothesis and led to a new definition of a gene, which is a continuous stretch of DNA that encodes a single enzyme or polypeptide.

This definition, however, did not last long. As the technology of DNA sequencing was developed, it soon became clear that the relationships between the DNA sequence and the protein sequence are quite complicated. One of the complications was the existence of overlapping genes. Overlapping genes are a pair of adjacent genes whose coding regions are partially overlapped. In other words a single stretch of DNA may encode two different polypeptides. Overlapping genes are ubiquitous and have been identified in prokaryotes, eukaryotes, and viruses. Overlapping genes make it possible to produce many different proteins from a limited length of DNA. A well-known example is the overlapping genes in the bacteriophage Φx174. This phage contains a relatively short circular DNA sequence but produces many more proteins than could be produced if the transcription occurred in a linear fashion, one gene after another.

Another important discovery of the complex nature of genes was that a large proportion of eukaryotic genes consist of exons and introns, and introns are spliced out before the translation of mRNAs (Fig. 6.1). However, when there are many introns and exons in a gene, some of the exons are sometimes spliced out together with introns when the mature mRNA is produced. In this case the exons spliced out are not always the same. This is called alternative splicing. For example, the α-tropomyosin in rats contains 12 exons including untranslated regions (UTRs). As a consequence,

Mutation-Driven Evolution. First Edition. Masatoshi Nei.

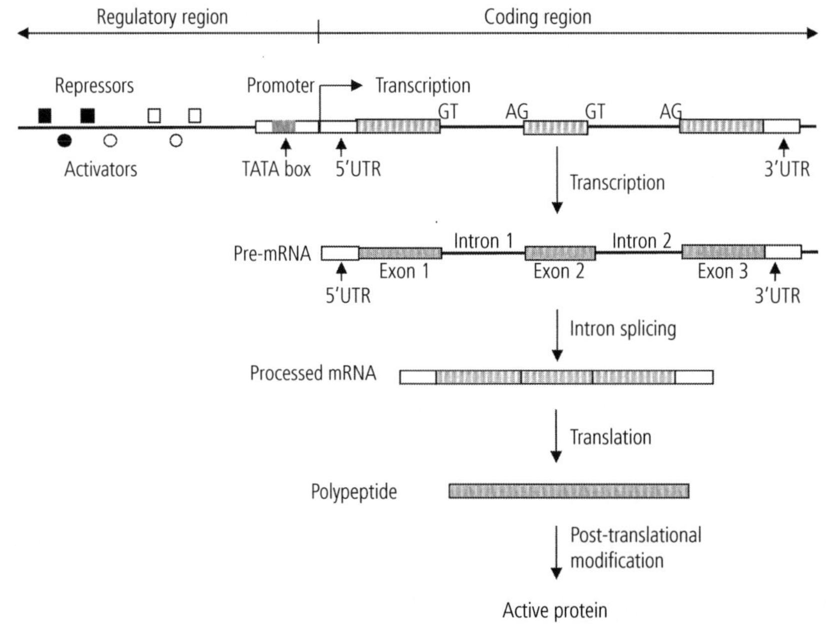

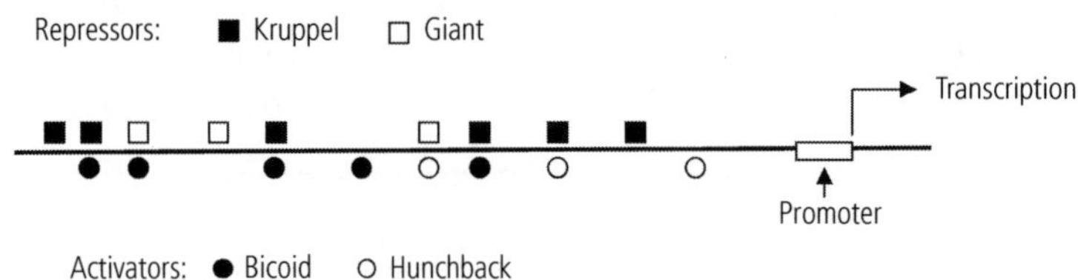

Fig. 6.1. Schematic structures of a gene. (A) A gene can be decomposed into the protein-coding region and the regulatory region. The coding region includes the 5′ and 3′ untranslated regions, exons, and introns, whereas the regulatory region includes the TATA box and the enhancer region that contains the activators and repressors. These activators and repressors are often called *cis*-regulatory elements or *cis*-elements, and they are the sites where transcription factors are attached. The transcription factors attached to the *cis*-elements form the transcriptional modules, and these modules interact with the RNA polymerase at the promoter region to control the level of the pre-mRNA. The introns in the pre-mRNA are spliced out to form a processed mRNA, which is then translated into a polypeptide. The expression of a gene is controlled by several other elements or factors such as microRNAs, transcribed small RNAs, and epigenetics. (B) *cis*-regulatory elements of the *Drosophila even-skipped* (*eve*) stripe 2 gene expression. There are two types of repressors and two types of activators.

seven different types of mRNAs are produced and generate seven different functional proteins (isoforms) (Fig. 6.2). It has been estimated that about 50 percent of human and mouse genes are alternatively spliced (Davuluri et al. 2008). Therefore, alternative splicing is an important mechanism for producing a large number of proteins from a rather limited number of genes in higher organisms. The current champion for making multiple proteins from the same gene appears to be the *Dscam* gene of *Drosophila* species. This gene encodes a membrane receptor protein involved in insect development and contains 24 exons. Theoretical consideration has suggested that this gene is capable of producing 38 016 different types of protein by alternative splicing. In fact, random searches for the proteins that could theoretically be produced indicated that a large proportion of them are actually produced

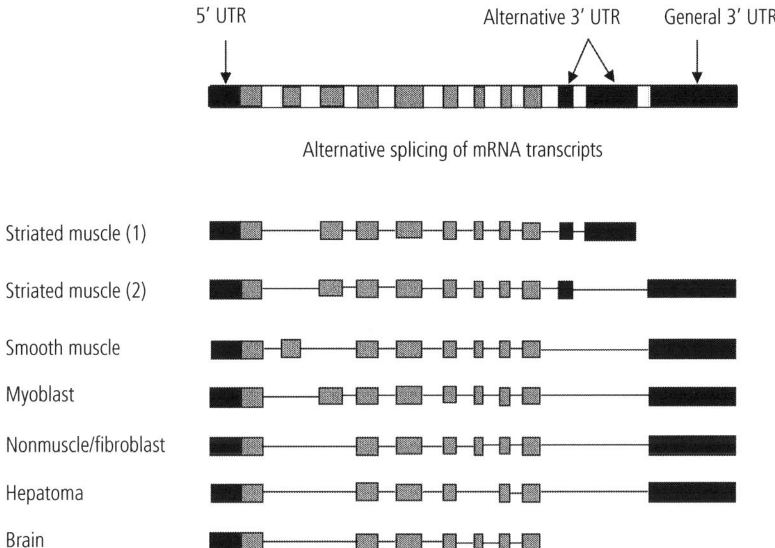

Fig. 6.2. A family of rat α-tropomyosin proteins formed by alternative RNA splicing. The α-tropomyosin gene is represented on top. The thin lines represent the sequences that become introns and are spliced out to form the mature mRNAs. Different mRNAs have different sets of exons. Therefore, a single gene produces seven different polypeptides. Modified from Breitbart et al. (1987).

(Gilbert 2006, pp 127–128). The *Drosophila* genome is currently thought to contain only 14 000 genes, but this genome can generate three times more proteins than the number of genes.

However, a more important change in the concept of a gene occurred when the mechanism of gene expression was discovered. As mentioned above, all cells in multicellular organisms have the same set of genes, yet some genes are expressed only in certain tissues at a given stage of development. This differential expression of genes is the cornerstone of development. This mechanism of differential gene expression was first studied by Jacob and Monod (1961), who discovered the *lac* operon in bacteria. This operon contained a region of DNA sequences encoding one or more proteins and a regulatory region. This regulatory region consisted of a promoter sequence that binds to RNA polymerase and a regulator sequence to which regulatory proteins bind. In the case of the *lac* operator, the regulator is turned on if the amount of lactose is lower than the level required for bacterial growth in a given environment, but if the amount exceeds the required level, it is turned off and no more lactose is produced. Later studies showed that there are other sequences that could affect various aspects of gene expression from transcription to post-translation modification (Fig. 6.1). Such sequences may exist within the protein-coding sequences as well as in the flanking regions of genes (Gilbert 2006). In some genes such as globin genes the regulatory elements may be found very far away from the protein-coding sequences. This has made the old concept of a gene as a compact genetic locus inapplicable. In fact, if we consider all regulatory elements of gene expression, a large part of the genome sequence may become a part of a gene (ENCODE Project Consortium 2007). For this reason, Gerstein et al. (2007) proposed the definition that *"a gene is a union of genomic sequences encoding a coherent set of potentially overlapping functional products."* However, this definition is too vague for non-specialists. I therefore would like to consider only basic aspects of gene structures and expression mechanisms.

Protein-Coding and Regulatory Regions of Genes

To understand the basic mechanisms of gene expression in eukaryotes, it is convenient to consider the protein-coding region and the regulatory region of genes separately (Fig. 6.1A). The protein-coding region is composed of the 5′ untranslated region (5′

UTR), exons, introns, and 3′ untranslated region (3′ UTR). In the formation of the pre-messenger RNA (pre-mRNA) all these regions are transcribed. However, this pre-mRNA undergoes RNA processing, and all introns are spliced out before the mature or processed mRNA is produced. The processed mRNA is then used for producing a polypeptide, which will be a component of a protein after post-translation modification. For example, in adult humans the α-globin polypeptide (chain) produced by the above process is inactive until the tetrameric hemoglobin molecule composed of 2 α-chains and 2 β-chains attached with a heme group is produced. The β-chain is produced by another gene. Formation of the functional molecule is called the post-translation modification.

However, to produce a particular protein in the right amount at the right time in development, various regulatory systems of gene expression are necessary. A well-studied regulatory system is the one mediated by so-called *cis*-regulatory elements, to which transcription factors bind. These transcription factors are divided into activators and repressors, and a proper combination of these transcription factors initiates the activity of RNA polymerase II attached to the TATA box and leads to the transcription of the protein-coding region (Gilbert 2006). It should be noted that *cis*-elements are not always confined to the 5′ DNA region of the gene but may be located at the 3′ side as well as in introns. The transcription factors (proteins) attached to these *cis*-elements are believed to function by forming a multiprotein complex.

The numbers and types of transcription factors vary from gene to gene. One of the most well studied regulatory regions of genes is the stripe 2 expression of the *even-skipped* (*eve*) gene in *Drosophila melanogaster*. The stripe 2 expression of this gene is controlled by 8 activators (5 *bicoid* and 3 *hunchback* proteins) and 9 repressors (6 *kruppel* and 3 *giant* proteins), and the regulatory region of the gene contains 17 *cis*-regulatory elements, where the transcription factors are attached (Fig. 6.1B). The *cis*-regulatory elements form a protein complex or a module and control the timing and level of gene expression. In this case, the classical notion that a gene is a clearly demarcated continuous DNA region no longer holds, because the transcription factors are generally encoded by genes that are located in separate genomic regions.

Gene Regulatory Networks

In the above section we considered the mechanism of production of only one protein. In any physiological or developmental process a large number of proteins are involved either as structural proteins (e.g. tubulin and ribosomal proteins) or as transcription factors. In morphogenesis the number of genes involved in production of proteins is small in the early stage of development, but as development proceeds, an increasing number of genes are used to produce both structural or transcription factor proteins. A functional unit composed of various transcription factors and *cis*-regulatory elements that interact with one another is called a gene regulatory network (GRN). The number of genes involved in a GRN is small in the early stage of development but may grow to thousands in a later stage (Davidson and Erwin 2006; Peter and Davidson 2011). There are also many different GRNs in the developmental process, and they may interact with one another. This indicates that gene interaction is an essential aspect of developmental processes.

It should also be noted that GRNs are often different in different organisms and developmental biologists are now studying the evolution of morphological characters by examining the differentiation of GRNs in different organisms (Davidson 2006).

Small RNAs that Control the Level of Gene Expression

In recent years several different types of small RNAs that affect the level of gene expression have been reported. Relatively well studied are microRNAs (miRNAs) that were first discovered by Lee et al. (1993) and Wightman et al. (1993). MicroRNAs are about 22-base-long noncoding RNAs that are key post-transcriptional regulators of gene expression in animals and plants. In mammals miRNAs are believed to control the activity of about 30 percent of all protein-coding genes (Filipowicz et al. 2008). By binding to the 3′ UTR region of mRNAs, miRNAs mediate translational repression. For example,

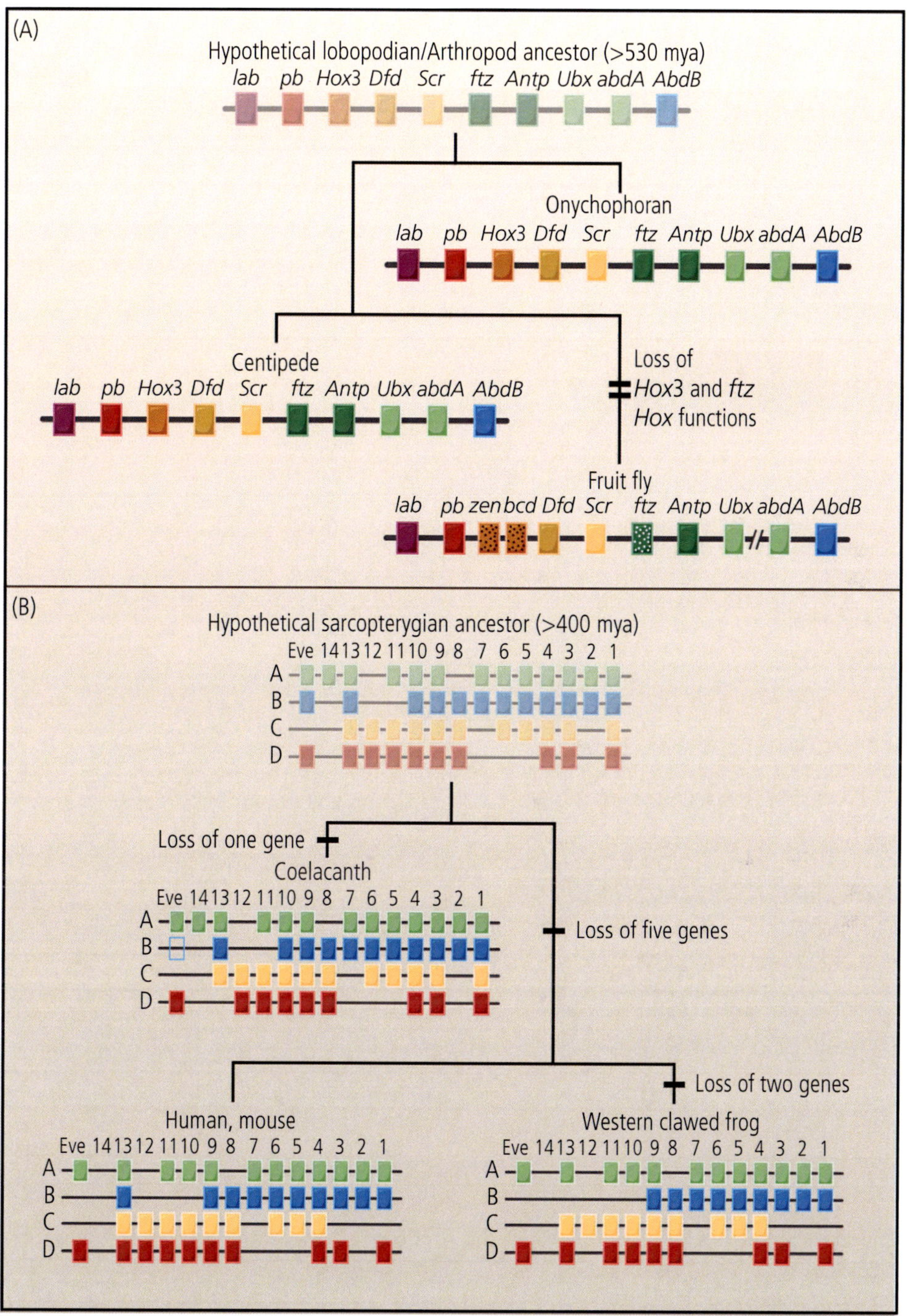

Plate 1. Ancestral complexity of *Hox* clusters and the lack of *Hox* gene duplications in arthropods and chordates. (A) Based upon the *Hox* gene complements of onychophora (velvet worms) and arthropods, a minimum of ten *Hox* genes must have existed in the common ancestor of lobopodians (ancestors of velvet worms) and arthropods. No new *Hox* genes arose in centipedes or insects while the *Hox3* and *ftz* genes were co-opted into new functions in certain insects (stippling). (B) No new *Hox* genes are known to have evolved since the divergence of tetrapods from a common sarcopterygian (lobe-finned fish) ancestor shared with coelacanths. Rather, gene loss has occurred in several lineages. From Carroll (2008). Reproduced with permission from Elsevier. See also Fig. 6.7.

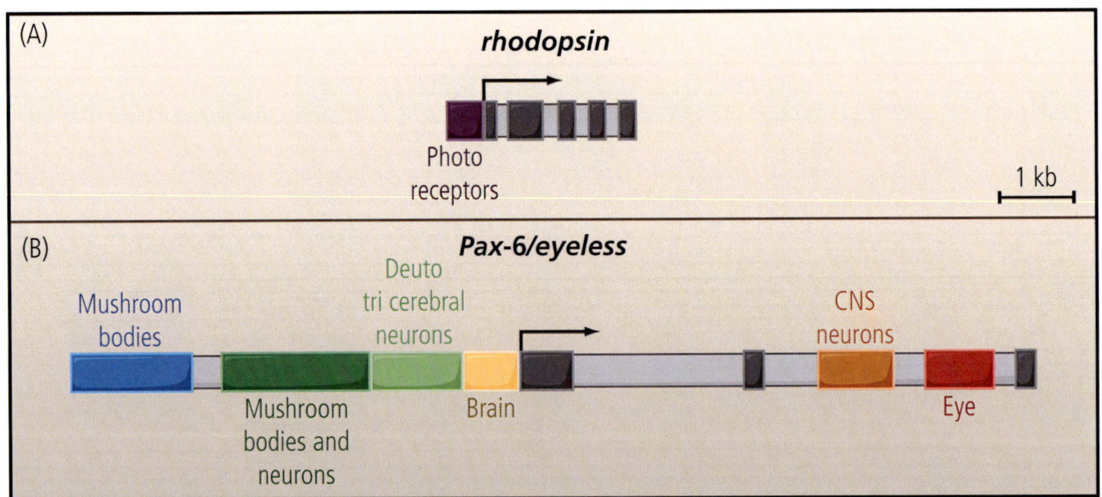

Plate 2. (A) Structure of the *rhodopsin* locus in the fruit fly *Drosophila*. Exons are shown in black, introns in gray, and the single *cis*-regulatory element (CRE) controlling gene expression in photoreceptor cells is shown in purple. (B) Depicted is the *rhodopsin* architecture with the locus encoding its chief regulator *Pax-6/eyeless*. Exons are in black, introns in gray, and the six distinct CREs governing gene expression in parts of the developing brain, central nervous system, and eyes are shown in various colors. From Carroll (2008). Reproduced with permission from Elsevier. See also Fig. 6.8.

Plate 3. Sea slug *Elysia chlorotica*, showing the highly branched digestive system. The body color of this organism is green. From Ma (2012). Reproduced with permission from Patrick Klug. See also Fig. 6.11.

(A)

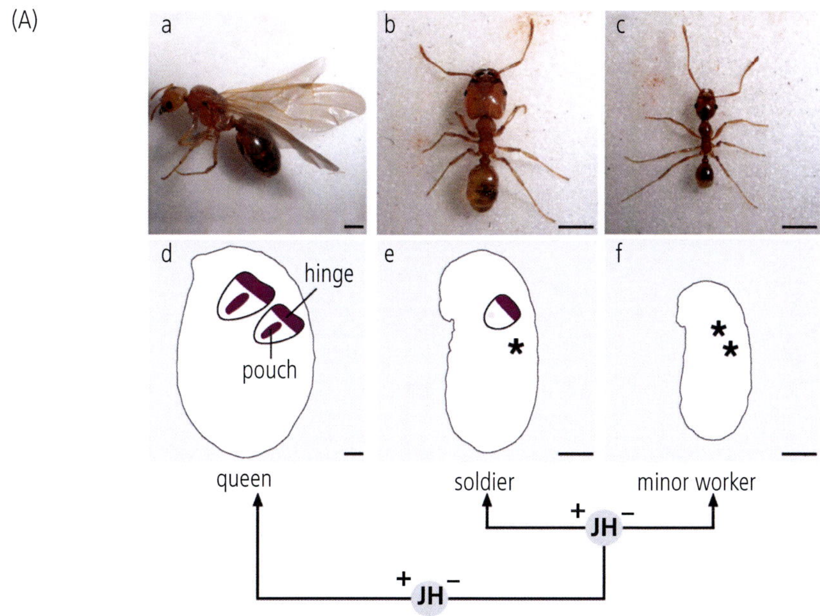

(B)

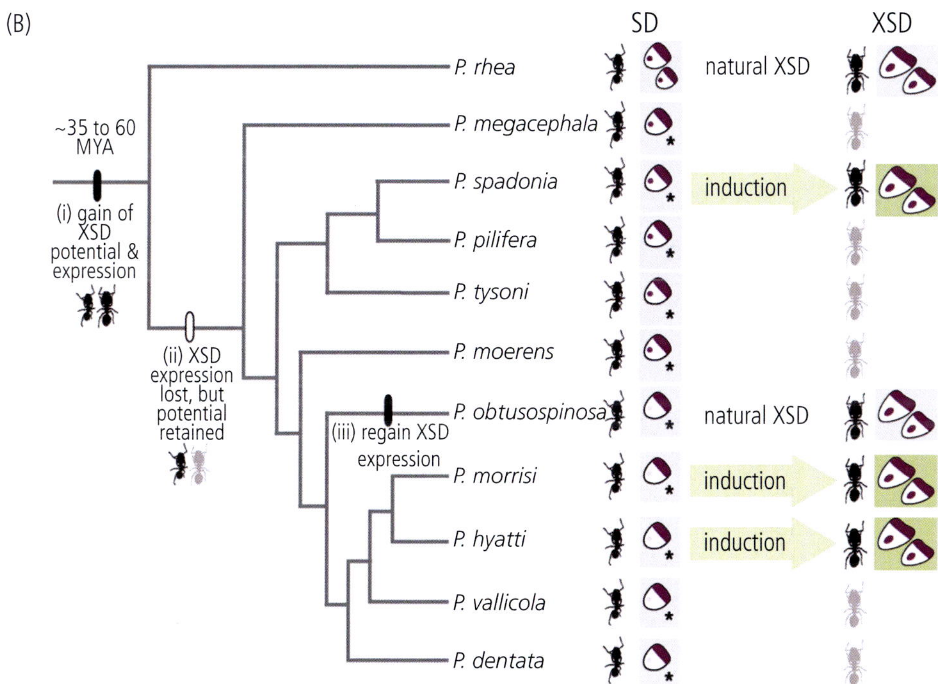

Plate 4. A. Wing polymorphism in *Pheidole morrisi*: the ability of a single genome to produce (a) winged queens and wingless (b) soldiers and (c) minor workers. Caste determination occurs at two JH-mediated switch points in response to environmental cues. (d) Wing discs in queen larvae showing conserved hinge and pouch expression of *sal*. (e) Vestigial wing discs in soldier larvae showing a soldier-specific pattern of *sal* expression, where it is conserved in the hinge but down-regulated in the pouch. Asterisks represent the absence of visible wing discs and *sal* expression in (e) soldier and (f) minor worker larvae. B. Evolutionary history of ancestral developmental potential and phenotypic expression of supersoldiers (XSDs). MYA, million years ago. Purple represents the pattern of *sal* expression; asterisks indicate the absence of vestigial wing discs and *sal* expression. Green arrows and boxes represent the induction of XSD potential. From Rajakumar et al. (2012). Reproduced with permission from the American Association for the Advancement of Science. See also Fig. 8.1.

(A)

(B)

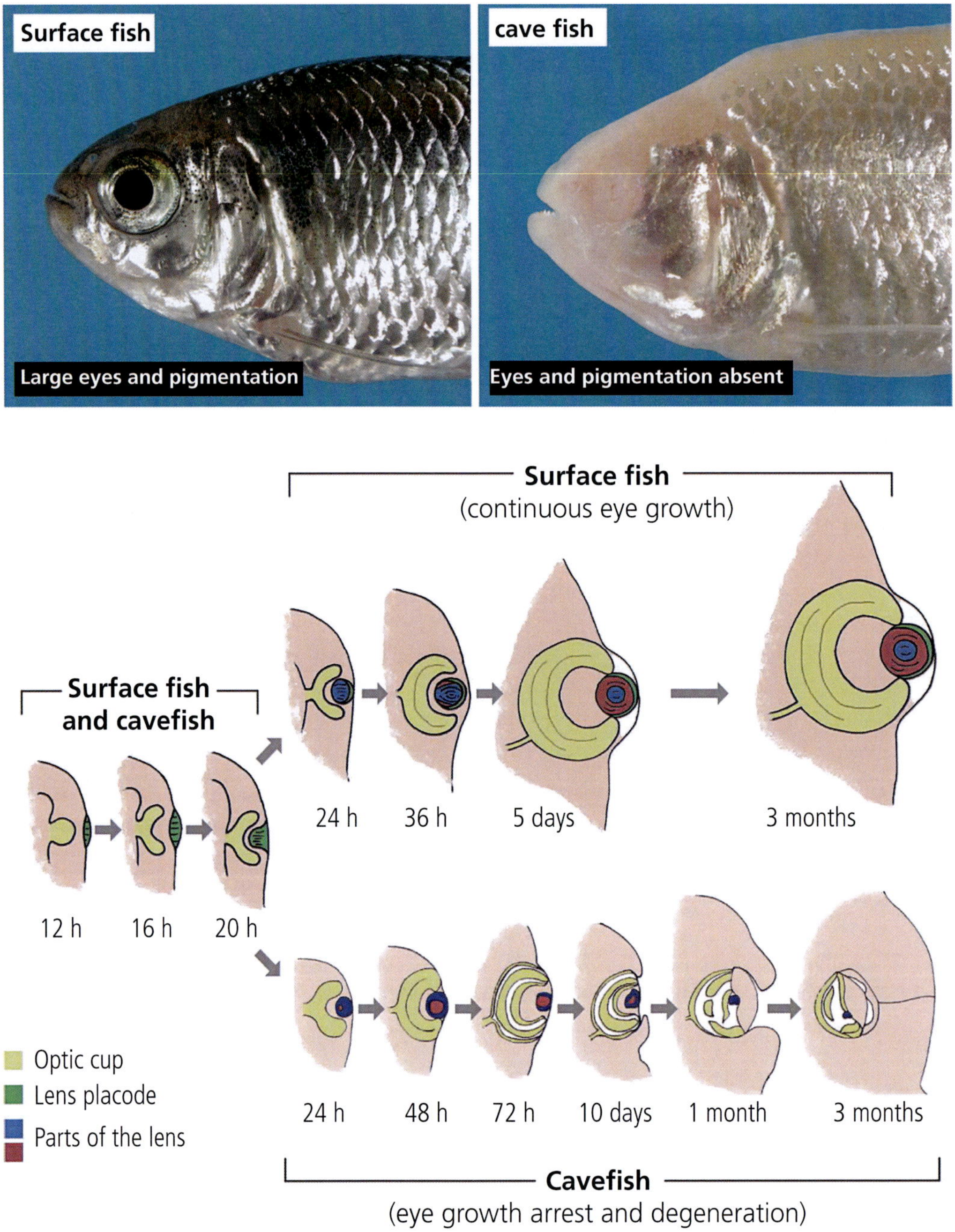

Surface fish
(continuous eye growth)

24 h
36 h
5 days
3 months

Surface fish and cavefish

12 h
16 h
20 h

Optic cup
Lens placode
Parts of the lens

24 h
48 h
72 h
10 days
1 month
3 months

Cavefish
(eye growth arrest and degeneration)

Plate 5. A. *Astyanax mexicanus* surface fish and cave fish. B. A diagram showing the events of *Astyanax* eye development and degeneration. Left: Early events of eye primordium formation are the same in surface fish and cave fish until approximately 1 day after fertilization. Top: In surface fish, the eye differentiates and the eye parts grow in concert with increased body growth. Bottom: In cave fish, the eye primordium grows for a while, then arrests, degenerates, and is internalized by overgrowth of the body. Reproduced with permission from Annual Review of Genetics. See also Fig. 8.2.

the miRNA miR-iab-4-5p that is encoded by the *iab-4* locus in *Drosophila melanogaster* controls expression of the *Ultrabithorax* (*Ubx*) gene, and induces the homeotic transformation of halteres (vestigial rear wings) to wings when this miRNA is ectopically (artificially) expressed (Ronshaugen et al. 2005). In normal flies the Ubx protein is abundantly produced, and this protein inhibits the development of the rear wings. However, in the presence of the miRNA encoded by the *iab-4* locus, the production of Ubx proteins is reduced, and this reduction transforms the rear wings into halteres. There are many other studies showing that deletion or overexpression of miRNA genes results in defective morphological characters such as cardiac morphogenetic defects, defective adaptive immunity, male sterility, etc. (Stefani and Slack 2008). It is now believed that miRNAs or other similar small RNAs such as Piwi-interacting RNAs are involved in the development of many morphological or physiological characters. For example, the rice miRNA, miR156, is known to generate differences in the height, tiller number, and panicle morphology in rice varieties when a single nucleotide difference exists at the target site (Jiao et al. 2010; Miura et al. 2010b).

In Chapter 5 we showed that the complexity of an organism is not necessarily correlated to the number of protein-coding genes in the genome but is correlated to the proportion of noncoding regions of DNA. A recent study suggests that the number of miRNA genes increases with increasing complexity of organism and therefore miRNAs are at least partially responsible for the evolution of complex organisms (Heimberg et al. 2008). This study was done mostly with animal species, and the number of miRNA genes examined is still small. Furthermore, studies showing that evolutionary changes of morphological characters occur by mutational changes of miRNAs or acquisition of new miRNAs are still rare. In addition to miRNAs, there are many other RNA elements that affect the level of gene expression and therefore the expression of phenotypic characters. For example, some small RNAs encoded by retrotransposons appear to be participated in the regulation of gene expression (Ponicsan et al. 2010). Many of these factors have not been well studied, but some of them will be discussed later.

Methylation and Epigenetics

Another mechanism of regulation of gene expression is DNA or protein methylation that occurs when a methyl group is added to specific nucleotides of DNA or specific amino acid residues. DNA methylation occurs when a methyl group is added to the cytosine pyrimidine ring. In adult mammals, this cytosine methylation occurs only when the cytosine (C) residue is followed by a guanine (G), and this methylation is observed in 60 to 90 percent of all CpGs. The methylated CpGs are usually present in the promoter regions of genes and prevent the transcription of genes. For these genes to be transcribed, the promoter region needs to be demethylated.

Protein methylation typically occurs for arginine or lysine amino acid residues in the protein sequence. This methylation often takes place in histone proteins. Histones are the major component of chromatin and form nucleosomes that wrap up many genes in a condensed form in the nucleus. If histones are methylated, they often act to repress gene expression epigenetically. For a gene to be expressed, methylated histones need to be acetylated to expose the DNA involved for transcription.

In many tissues most genes are in an inactive state because of the methylation of histones in the nucleosomes. Once a portion of histones is demethylated, the genes involved are in an active state. Yet, the transcription of a gene may not start if the promoter region of the gene is methylated. For example, the β-globin gene cluster consists of the ε, γ_A, γ_G, δ, and β genes (Fig. 6.3). The ε gene is expressed in the embryonic stage, the γ_A and γ_G genes in the fetal stage, and the δ and β genes in the adult stage. In blood cells, this cluster of genes is exposed from the nucleosomes and is in an active state. How then are different genes expressed in different developmental stages? The answer to this question appears to be the differential methylation of the promoter of the genes. In the embryonic stage the promoters of the γ_A, γ_G, δ, and β genes are apparently methylated, so that only the ε gene is expressed. In the fetal stage, however, the promoter regions of only the γ_A and γ_G genes are unmethylated to allow the initiation of transcription. Similarly, in the adult stage only the promoters of the δ and β genes are unmethylated (Fig. 6.3).

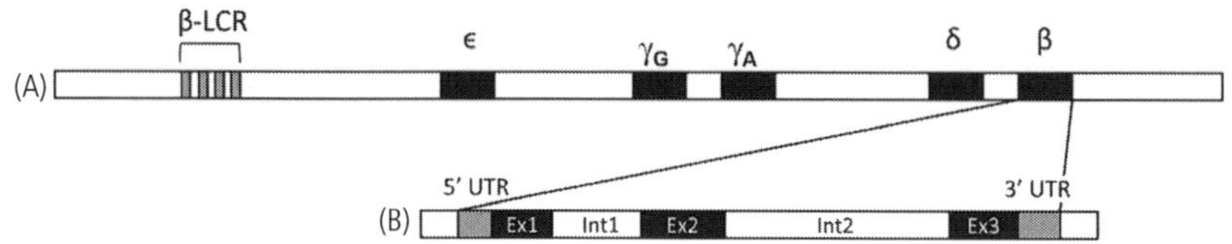

Fig. 6.3. (A) Schematic structure of the β-globin gene cluster in humans. (B) An extended figure of the β-globin gene. β-LCR: β-locus control region.

However, how is this differential methylation and demethylation of genes controlled? This control is mediated by a DNA sequence called the locus control region (LCR) that exists far away from the gene cluster in the 5′ side (Fig. 6.3). Of course, the detailed mechanism of this control is still unknown. This type of LCR has been identified for the differential expression of α-globin cluster genes, which are located on a different chromosome. A similar LCR is believed to control the expression of MHC class II genes (Masternak et al. 2003), HOX genes (Lee et al. 2006), and some other multigene families.

For certain types of genes, methylation or demethylation of the promoter regions of genes are caused by environmental factors, and this is called epigenetic control. For example, some groups of plants such as wheat require exposure to cold temperatures for an extended period of time (in winter) to initiate flowering, which is called vernalization. In *Arabidopsis*, the response to vernalization is mediated by a repressor gene called *FLC*. This gene functions as a repressor of flower development by inhibiting the activation of a set of genes controlling the transition of plants from vegetative growth to the formation of reproductive organs. Therefore, if the expression of this gene is suppressed, flower formation is initiated. This suppression of the *FLC* gene is achieved by methylation of histones in the nucleosomes, which is generated by vernalization (Bastow et al. 2004). In other words, the inactivation of the *FLC* gene is controlled by an environmental factor.

The study of this type of change of gene expression by external or environmental factors is called epigenetics. There are many other examples of morphogenesis controlled by external factors. Well-known examples are the sex determination of some reptiles (e.g. turtles) by temperature, seasonal changes of wing patterns of butterflies, and others. Strictly speaking, most morphological characters are controlled by both genetic and environmental factors, as will be discussed in Section 6.4. Therefore, the study of epigenetics in developmental biology is very important. At the present time, however, the molecular basis of epigenetics is poorly understood.

Signaling Pathways and Gene Interaction

In the above section we discussed relatively simple mechanisms of controlling gene regulation. In practice, gene expression is regulated by a large number of genes, which often interact with one another. One such mechanism is the signal transduction pathway. Previously, we mentioned that mammalian organisms have a large number of odorant receptors and these receptors are responsible for perceiving various kinds of odor. Biochemically, these odorant receptors are called G protein coupled receptors (GPCRs) and are located in the cell membrane as 7 transmembrane receptors. An airborne or water-soluble odorant binds to the extracellular portion of the GPCR, and the odor signal then goes through the GPCR and is transferred to the G protein (guanine nucleotide-binding protein) located inside the cytoplasm. The G protein activates a cascade of signal transduction compounds and finally transfers the odor signal to the brain.

This signaling pathway is only one of many important G protein pathways that operate in eukaryotic cells. The signals from color vision, hormones, neurotransmitters, and other signaling factors are all processed by G protein pathways, though the detail varies considerably with signal (Gerhart and Kirschner 1997, chapters 2 and 3). G protein pathways are prevalent in eukaryotic cells, but G proteins themselves are highly conserved. Yet, they can be classified into highly differentiated functional groups. Actually, G protein pathways are known to be diversified with respect to receptors as well as target molecules, generating potentially rich networks of information transfer. A single receptor may interact with several types of G proteins, and one G protein may interact with many receptors, as is in the case of odorant receptors. This makes the G protein pathway highly divergent and of multiple utility (Figs 6.4, 6.5).

However, it should be noted that the G protein pathway is only one of many signaling pathways used by eukaryotic cells. There are many other signaling pathways that are required for the formation of phenotypic characters. Important ones in fruitfly embryos are the TGF-β, wingless (wnt), notch, hedgehog, Toll, and FGF families (Carroll 2005a, p. 44). All of these pathways are known to have at least one signaling ligand, at least one receptor spanning the cell membrane, and at least one transcription factor that responds to signaling inputs by binding to target genes. These pathways are essential for morphogenesis.

In Chapter 5 we mentioned that some genetic systems such as the adaptive immune system work in the presence of interaction of many component gene families. Actually each of these gene families has its own signaling pathways. For example, the immunoglobulin and MHC class I gene families in

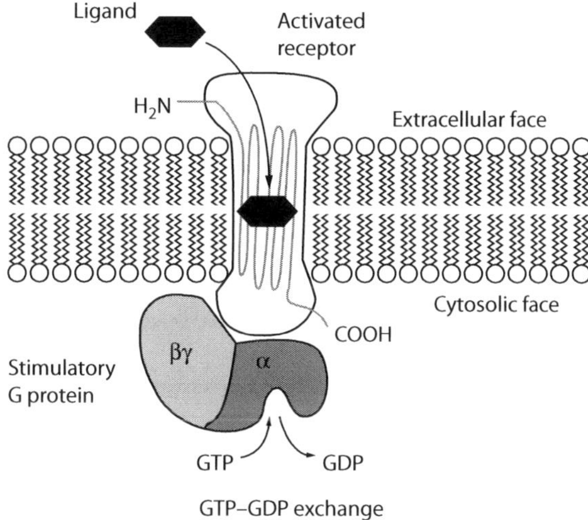

Fig. 6.4. Signal send-off. Ligand binding to a GPCR's extracellular region triggers changes in the protein's transmembrane region. This causes the release of guanosine diphosphate (GDP) and the uptake of guanosine triphosphate (GTP) from the G protein, spurring activation of predefined signaling pathways. Reproduced with permission from the American Chemical Society.

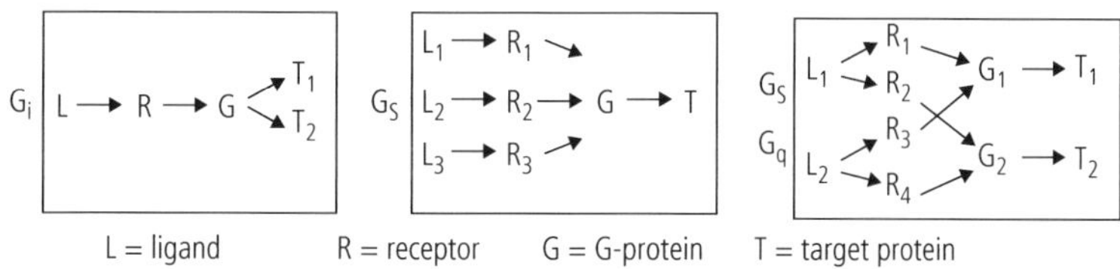

Fig. 6.5. Convergence, divergence and interaction in G protein signaling, Example pathways for three different types of Gα subunits are shown, but note that all types of G proteins can participate in all types of pathways. Left, G_i. A single ligand activates a single receptor linked to G_i, generating $G\alpha_i$GTP (G). This activates more than one target in the cell. Center, G_s. Several ligands activate several different receptors, which in turn converge on one $G\alpha_s$GTPs (G) which activates a single target. Right, G_s and G_q. There is cross-talk among ligands and receptors linked to G_s and G_q, which results in the activation of two targets (see text for examples).

the adaptive immune system are known to have their own signaling pathways. The presence of these pathways indicates that in the development of physiological and morphological characters an interaction of a large number of genes is required. It is also known that many genes controlling phenotypic characters are pleiotropic.

6.2. Evolution of Physiological and Morphological Characters

In the study of evolution, phenotypic characters are often separated into physiological and morphological characters for convenience. Strictly speaking, however, it is difficult to distinguish between the two types of characters, because the formation of morphological characters depends on various physiological processes in the developmental process and the function of physiological characters is dependent on the anatomy or morphology of the organism. For example, coat color of mammals is a morphological character in the sense that individuals with different color patterns can be easily distinguished, but it is also a physiological character because it affects heat sensitivity or behavioral pattern. Nevertheless, it is convenient to treat the evolution of physiological and morphological characters separately, because the former characters are concerned primarily with adult life and the latter are products of morphogenesis in the developmental process. For example, the transportation of oxygen from the lungs to various tissues in vertebrates is carried out primarily by hemoglobins and myoglobins. Therefore, examining the molecular structures and expression patterns of these proteins from different organisms, one can study the mechanism of evolution of oxygen transportation to some extent. By contrast, to understand the evolution of morphological characters, we must study the evolutionary change of morphogenesis, which depends on the complicated molecular and cellular processes carried out by a large number of genes. Furthermore, some developmental biologists (e.g. Carroll 2005a, b) have proposed that the evolution of physiological characters depends primarily on amino acid substitution in protein-coding genes whereas the evolution of morphological characters is controlled mainly by changes in the regulatory regions of genes. Although this view was criticized by Hoekstra and Coyne (2007), there are some empirical data to support this idea. I therefore consider this issue before we discuss morphological evolution.

Changes in the Protein-Coding Regions of Genes

The study of molecular evolution started with interspecific comparison of protein molecules concerned with various physiological functions (e.g. hemoglobin, cytochrome c, and insulin). This type of study soon uncovered that most amino acid substitutions in proteins are more or less neutral (King and Jukes 1969) and the functional change of proteins is generally caused by a small number of crucial amino acid substitutions occurring in the active sites of proteins, as mentioned in Chapter 4. Furthermore, for a protein to maintain its function there are generally functional or structural constraints in the amino acid sequence, and many random mutations are eliminated by purifying selection. For this reason, most proteins evolve conservatively (Fig. 4.3). This is a general principle of evolution of proteins controlling physiological characters (Kimura 1983; Nei 1987, 2005). Since we have already discussed this issue in Chapter 4, I shall not repeat it here.

In the last two decades the evolutionary changes of protein-coding regions of genes involved in morphological evolution (called morphogenes, Liao et al. 2010) have also been studied, and it appears that the pattern of amino acid substitution is similar to that for the genes involved in physiological characters (physiogenes). That is, the functional change of a gene generally occurs by a small number of amino acid changes and the remaining changes are more or less neutral (see Tables 4.2 and 6.1). In the case of morphogenes, however, there are many exceptions, and the functional change is often caused by nucleotide deletion/insertion, exon deletion and transposon insertion. One of the genes listed in Table 6.1 is *SBE1*, which causes wrinkled seeds in peas. This mutation is caused by insertion of a transposon. Gregor Mendel used this wrinkled seed character in his famous genetic experiments without knowing its mutational origin. Another character Mendel used in his experiments is plant height. The short plant is caused by one amino acid substitution in the *Le*

Table 6.1. Examples of changes in morphological characters caused by nondeleterious mutations in the protein-coding regions of genes. Similar examples for physiological characters are presented in Table 4.3.

Protein/Gene	Organism	Amino Acid or Nucleotide Changes	Character Involved
MC1R	Beach mouse	1 aa	Coat color
HOXD13	Human	1–4 aa	Limb
Oca2	Cave fish	Exon deletion	Albinism
MCR1R	Rock mouse	4 aa	Coat color
MSH-R	Leopard	1 aa	Coat color
MSH-R	Cattle	1 aa	Coat color
MC1R	Pig	1 aa	Coat color
ASIP	Jaguarundi	2bp deletion	Coat color
MC1R	Jaguarundi	5/8 deletion	Coat color
ABCC11	Human	1 aa	Earwax
Le	Pea	1 aa	Stem length
SBE1	Pea	Transposon insertion	Wrinkled seed
I	Pea	6bp insertion	Cotyledon color
A	Pea	Splice site	Flower color
VRS1	Barley	1 aa	Six-rowed spike
Romosa2	Maize	1 aa	Flower shape
Div	Snapdragon	4bp deletion	Flower symmetry
Apelata 1–2	*Arabidopsis*	1 aa	Flower development
Agamous	*Arabidopsis*	Transposon insertion	Flower development

Data obtained from various sources including Hoekstra and Coyne (2007) and Reid and Ross (2011).

locus, which controls stem length. In general, however, many morphogenes appear to be subject to various forms of structural changes such as exon shuffling (Xu et al. 2009, 2012).

The evolution of both physiological and morphological characters is affected by mutational changes of the regulatory regions of genes that include promoters and enhancers surrounding the coding regions. As mentioned above, the β-globin gene family in humans consists of a cluster of duplicate genes ε, γ_A, γ_G, δ, and β (Fig. 6.3). The gene ε is expressed in the early embryonic stage, γ_A and γ_G are in fetal liver, and δ and β are in adult individuals. The regulatory region of each globin gene determines the successive activation and suppression of expression of β-family genes in development. How this complex system of gene expression evolved is unclear, but this type of gene expression control would make the evolutionary changes of regulatory regions conservative because all the regulatory mechanisms must be maintained in the evolutionary process.

Furthermore, the expression of every gene is controlled by many other mechanisms such as *cis*-regulatory systems, gene regulatory networks, protein-protein interactions, small RNAs, etc. Operation of these regulatory systems must be coordinated with delicate balance. Therefore, the gene regulatory systems are also expected to evolve slowly. Many investigators have examined the rate of nucleotide substitution in the 5′ flanking regions of genes where *cis*-regulatory elements are located and showed that the rate is generally lower than the rate of synonymous substitution in the protein-coding region but higher than the rate of nonsynonymous substitution (Purugganan 2000; Miyashita 2001; de Meaux et al. 2005; Keightley et al. 2005; de Meaux et al. 2006). If we consider the possibility that the nucleotides outside the *cis*-elements would evolve at neutral rates, this observation suggests that the regulatory elements evolve under some functional constraints.

Any specific morphological characters or organs such as animal eyes, hearts, or limbs, plant flowers, etc. are products of complex processes of temporal and spatial expression of many interacting genes in the developmental process. This observation suggests that regulatory regions of genes play an important role in morphological evolution. In fact, many developmental biologists suggest that the major factor of morphological evolution is the change of *cis*-regulatory elements (CRE) and the changes in protein-coding regions play only minor roles (Gerhart and Kirschner 1997; Wilkens 2002; Carroll 2005b, 2008; Davidson 2006). Let us call this view the gene regulation hypothesis.

By contrast, other traditional evolutionists believe that changes in both protein-coding and regulatory regions of genes are important but that only minor proportions of mutations in these regions affect morphological evolution. Nei (1987) called this view the major gene effect hypothesis. This view is based on the idea that a majority of mutational changes are more or less neutral and do not affect morphological evolution appreciably. The importance of both coding regions and regulatory regions of genes were also emphasized by Hoekstra and Coyne (2007). Here let us consider these two different views separately.

Gene Regulation Hypothesis

It is well known that Ohno (1967, 1970) emphasized the importance of genome duplication in generating new genes, but it is less well known that in 1972 he discussed the inefficiency of gene duplication for producing new genes and advocated the gene regulation theory (Ohno 1972b). He first argued that natural selection operating at individual loci is a conservative evolutionary force and does not create innovative characters and that gene duplication certainly creates new genes but new paralogous duplicate genes maintain a function similar to that of the original gene.

Extending Jacob and Monod's (1961) idea, he then suggested that truly innovative characters are generated by evolutionary changes in gene regulatory systems. Thus, he stated: "*Drastic evolutionary changes in organisms' appearances are usually due to changes in regulatory systems rather than in structural genes. Man uses all five digits while the horse stands on its middle toes. Nevertheless, a digit is a digit. The same set of structural genes are mobilized for the formation of human fingers and equine cannons. It would be safe to say that the creating of additional regulatory systems contributed more to big evolutionary changes than did the creation of new structural genes.*"

King and Wilson (1975) noticed a conspicuous difference between protein evolution and morphological evolution in their study of evolution of humans and chimpanzees. They estimated that the number of codon differences detectable by electrophoresis is 0.62 per locus between humans and chimpanzees, using Nei's (1972) genetic distance measure. This value was similar to the genetic distances observed between many sibling species of *Drosophila* and rodents. They then suggested that this degree of protein differences is too small compared with the conspicuous morphological differences observed in such characters as brain size and the anatomy of the pelvis, foot, and jaws. To explain this difference, they turned to Ohno's (1972b) gene regulation hypothesis mentioned above. Ohno's idea (often erroneously credited to King and Wilson 1975) has recently been refined by many developmental biologists, who have studied the molecular basis of gene regulation (Gerhart and Kirschner 1997; Carroll et al. 2005; Davidson 2006), and it has become a popular explanation for morphological evolution (Carroll 2005a, 2008).

One recent example showing the importance of *cis*-regulatory elements has been obtained with Darwin's finches in the Galapagos Islands. There are 14 species of finches in the Islands, and they are often used as a textbook example of adaptive radiation of morphological characters. One character that has been studied well is the beak shape of the birds living on different islands. Several species eat insects and flowers of cactuses, while some others feed on seeds dropped on the ground. Cactus finches generally have long and pointed beaks, whereas ground finches have broad and thick beaks used for crushing seeds. Abzhanov et al. (2004) found that there is a high correlation between the extent of beak breadth and the expression level of bone morphogenic protein (BMP4), in the frontal part of the beak in the embryonic stage.

Later they searched for other genes affecting the beak shape and showed that calmodulin (CaM), a protein involved in mediating calcium signaling, is expressed at higher levels in the long and pointed beak of cactus finches than in more broad beaks of ground finches (Abzhanov et al. 2006). Therefore, it appears that the breadth and length of finch's beaks are controlled primarily by the expression levels of genes *Bmp4* and *CaM*, respectively. Darwin's finches are believed to have originated from finches in South or Central America about 2 MYA (Sato et al. 2001) through a mild bottleneck of population size. It is therefore likely that the beak shape of the finches evolved by new regulatory mutations and natural selection that occurred during the last 2 million years. In this case many different regulatory mutations appear to have occurred, because there is continuous variation in the beak shape and the expression levels of BMP4 and CaM among different species of Darwin's finches.

Another example of morphological evolution by regulatory mutations is that of freshwater stickleback fish living in lakes near the northern Atlantic and Pacific. They were apparently derived from the oceanic marine sticklebacks about 12 000 years ago when glaciers started to retreat and fresh water lakes were formed. Oceanic sticklebacks have relatively long pelvic (rear) fins, but the fins are almost absent or substantially reduced in freshwater sticklebacks (Fig. 6.6A). It has been shown that the presence of pelvic fins is associated with a high level of expression of transcription factor gene, *Pitx1*, in the pelvic region of the embryo (Shapiro et al. 2004). By contrast, freshwater sticklebacks showed no or low levels of expression of the gene. Study of the PITX1 proteins from oceanic and freshwater sticklebacks showed that there were no amino acid differences between them. From these observations, they concluded that the formation of pelvic fins is initiated by the expression of *Pitx1* and the evolutionary change of the regulatory region of the *Pitx1* gene is responsible for the reduction of pelvic fins. There are many other examples of *cis*-regulatory mutations that have generated morphological changes (Carroll 2005a; Wray 2007). Figure 6.6B shows a case where a dark spot was generated in a wing of the fruitfly *Drosophila biarmipes* by addition of the "wing spot" regulatory elements. This represents a case of gain of function mutation. Therefore, this form of mutation seems to play important roles in phenotypic evolution.

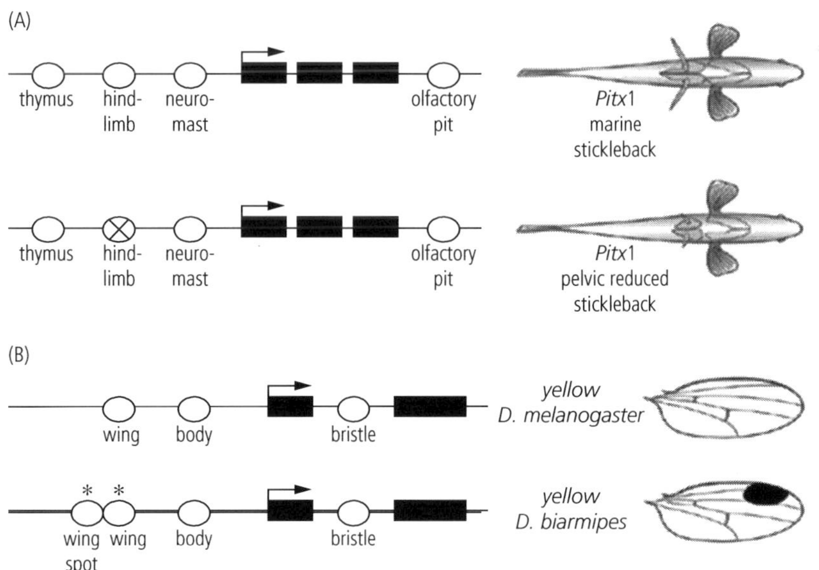

Fig. 6.6. Morphological changes apparently caused by regulatory mutations. (A) "Conjectured" loss of function by deletion of the "hind-limb" elements of the regulatory regions of the *Pitx1* gene in freshwater stickleback. (B) Gain of function by addition of the "wing spot" regulatory elements in the fruitfly *Drosophila biarmipes*. Modified from Carroll (2005b).

However, these case studies are not sufficient to establish the rule that the gene regulation hypothesis is the general explanation of morphological evolution, because morphological evolution may occur in many different ways. Carroll (2008) is fully aware of this criticism and presents more general arguments. According to him, there are eight general observations that support the gene regulation hypothesis. They are: (1) abundance of pleiotropic gene expression; (2) ancestral genetic complexity;

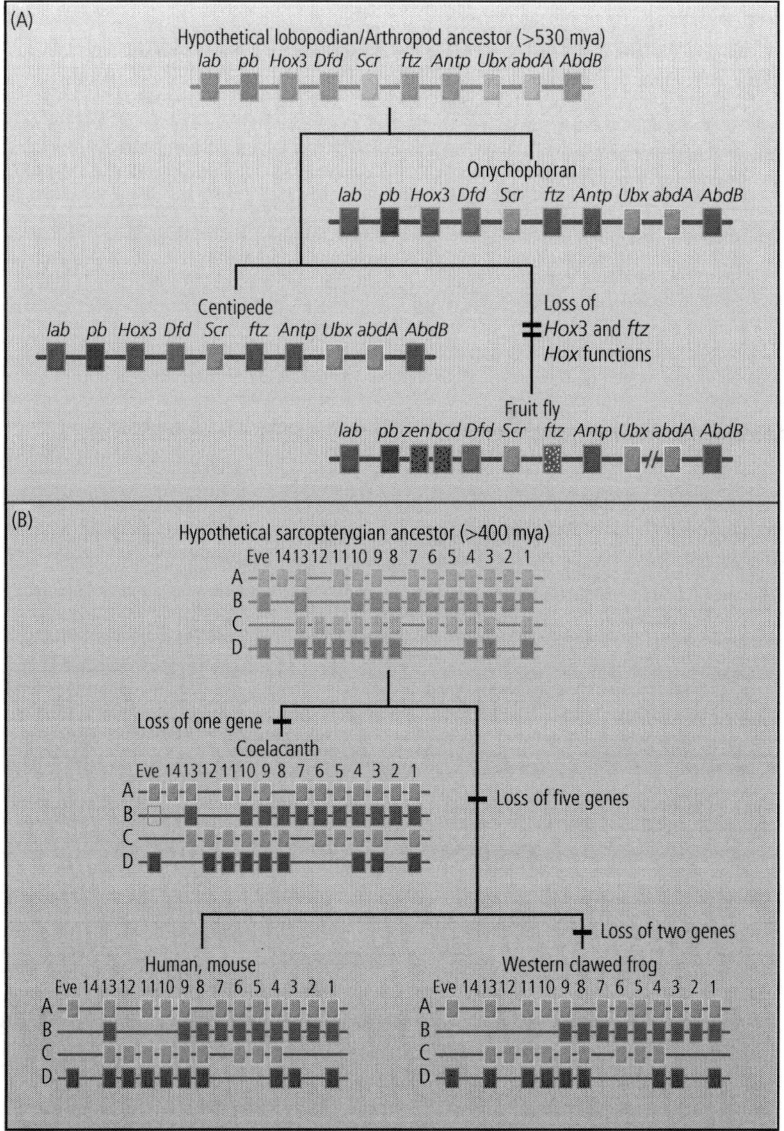

Fig. 6.7. Ancestral complexity of *Hox* clusters and the lack of *Hox* gene duplications in arthropods and chordates. (A) Based upon the *Hox* gene complements of onychophora (velvet worms) and arthropods, a minimum of ten *Hox* genes must have existed in the common ancestor of lobopodians (ancestors of velvet worms) and arthropods. No new *Hox* genes arose in centipedes or insects while the *Hox3* and *ftz* genes were co-opted into new functions in certain insects (stippling). (B) No new *Hox* genes are known to have evolved since the divergence of tetrapods from a common sarcopterygian (lobe-finned fish) ancestor shared with coelacanths. Rather, gene loss has occurred in several lineages. From Carroll (2008). Reproduced with permission from Elsevier. See also Plate 1.

(3) functional similarity of duplicate genes; (4) deep homology; (5) infrequent duplication of basic regulatory genes; (6) spatially different gene expression (heterotopy); (7) modularity of *cis*-regulatory elements; and (8) vast regulatory networks. Observations (2), (3), (4), and (5) are all concerned with the antiquity and conservation of regulatory genes, whereas the other observations are related to the importance of gene interaction and the versatility and flexibility of gene regulatory systems. The first of the above two features may be exemplified with the HOX homeotic gene clusters that specify body segmentation of a wide variety of animals. The mammalian genome has four sets of HOX gene clusters, but the fruitfly genome has only one set of clusters.

Because the HOX genes are now known from many different organisms, it is possible to reconstruct the evolutionary history of the HOX gene clusters for the last 540 million years (e.g. Hoegg and Meyer 2005). A simplified picture of the history is presented in Fig. 6.7. This figure suggests that the majority of HOX regulatory genes were present in the ancestor of lobopodians and arthropods about 530 MYA, and that they have not changed very much and are similar to those of the present centipedes and fruitflies. In the vertebrate lineage the four clusters of HOX genes were generated more than 400 MYA, and the current genomic structures of the clusters are only slightly different from the ancestral clusters. The HOX genes composing the HOX cluster are known to have been generated by gene duplication or deletion (Zhang and Nei 1996; Gehring et al. 2009). However, almost all mammalian species are known to have the same set of genes.

These findings suggest that regulatory genes have been highly conserved in a diverse group of animals and therefore the evolutionary changes in these regulatory genes would not be able to explain various innovative changes of morphological characters. How can we explain the evolution of innovative characters and how does the diversification of organisms evolve? Carroll's answer to these questions is the second feature of developmental biology mentioned above, that is, the changes in regulatory regions of genes. He argued that mutational changes in protein-coding regions may change the function of the protein encoded but the extent of change is limited. As Sturtevant (1925), Bridges (1935) and Ohno (1970) suggested, gene duplication may generate genes with new functions but the paralogous duplicate genes show functions which are not very different from that of the original genes. Carroll proposed, as Ohno (1972b) did earlier, that changes in regulatory regions of genes may introduce innovative changes of morphological characters.

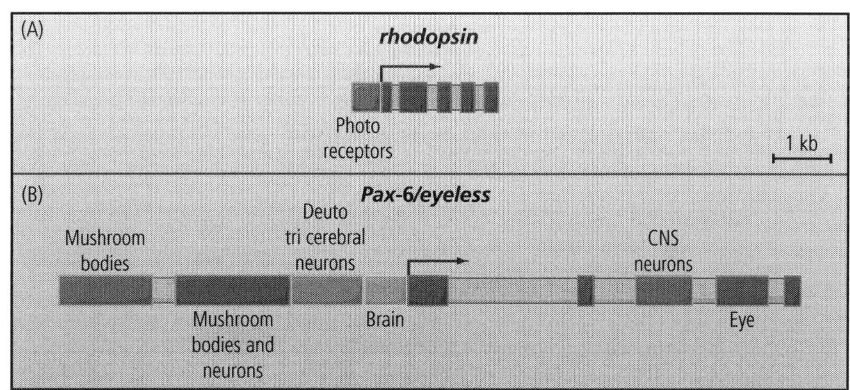

Fig. 6.8. (A) Structure of the *rhodopsin* locus in the fruitfly *Drosophila*. Exons are shown in black, introns in gray, and the single *cis*-regulatory element (CRE) controlling gene expression in photoreceptor cells is shown in purple. (B) Depicted is the *rhodopsin* architecture with the locus encoding its chief regulator *Pax–6/eyeless*. Exons are in black, introns in gray, and the six distinct CREs governing gene expression in parts of the developing brain, central nervous system, and eyes are shown in various colors. From Carroll (2008). Reproduced with permission from Elsevier. See also Plate 2.

In fruitflies, the morphologies of larval and adult individuals are so different despite the fact that the entire set of genes in the genome is the same. This suggests that the difference in gene regulation alone may produce drastically different morphological characters. Carroll then argued that the *cis*-regulatory regions for transcription factor genes are more complex than those of most genes encoding cell-type specific genes engaged in the chemistry of physiological processes. For example, the rhodopsin gene encoding vision-related protein (photoreceptor) in *Drosophila* is known to have only one *cis*-regulatory element (CRE) (Fig. 6.8A). In striking contrast to this simple system is the *cis*-regulatory region of the *Eyeless* (*Ey*) locus, which is required not only for eye development but also for patterning parts of the developing brain and central nervous system. The *Ey* gene is known to be orthologous to the *Pax–6* gene, which is essential for eye formation throughout the animal kingdom (Gehring and Ikeo 1999). There are six distinct CREs, averaging about 1kb size, so that each drives *Ey* expression in a particular special pattern in the eye and larval and adult brains, etc. (Fig. 6.8B). This complex special pattern of *Ey* expression facilitates the independent activities of multiple modular CREs.

In Carroll's view, the arrays of CREs that independently govern individual regulatory gene expression at different times and different places during development present a new picture of gene organization that is needed for phenotypic evolution. He stated: "*First, the multiple CREs are plain evidence of how gene function has expanded and diversified without duplication of coding sequences. Second, mutations in one CRE will not affect the function of other CREs or of the protein.*" (Carroll 2008, p. 30). If each CRE evolves independently and exerts its effects independently, his view will be the same as the classical view of independent evolution though we have to treat each CRE as though it is a classical gene locus. However, this form of multifunctional CREs is expected to have a wide range of pleiotropic effects and gene interaction.

Carroll also considers the existence of vast regulatory networks, recently identified, as support of his gene regulation theory. According to him, new techniques are revealing that transcription factors typically regulate tens to hundreds of target genes, and this will generate pleiotropic effects on an enormous scale that is not yet widely appreciated. One study has shown that there are on the average 124 target genes for each of the 67 *Drosophila* transcription factors examined (Stark et al. 2007), whereas in another study about 500 target genes have been identified for the *Drosophila* Twist transcription factor (Sandmann et al. 2007). These results suggest that GRNs are quite complex and evolutionary changes of morphological characters may be achieved by changes in GRNs alone.

For these reasons, Carroll and some other developmental biologists believe that morphological evolution is caused primarily by changes in regulatory regions of genes. While these arguments appear to support the gene regulation hypothesis, they do not really exclude the possibility that changes in coding sequences also contribute to morphological evolution.

Major Gene Effect Hypothesis

Three years before King and Wilson's (1975) paper appeared, Nei and Roychoudhury (1972) had reported similar observations about the discrepancy between protein and morphological variation in the comparison of the American black, Caucasian, and Japanese populations. They showed that the electrophoretically detectable codon differences between the three major human populations are only about 10 percent of the codon differences between individuals within the populations despite the fact that the morphological differences between populations are much greater than those within populations. Nei and Roychodhury then proposed that the protein differences are caused primarily by neutral or nearly neutral mutations whereas morphological differences were magnified by natural selection.

Later Nei (1987) explained this result by the hypothesis that a small proportion of mutations causes large phenotypic effects and they are subject to natural selection, whether they are due to coding gene mutations or regulatory gene mutations. He noted that transcription factors are generally proteins and therefore the mutations occurring in both regions should have similar effects. A large proportion of mutations affecting *cis*-elements also appear to be more or less neutral (see below). Originally, he did not particularly distinguish between physiological and morphological characters. In 2007, however, he proposed that the genes expressed in early stages of development are generally conserved whereas those which are expressed in later stages of development are less conserved, whether they are concerned with physiological or morphological characters (Nei 2007). Hoekstra and Coyne (2007) criticized the regulatory gene hypothesis by arguing about the difficulty of distribution between physiological and morphological characters. Their view is essentially the same as Nei's (1987).

Actually, there are many examples showing that changes in coding sequences alone cause morphological changes (Table 6.1). As in the case of genes controlling physiological characters, the genes controlling morphological characters are also often caused by amino acid substitutions or nucleotide deletion/insertion in the protein-coding regions of genes (Table 6.1). One example is the *Le* gene, which controls the plant height (or stem length) in the pea plant. A mutation of this gene shortens plant height, and this mutant form is caused by a single amino acid difference, as mentioned earlier in this section.

One of the commonly observed morphological variations within and between species is that of pigmentation of the hair, skin, and eyes of mammals and birds (Table 6.1). Many mammalian polymorphisms of black coat color (caused by pigment eumelanin) and reddish or yellowish color (caused by pigment phaeomelanin) are controlled by proteins called melanocortin-1 receptor (MC1R) and Agouti (Bennett and Lamoreux 2003; Carroll 2005a). The wild-type coat color of jaguars and jaguarundis of the cat family is reddish or yellowish and is determined by phaeomelanin. However, there are mutant genotypes with black coat color. This color is dominant to the wild type and is caused by deletion of several nucleotides as well as amino acid substitutions in the *MC1R* and *Agouti* genes (Eizirik et al. 2003). Jaguars and jaguarundis live in the jungles of Central and South America, and the selective advantage or disadvantage of the black form over the wild type is unclear (Carroll 2005a). It is possible that the mutant black form has spread through the population largely by genetic drift.

However, there are cases in which coat color is clearly related to the adaptation of organisms. In the Pinacate region of southwest Arizona, the rock pocket mouse, *Chaetodipus intermedius*, inhabits both dark and sandy rocky areas in the region. Dark areas have been formed by lava flow from a volcanic eruption that occurred more than one million years ago (MYA). Rock pocket mice are generally light-colored, but in the lava areas dark-colored individuals are observed. Nachman et al. (2003) showed that there are four amino acid differences in MC1R between dark-colored and light-colored individuals in this region. Since dark-colored mice were derived from light-colored mice by mutation, the former were apparently adapted to the dark environment to avoid attack from predators such as birds and large mammals. Similar adaptation to new environments caused by a single amino acid substitution in MC1R has been reported in the beach mouse, *Peromyscus polionotus*, in Florida (Hoekstra et al. 2006). These examples suggest that new mutations are responsible for adaptation (or preadaptation) to new environments and they have spread through the population primarily by natural selection. At this stage, it should be noted that there are more than 150 genetic loci that control the coat color of mammals, and the *MC1R* gene is only one of them (Bennett and Lamoreux 2003). These genes interact with one another, and the identification of the gene or genes controlling a particular polymorphism of color pattern is not always easy.

These examples show that morphological characters can be changed by a few amino acid substitutions, but it should be noted that as in the case of physiological characters most amino acid substitutions do not affect them appreciably. In the case of MC1R, there are 63 amino acid differences (out of 315 shared sites compared) between the wild-type mice and the wild-type rock pocket mice, but the two species have essentially the same coat color, indicating that only a few specific mutations can change coat color (Nachman 2005).

Carroll (2008) indicates that a large portion of coat/plumage color variation in mammals and birds is caused by changes in coding gene sequences. However, he considers this form of variation exceptional and has proposed the hypothesis that it occurs when the pleiotropic effects of genes involved are of minor importance. Yet, he has not really explained why the genes involved in coat color variation should have a lesser extent of pleiotropy. Actually, Sewall Wright studied the coat color variation of guinea pigs and concluded that the genes involved show extensive pleiotropic effects. Therefore, it is still unclear whether Carroll's hypothesis is correct or not.

According to the major gene effect hypothesis, this type of observation is expected to occur for genes that are expressed in later stages of development, whether the character is physiological or morphological. Because coat color is determined in a relatively late stage of development, this hypothesis is capable of explaining the evolution of coat color. Of course, at the present time, the detailed mechanism of coat color development has not been studied, so that it is premature to derive a definitive conclusion.

Another problem with the gene regulation hypothesis is the assumption that the expression of a gene is controlled by a complex regulatory system as in the case of the *Pax–6/eyeless* gene. If the number of CREs is large, it would impose a large amount of genetic load (Chapter 2), and the survival of the species may be threatened, as argued by Ohno (1972b). The genetic load argument is very crude and may apply only to large organisms like mammals, yet it cannot be neglected. For this reason, the number of genetic elements involved in the expression of a gene cannot be very large. I believe this is a good cautionary note.

Gene Regulatory Networks and Morphological Evolution

If we compare different phyla or classes of organisms, we are deeply impressed with the enormous amount of phenotypic diversity. For example, sea urchins and starfish, which belong to different classes of the phylum *Echinodermata* and diverged more than 540 MYA, show strikingly different morphologies, and they are apparently well adapted to different environments. However, studies of the early stage of embryonic development have shown that sea urchins and starfish have similar morphologies and developmental patterns and there is a common form of gene regulatory network (GRN) consisting of about 6 transcription factor genes (Davidson and Erwin 2006; Peter and Davidson 2011). This basic core GRN is specific for the early development of echinoderms and has not changed for the last 540 million years. However, as their development proceeds, the GRN in each of the two species expands into a more complex form including a large number of genes for transcription factors, signaling proteins, and structural proteins. In this process of expansion of the GRN, different genes are added in the two species so that their GRNs are gradually differentiated. This gradual differentiation of GRNs is responsible for the formation of the very different morphologies of sea urchins and starfish. The basic core GRN is highly conserved, and any significant change in the core results in deformation of the organism. This is also true with the GRNs functioning in successive developmental stages, but the extent of developmental constraint gradually becomes weaker as the development proceeds. The idea of GRNs has now been shown to work at the genomic level (Oliveri et al. 2008; Nam et al. 2010).

This property appears to apply to many different animal phyla, and new species in each phylum are generated by modifying the GRNs in later stages of development (Davidson and Erwin 2006; Peter and Davidson 2011). In other words, evolution occurs by modifying the old form of organism by mutation in later stages of development. In fact, the evolution of eye spots in the wings of some butterflies or the evolutionary changes of the number and form of body segments in insects and vertebrates have occurred by modification of GRNs in late stages of development (Brakefield et al. 1996; Carroll et al. 2005; Davidson 2006). In this view, the evolution of phenotypic diversity of different phyla has occurred not by positive Darwinian selection but by novel mutations and elimination of pre-existing less fit genotypes (Nei 2007). Evolutionists have proposed various mechanisms by which evolution can occur so fast that enormous amounts of phenotypic diversity among organisms can be explained (Fisher 1930; Muller 1932; Wright 1932). In reality, evolution is an intrinsically slow process, and the current phenotypic diversity has been generated only because there has been a long evolutionary time, more than 3 billion years.

6.3. Evolution of Gene Regulatory Systems

We have seen that in the formation of physiological and morphological characters various regulatory systems of gene expression are involved. In the past the majority of noncoding DNA was considered to be junk or virtually nonfunctional. Recent studies have raised serious doubts about this conception, suggesting that a substantial portion of the noncoding DNA is involved in gene regulation. Even some transposons and pseudogenes are now believed to encode small RNAs and contribute to gene regulation and genetic diversity. In this section, let us consider how such genetic elements have evolved. Actually, many regulatory systems involving small RNAs apparently existed from the very early stage of evolution, even from the time of the RNA world. However, such ancient events of evolution are difficult to study and remain highly speculative. Therefore, the early events of evolution will not be considered here.

Cis-Regulatory Elements

A number of authors have studied the evolutionary change of gene regulatory systems using relatively closely related species. The *Drosophila* homeotic gene *even-skipped (eve)* is known to produce seven transverse stripes along the anterior-posterior axis of the early embryo. Expression of each of these stripes is regulated by more than a dozen *cis*-regulatory elements (CREs) in the enhancer (activator and repressor) region. Ludwig et al. (1998) studied the CREs of stripe 2 enhancer of the *eve* gene from six different *Drosophila* species and showed that the CREs of this gene are generally highly conserved but change gradually as time goes on. Figure 6.9 shows that the evolutionary changes in CREs have occurred by deletion, insertion, or nucleotide substitution in 5 species of *Drosophila*. CREs *bicoid–3* and *hunchback–1* were apparently added to the *melanogaster* group of species after these species diverged from the *D. pseudoobscura* lineage, whereas part of CRE *giant–3* was deleted. CRE *bicoid–5* is highly conserved, and no nucleotide substitution is observed among the 5 species compared. By contrast, *bicoid–3* shows a fair number of nucleotide substitutions even among the *melanogaster* group species. Therefore, it is quite possible that a new CRE can be generated by nucleotide substitutions alone (Carroll 2005b).

Nevertheless, when the genetic constructs of enhancers and coding regions from different species were examined, all of them showed essentially the same stripe 2 expression. This indicates that the genomic positions of CREs can change in the evolutionary process, but as long as proper numbers of CREs to act as activators and repressors are retained, the regulatory system remains unchanged. Apparently, what is more important is to form a proper complex of transcription factors that control the activity of RNA polymerase. Therefore, the CREs can change with mutations though they are certainly more conserved than non-CREs. A similar but

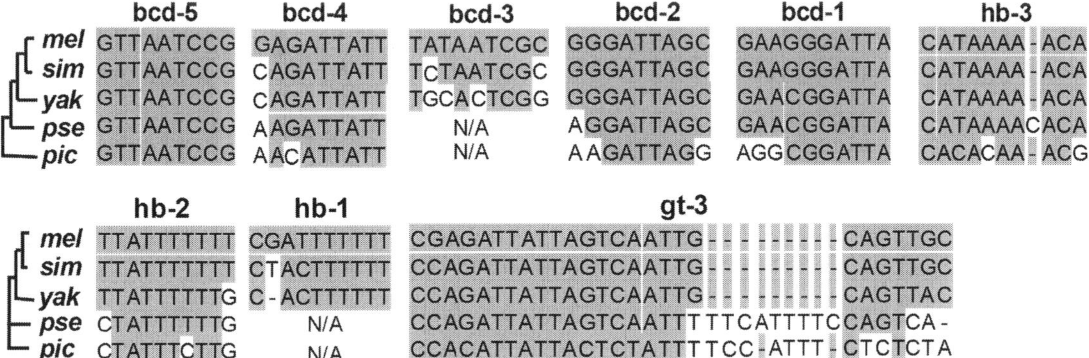

Fig. 6.9. Nucleotide sequence of 9 *cis*-regulatory elements of the *even-skipped* stripe 2 gene expression in 6 *Drosophila* species. *Drosophila* species: *mel, melanogaster*; *sim, simulans*; *yak, yakuba*; *erec, erecta*; *pse, pseudoobscura*; *pic, picticornis*, *cis*-regulatory genes: *bcd, bicoid*; *hb, hunchback*; *gt, giant*. Modified from Ludwig et al. (1998).

more complicated regulatory region of a gene has evolved with respect to the mating type genes, *MATa* and *MATα*, in the ascomycete yeast lineages (Tsong et al. 2006).

These results suggest that most of the nucleotide substitutions in the regulatory region evolve in a more or less neutral fashion as in the case of the protein-coding region. Therefore, the evolutionary change of gene regulation is apparently controlled by major gene mutations including deletions/ insertions.

Evolutionary Change of MicroRNAs and other Small RNAs Controlling Gene Expression

In recent years many studies have shown that the level of gene expression is controlled by various types of small RNAs (21 to 28 nucleotides). Among others, microRNAs (miRNAs) appear to play important roles in development, as mentioned earlier. A miRNA is transcribed as a primary miRNA, which forms a hairpin structure and undergoes several steps of processing to produce a mature miRNA with about 21 nucleotides. This mature miRNA interacts with the transcripts of the target genes. In animals, this interaction occurs between the seed sequence of about 7 nucleotides of the miRNA and the target site in the 3′ UTR region. In plants, however, the entire mature miRNA is used for recognizing the transcript with near perfect base-pairing in the protein-coding region (Chen 2005; Axtell and Bowman 2008; Bartel 2009). When a miRNA recognizes its target sites on a transcript, this transcript is generally degraded.

Many miRNAs are encoded by genomic regions which are located primarily in introns and intergenic regions (Stefani and Slack 2008). The transcription of intronic miRNAs is believed to be controlled by the same regulatory system as that for the expression of the gene itself. By contrast, the expression of intergenic miRNAs is apparently controlled by its own promoter. Therefore, creation of new intergenic miRNAs must be more complicated than that of intronic or exonic miRNAs.

In general, miRNAs are highly conserved, and new miRNA loci are often produced by gene duplication of pre-existing miRNA genes or from random hairpin structures of DNA (Tanzer and Stadler 2004; Tanzer et al. 2005; Nozawa et al. 2010). In plants there is evidence that some miRNA genes are generated from inverted duplicates of protein-coding genes (Allen et al. 2004). Because a duplicated gene has a high degree of sequence similarity to the original gene, a new miRNA locus that matches the 3′ UTR region of the original gene may be easily produced. However, miRNAs may also be generated by mutations. The probability of generating a miRNA locus from a random sequence by mutation is very small, but the number of DNA segments that can be mutated is so large that a fair number of miRNA loci may be generated over a long evolutionary time (Lu et al. 2008). If these miRNA loci are generated in intronic regions, the probability of producing new miRNA genes is quite high because there is no need to produce a new transcription system for this gene. In this case the transcription system for the gene encompassing the introns may be used.

Of course, the new miRNA loci created in this way may deteriorate unless they find proper target genes. Because most new miRNAs would not satisfy this requirement, most new loci are unlikely to survive very long. In fact, most newly arisen miRNA genes are disintegrated relatively quickly (Berezikov et al. 2006; Lu et al. 2008; Nozawa et al. 2010; Nei and Nozawa 2011). Furthermore, the mutation of a miRNA sequence may result in a nonfunctional miRNA. In general, miRNA loci are subject to birth-and-death evolution, and the turnover of many miRNA gene families may occur relatively fast (Nozawa et al. 2010).

At the present time, miRNAs are believed to exist only in animals and plants. However, their related RNAs, small interference RNAs (siRNAs), are also involved in the degradation of mRNAs. Actually, siRNAs are observed in all kingdoms of eukaryotes, and therefore RNA interference appears to have evolved earlier than the evolution of miRNAs.

RNA interference (RNAi) appears to have evolved originally for protecting the host plants and animals from exogenous viruses and endogenous transposons. In fact, RNAi was first discovered as an immune system for protecting plants from viruses (Hamilton and Baulcombe 1999). MicroRNAs are endogenously produced, but their basic function is to control the level of production of mRNAs at the post-transcriptional level. MiRNAs have more sophisticated functions to regulate gene expression by controlling the tissue specificity, timing and level of gene expression (Heimberg et al. 2008). At the present time, however, the detailed aspects of the origination of miRNAs are not well known.

As mentioned in Section 6.1, many noncoding RNAs appear to be involved in the regulation of gene expression. A group of small RNAs with 20–300 nucleotides are involved in the modification of target RNAs, synthesis of telomeric DNA, chromatin structure dynamics, etc., whereas medium and large RNAs (300–10 000 nucleotides) participate in X chromosome inactivation, DNA demethylation, gene transcription, epigenetics, etc. (Costa 2007). There are also many different small RNAs whose functions have not been well understood. It is also known that some pseudogenes are transcribed and certain pieces of RNA transcripts are apparently used for regulating the expression of other genes (Wen et al. 2011). The functions of these RNAs and their mechanisms are currently under investigation. How these types of control of gene expression evolved is unclear. This could be an example of Jacob's (1977) theory of evolution by tinkering. That is, innovation of a phenotypic character of gene function occurs by using many pieces of genetic material, whatever is available at the time of necessity.

6.4. Epigenetics and Phenotypic Evolution

Any organism lives in a given environment, and therefore the life of an organism must be affected by environmental factors. If the environmental factors are different for different individuals, these individuals may develop different phenotypes. This occurs even for the same genotypes, because the pattern of gene expression is modified by environmental factors. This modification of gene expression by environmental factors is called epigenetics. In practice, all individuals grow in different environmental conditions, so that epigenetics is an important component of development and morphogenesis. However, the molecular mechanism of epigenetics is not well understood, and it is now under intensive study. Here I would like to discuss

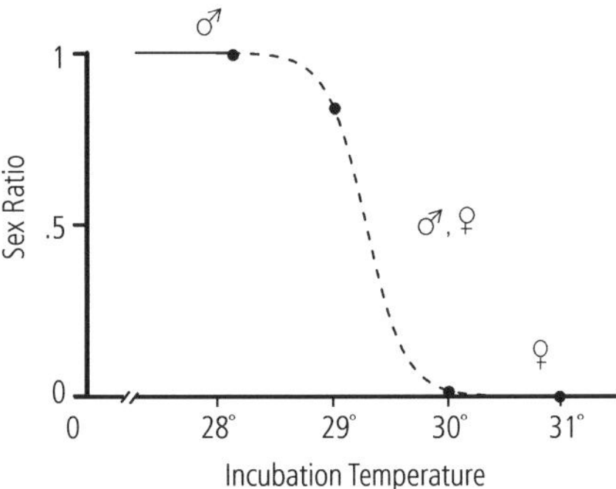

Fig. 6.10. Sex ratio (proportion of males) as a function of incubation temperature in the Ouachita map turtle. Modified from Bulmer and Bull (1982). Reproduced with permission from John Wiley & Sons.

two well-known examples of epigenetic development of characters without going into the detailed mechanism.

Environmental Sex Determination

The first example I would like to discuss is the temperature-dependent sex determination (TSD) in some alligators and turtle species (Bull 1983; Ramsey and Crews 2009; Shoemaker and Crews 2009). In these organisms sex is determined by the temperature in about one third of the period (14–15 days) of embryonic development (incubation time), which is called the temperature sensitive period (TSP). In the Ouachita map turtle a low temperature (22–28°C) in the TSP generates males only, whereas a high temperature (30°C or higher) produces females only (Fig. 6.10). In the intermediate temperatures (28–30°C), both males and females are born, so that the sex ratio (proportion of males) is between 0 and 1. This occurs because female hormone estrogens trigger ovarian development in a high temperature environment, whereas male hormone androgens produce testes in low temperatures.

However, the reason why the low temperature generates androgens and the high temperature causes the production of estrogens is unknown. A number of investigators (e.g. Lance 2009; Ramsey and Crews 2009; Shoemaker-Daly et al. 2010) have attempted to clarify the molecular basis of gonad formation in reptiles with TSD, following the mammalian system in which the molecular pathway of forming reproductive organs is known. So far, no clear answer has been obtained. In nature, of course, the incubation temperature varies from individual to individual depending on the environment, and the sex ratio is not far from 0.5 (Bulmer and Bull 1982). In practice, however, the incubation temperature may vary with year or location, so that the sex ratio may be distorted substantially in some years or locations.

Evolution of Temperature-Dependent Sex Determination

How did the complicated TSD system evolve? This is part of a larger question about the evolution of sex determination, which will be discussed again in Chapter 8. Actually, there are hundreds or thousands of different ways of sex determination even if we consider the animal kingdom alone. The reproductive system varies with species even within reptiles alone. In some species such as snakes and several lizards sex is determined by heteromorphic chromosomes, as in the case of mammals and birds (Sarre et al. 2011). In these species environmental factors do not seem to be important. Even if the same TSD system is used, the detailed aspects such as the temperature-sensitive period varies from species to species (Shoemaker and Crews 2009). In American alligators the very low or high temperatures produce females, and the intermediate temperatures produce primarily males only.

If we consider non-reptile animals, the diversity of sex-determining mechanisms becomes much higher (Bull 1983). In some marine worms such as *Bonellia viridis*, the free-living planktonic larvae are not sexually differentiated. Larvae which land on unoccupied sea floor mature into adult females. Those that settle on adult females become males. This is caused by the chemical called bonellin, which is produced by females. Males cling to the female's body or are sucked inside the female by the feeding tube and spend their remaining life inside her genital sac. The purpose of the male is to produce sperm to fertilize the female's eggs. Therefore, the sex is determined by bonellin in this organism. A somewhat similar sex determination has recently been reported in another group of marine worms (genus *Osedax*) that feed on the bones of dead whales (Rouse et al. 2004).

The above examples are rather clear cases where the same genotypes express different phenotypes because of different environmental factors. However, epigenetic effects of phenotypic expression must exist whenever genotypes and environmental factors interact with each other, though the actual molecular mechanism may not be known. Identical twins in humans contain essentially the same genotype, but close examination of their phenotypes indicates that they are not identical and there are always small differences. These small differences are generated by epigenetic effects of environmental factors.

Vernalization and Flowering in Plants

The second example of epigenetic control of gene expression is the effects of vernalization on plant flowering. It has long been known that flowering of wheat and barley can be accelerated by prolonged exposure to cold temperatures, which is called vernalization. Vernalization occurs during winter when the temperature is low (0–10°C). A few weeks of exposure to cold temperature is usually sufficient to promote flowering, but longer periods are more effective than shorter periods.

In wheat and barley three genes, *VRN1*, *VRN2*, and *FT*, are important for the regulation of flowering through vernalization (Trevaskis et al. 2007). The expression of gene *VRN1* is induced by vernalization, and the gene then promotes the transition of vegetative growth of the shoot apex to the production of reproductive organs. Gene *FT* is induced by long days and accelerates the transition of vegetative growth to reproductive growth of the apex. *VRN2* is a floral repressor and represses the activity of the *FT* gene until plants are vernalized. Only after the repression of the *FT* is removed is flowering permitted. Vernalization accelerates flowering by inducing the expression of the *VRN1* gene, a promoter of a MADS-box transcription factor. Before vernalization is applied, the *VRN1* gene region of chromatin is methylated, but this methylation is reduced by vernalization, and the level of expression of *VRN1* is enhanced (Kim et al. 2009; Oliver et al. 2009b).

It should be noted that there are varieties of wheat and barley in which flowering occurs without vernalization. In these varieties the expression level of the *VRN1* gene increases during flower development. This suggests that a high level *VRN1* gene is required even without vernalization. Note also the effect of vernalization is quantitative rather than qualitative and varies with genotype considerably. The epigenetic mechanism of vernalization in *Arabidopsis* is somewhat different from that of cereals, and a different set of genes is involved. Therefore, it seems that the evolution of flowering mechanisms occurred by Jacob's (1977) scheme of evolution by tinkering mentioned in Section 5.5.

It should also be noted that vernalization is merely one way of controlling flower formation or the switch from vegetative growth to flower development. Actually, a more important factor in this respect is photoperiodism, which is a universal factor for plants in temperate climates. Furthermore, there are various molecular mechanisms such as RNA silencing and the mobile flowering promotion signal (florigen) (Baurle and Dean 2006). Therefore, a large number of genes with their epigenetic expression are involved in the determination of flower formation and flowering time. Our understanding of the flowering process still remains very poor.

In this section I discussed epigenetic gene expression using two well-known examples, that is, environmental sex determination in reptiles and vernalization in plants. However, our ignorance of the interaction between genes and environments is profound. This problem will be an important issue in evolutionary biology in the next few decades.

6.5. Gene Co-Option and Horizontal Gene Transfer

As discussed above, the evolution of genes usually occurs through nucleotide changes or gene duplication in the protein-coding or regulatory regions of DNAs. In some cases, however, new functional genes may evolve by recruiting or co-opting other genes for a new function. This type of evolution is called gene co-option or gene recruitment. Occasionally genes with new functions may also be developed by adopting foreign genes that come from outside the individual. This phenomenon is called horizontal or lateral gene transfer. In this section I would like to present a brief discussion about this type of evolution.

Gene Co-Option

In Chapter 5 we discussed that the adaptive immune system in vertebrates probably evolved by recruiting different gene families which were originally developed for other purposes. There are many such examples of evolution, particularly with morphological characters. A well-cited example is the evolution of bird feathers, which are believed to have evolved initially for heat regulation but later co-opted for body-color display and finally for bird flight. Similarly, the swimbladder of teleost fish may have been co-opted into the lungs of terrestrial vertebrates, as was argued by Darwin (1872, p. 147). Co-option is certainly important for adaptive evolution, but the very initial step could have been more or less neutral. An interesting case of neutral gene co-option is observed with the gene sharing first discovered by Piatigorsky et al. (1988). Gene sharing refers to the event in which the same protein shows two different functions. For example, the protein argininosuccinate lyase is known to function as an enzyme in many different tissues of vertebrates, but it is also used as the lens structural protein δ-crystalline that is water-soluble and maintains lens transparency. Another protein, small heat-shock protein, has also been co-opted as a lens structural protein as well as a protein that regulates the folding and unfolding of other proteins in vertebrates. Similarly, other crystallines (e.g. β- and γ-crystallines) are known to be bifunctional. In general, these bifunctional proteins are not duplicate genes, and they are produced by changes in gene expression in different tissue. Therefore, there are no amino acid differences between the proteins expressed in different tissues.

In recent years many such genes have been discovered (Piatigorsky 2007), and these are often called moonlighting genes, because they have a primary and a secondary job (function) (Jeffery 2003). For example, cytochrome c has a primary function in energy metabolism and a secondary function of apoptosis, whereas enolase in fungi has a primary function as a glycolytic enzyme and a secondary function of mitochondrial tRNA import. According to Piatigorsky (2007), most proteins are multifunctional. For example, a number of enzymes used for glycolysis bind to cytoskeletal proteins such as actin and tubulin, and therefore they have a role as structural proteins. As another example, serum albumins are abundant structural proteins and are important for transporting fatty acids and for binding toxic metabolites. Albumins are also known to have a number of catalytic activities such as oxidation of nitric oxide and formation of s-nitrosothiols. Therefore, gene sharing or moonlighting is not a rare event.

Another type of gene co-option is observed in the formation of antifreeze proteins (AFPs) that exist in many different species living in frigid areas. AFPs circulate in the blood and bind to ice crystals and prevent their growth. The best studied AFPs are those from the perch-like fish belonging to the suborder Notothenioidei living in the seas surrounding Antarctica. These notothenioid fish live at seawater temperatures between –2°C and 4°C without being frozen. There are four or five different types of AFPs, and these proteins have evolved independently from different proteins. One of them is the antifreeze glycoprotein (AFGP), and this protein is apparently derived from pancreatic trypsinogen-like protease. However, the genomic structure of this gene is quite complex and is composed of a polyprotein multigene family (Cheng and Chen 1999). Molecular dating has suggested that this gene family originated 22–42 million years ago, during which a global cooling is known to have occurred (Near et al. 2012). A member of the AFGP gene family contains Ala-Ala-Thr repeats generated by duplication and amplification of conserved, partially noncoding, nine-nucleotide sequences. Most AFGPs have lost exons 3–5 of the trypsinogen gene but have retained exons 1 and 6. Interestingly, a chimeric AFGP-trypsinogen gene, which contains exons 2–6 of the trypsinogen gene was also discovered in some species. The evolutionary history of the AFGP gene is quite complicated (Chen et al. 1997).

AFPs are also observed in the teleost fish (e.g. northern cod) living in the northern Arctic region. They are again classified into several types and are composed of multigene families. However, they originated independently of the AFPs from the Antarctic areas. For example, AFGPs have the same amino acid sequence motif (Ala-Ala-Thr) as that of the Antarctic proteins, but the origin of these genes is unknown (True and Carroll 2002). These examples of AFP genes are interesting, because they show that proteins with essentially the same function can evolve independently from different proteins. This finding is against the general rule of protein evolution, which predicates that protein functions almost always diverge and convergent evolution is very rare. Yet, this rare event happens in evolution when the proper genetic material is available and the proper ecological condition is met. This type of evolution by gene co-option is similar to Jacob's (1977) idea of the evolution by tinkering.

Horizontal Gene Transfer

Horizontal gene transfer (HGT) refers to the transfer of genetic material between different organisms by a process other than the inheritance of genes from parents to offspring. The evolutionary significance of HGT was not well understood until the genomic study of bacterial phylogenies was conducted in the 1980s, though the occurrence of symbiosis between different organisms had been known for some time (Ochiai et al. 1959; Syvanen 1985). When HGT was shown to occur frequently in bacteria, it was regarded as a factor that complicated the study of phylogenetic relationships of different species (Doolittle 1999). However, HGT is also an important mechanism for the evolutionary change of phenotypic characters.

It is well known that the chloroplast in plant cells originated from cyanobacteria about two billion years ago. At present, the number of genes in the cyanobacterial genome ranges from 1700 to 7000, but the number of genes in the chloroplast genome appears to be 50 to 200 (Martin et al. 2002). Therefore, a majority of the original cyanobacterial genes have been lost or transferred to the host nucleus. Acutally, Martin et al. (2002) showed that a large proportion of genes are now located in the host nucleus and that about half of these genes are targeted at the chloroplasts but the remaining half are now used for various purposes of the host organism including metabolism, biosynthesis, transcription, and cell division. These results indicate that the genes that have been transferred from an entirely different species can be used for the purpose of evolution of the host organism. In fact, symbiosis and

HGT have been important factors for the evolution of multicellular organisms from unicellular organisms (Rokas 2008).

In recent years, the number of reports about evolution by HGT is increasing even in eukaryotes (Hall et al. 2005; Keeling and Palmer 2008). One of the best studied cases is the transfer of the genes of *Wolbachia* bacteria to insect species (Dunning Hotopp et al. 2007). The bacteria *Wolbachia pipientis* is a maternally inherited endosymbiont that infects a wide range of arthropods and nematodes. It is present in developing gametes and therefore provides conditions favorable for heritable transfer of bacterial genes to the eukaryotic hosts. Dunning Hotopp et al. (2007) examined the gene transfer from *Wolbachia* bacteria to 26 species of fruitflies (*Drosophila*) and other insects and found that 11 species contained Wolbachian DNA, the size of which ranged from a small piece to almost the entire region of the genome. Generally speaking, HGT occurs from prokaryotes to eukaryotes, but it is also known to occur from eukaryotes to prokaryotes as well. HGT of nuclear genes between different eukaryote species seems to be rare, but there are a number of cases reported (Moran and Jarvik 2010; Yoshida et al. 2010).

Photosynthetic Animals

Generally speaking, photosynthesis takes place only in plants, algae, and some bacteria, and animals which are incapable of photosynthesis live on the food produced by photosynthetic organisms. However, there are exceptions. Some animals such as giant clams, sponges, corals, and flatworms are capable of capturing photosynthetic products through the formation of symbiotic associations with intact unicellular algae or cyanobacteria (Rumpho et al. 2011). In these cases algae or cyanobacteria act as an autonomous photosynthetic factory.

In some animals such as sea slugs, however, the ability of retaining only the functional plastids from their algal prey has evolved. In this case plastids are retained intracellularly in cells lining the host animal's digestive system and remain photosynthetically active for an extended period of time. This photosynthesis was first discovered by Kawaguti and Yamasu (1965) in the sea slug *Elysia atroviridis*, and extensive studies have been made into the mechanism of photosynthesis in the species of the genus *Elysia*. Here I would like to discuss Rumpho et al.'s (2011) study of *E. chlorotica*.

E. chlorotica is a green marine animal and is found in the eastern coastal marshes of the United States. It is usually 2–3 cm long but can grow as long as 6 cm (Fig. 6.11). At the juvenile stage, the sea slug eats the yellow-green alga *Vaucheria litorea*. The sea slug is brown-colored before feeding on the algae but become green-colored after feeding (Ma 2012). Upon feeding on *V. litorea*, the slug breaks down the unicellular filamentous algae and sucks the contents, accumulating chloroplasts in their branched digestive system distributed throughout the body (Fig. 6.11). The feeding continues until the number of chloroplasts saturates the digestive system. The sea slug can then live as a photosynthetic organism for several months without eating.

The relationship between the sea slug and the alga is different from the ordinary endosymbiosis, because the alga is broken down and is not a living entity. However, the relationship is also different from the ordinary HGT, because the algal genes transferred are functional only for one generation and they are not transmitted to the next generation. If algal chloroplasts were incorporated into the slug

Fig. 6.11. Sea slug *Elysia chlorotica*, showing the highly branched digestive system. The body color of this organism is green. From Ma (2012). Reproduced with permission from Patrick Klug. See also Plate 3.

germ cell and remained functional, a true photosynthetic animal would be generated.

In this section we have considered several issues of gene co-option and horizontal gene transfer and showed that these mechanisms play important roles in evolution. It is interesting to note that different genes which originally evolved for different purposes can come together and form a new genetic system, and that even genes which evolved in distantly related organisms can be combined to start new evolutionary lineages.

6.6. Summary

To understand the evolution of phenotypic characters, it is important to know how these characters are formed by interaction of many genes in the developmental process. In this case different genes are expressed in different tissues at different developmental times. This form of time-dependent gene expression must be coordinated by some molecular mechanism. For this reason, it is important to know the mechanism of gene regulation in the developmental process.

In general, a gene is composed of the protein-coding region and the gene regulatory region. The protein-coding region is for producing a protein with a given amino acid sequence, whereas the gene regulatory region is responsible for initiating the transcription of messenger RNAs (mRNAs) and consists of the transcription promoter (TATA box) and the *cis*-regulatory region. The *cis*-regulatory region is for the attachment of transcription factors.

Evolutionary changes of phenotypic characters are achieved by changes in either or both of the protein-coding and the regulatory regions. Changes in the protein-coding region result in alteration of the amino acid sequence of the protein encoded. This alteration sometimes changes protein function, and this change may affect phenotypic characters. For example, the difference between red and green color vision in humans is caused by two amino acid substitutions in the vision pigment. It is also known that changes in the *cis*-regulatory region often result in evolutionary changes of phenotypic characters. This seems to be particularly important for morphological characters. Here too, however, only a small proportion of changes seem to be important, and the remaining changes appear to be more or less neutral. There are many other genomic changes that affect phenotypic evolution. DNA methylation and histone methylation are known to control the initiation or prevention of gene expression in particular tissues at a given time. MicroRNAs (miRNAs) and other small RNAs encoded by the noncoding regions of DNA are also known to control the level of protein production by degrading mRNAs. Changes in these small RNAs may then alter the expression of phenotypic characters.

Development of a phenotypic character is controlled by a large number of interacting genes, and it is usually initiated by signal molecules that activate the signaling pathways. For example, the SRY protein in mammals is a signal protein that activates the function of the SRY signaling pathway for the formation of the male phenotype. There are many signaling pathways involved in the development of a phenotypic character. The transcription factors generated by different pathways often interact with one another in the developmental process.

Recent studies have shown that the genes controlling phenotypic characters that are expressed in the early stage of development are highly conserved and that recently evolved characters are expressed in a later stage of development. Even the genes controlling the latter characters are generally conserved, but there is a large component of neutral or nearly neutral genetic variation within and between species.

In recent years significant progress has been made in the study of the effect of environmental factors on the formation of phenotypic characters at the molecular level (epigenetics). We now know the molecular mechanism of vernalization and photoperiodism in plants, but there are many unsolved problems in this area. It seems that this is one of the most important problems in the current study of phenotypic evolution. Evolution does not always occur phyletically. Different genes originally evolved for different purposes may come together and form a new genetic system for a new function. Genes generated in different organisms can also come together by horizontal gene transfer, and previously unforeseen organisms may be generated. Symbiosis, which has been important in generating multicellular organisms, is also a form of horizontal gene transfer.

CHAPTER 7

Mutation and Selection in Speciation

In this chapter I would like to discuss the genetic and molecular basis of speciation or formation of new species, which has been controversial for the last 150 years. In his book *The Origin of Species* Charles Darwin was primarily concerned with the evolution of new species by means of natural selection operating on continuous variation. He was aware of the fact that the hybrids between different species are often inviable or sterile, but he had difficulty explaining it by natural selection. During his time some authors suggested that the hybrid sterility between different species might be enhanced by natural selection because the mixing of two incipient species by hybridization is disadvantageous for the formation of new species. Darwin rejected this idea after examination of various cases of species hybridization and concluded that "*hybrid sterility is not a specially acquired or endowed quality but is incidental on other acquired differences*" (Darwin 1859, p. 245). However, he did not present any concrete theory for explaining the development of reproductive isolation. Because Darwin maintained the view that adaptive evolution occurs very slowly by means of natural selection, he also believed that speciation occurs as a very slow process.

As mentioned in Chapter 1, de Vries (1901–1903, 1909) proposed a diametrically opposed view called the mutation theory, in which new species or elementary species (meaning incipient species) are produced spontaneously by single mutational events. According to this theory, new incipient species are instantly produced, and they are reproductively isolated from the parental species. Because this theory was based on experimental studies conducted with the evening primrose *Oenothera lamarckiana*, it was accepted by many biologists when it was proposed (Allen 1969). About two decades later, however, the mutation theory was almost abandoned mainly because *O. lamarckiana* was found to be a heterozygote for chromosomal complexes and the mutant forms he discovered were mostly caused by chromosomal rearrangements derived from this unusual parental species (Davis 1912; Renner 1917; Cleland 1923). The fact that a number of *Oenothera* species contained these chromosomal complexes was a new discovery in genetics at that time, and much attention was given to this discovery rather than to de Vries's mutation theory. At the time of de Vries the genetic cause of mutations was not known, and he regarded any heritable changes of phenotypic characters as mutations. Later studies showed that at least one of the elementary species he discovered was a tetraploid (see Section 7.1), and it established itself as a new species in self-fertilizing evening primrose. Therefore, he was right in his proposal of mutation theory. In fact, recent genomic data abundantly support his theory of origin of species by chromosomal changes.

In general, however, the formation of new species by chromosomal mutations appears to be rare, and most speciation events are assumed to occur by the establishment of genic sterility or inviability of hybrids between different species. The evolutionary mechanism of genic speciation is complicated, and there are many different ways of generating hybrid weakness. In these cases many investigators have emphasized the importance of natural selection rather than mutation (e.g. Presgraves et al. 2003; Coyne and Orr 2004; Wu and Ting 2004; Maheshwari et al. 2008). Some authors implied that adaptive evolution of incompatibility genes is important in speeding up speciation. In my view, the crucial event of speciation is the development

Mutation-Driven Evolution. First Edition. Masatoshi Nei.

of reproductive barriers between species, and this is accomplished mainly by mutation.

In this chapter, we first discuss the roles of chromosomal variation in speciation in the light of recent genomic data and then discuss various mechanisms of speciation by means of genic mutation and selection. We will consider both theories and experimental data that support or do not support a particular speciation model. We will be concerned only with the case of allopatric speciation, in which the populations to be differentiated are geographically or ecologically isolated. My primary purpose is to clarify the roles of mutation and selection in the evolution of reproductive isolation and show that the molecular basis of speciation is more complicated than generally thought at present.

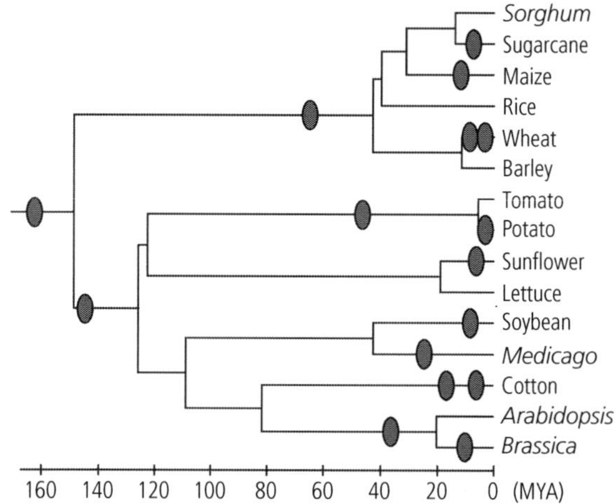

Fig. 7.1. Inferred polyploidization events during the evolution of angiosperms. Circles indicate suspected genome duplication events. Approximate time scale is shown below the tree. Modified from Adams and Wendel (2005). Reproduced with permission from Elsevier.

7.1. Speciation by Chromosomal Mutations

Formation of New Species by Polyploidization

Soon after de Vries reported various mutants derived from *O. lamarckiana*, a number of investigators studied their chromosomal numbers and chromosomal segregation at meiosis (see Cleland 1972). They found many aneuploids and trisomics, but there was one elementary species (*O. gigas*) which was bigger and more vigorous than *O. lamarckiana*. This was later shown to be a tetraploid (Lutz 1907; Gates 1908; Davis 1943). Furthermore, cytogenetic studies of flowering plants (angiosperms) in the mid-20th century showed that 20–40 percent of the species had experienced polyploidization in their origin (Stebbins 1950; Grant 1981). At this stage, it was clear that the chromosomal mutation called polyploidization is an important mechanism of creating new species in angiosperms. As is well known, polyploid plants establish a sterility barrier from their parental species immediately after their occurrence, because the hybrids between them have an abnormal segregation of chromosomes at meiosis and consequently they are sterile. Yet, many plant geneticists did not realize the importance of polyploidization in evolution. For example, Stebbins (1966, p129) stated "*the large amount of gene duplication dilutes the effects of mutations and gene combinations to such an extent that polyploids have great difficulty evolving truly new adaptive gene complexes.*"

In recent years, our knowledge of polyploid evolution has expanded enormously because of the availability of genomic sequences of many different organisms. Statistical analyses of these sequences have shown that polyploidization or genome duplication has occurred quite often particularly in flowering plants. Doyle et al. (2008) state that the genomes of flowering plants are fundamentally polyploid and most species in plants have experienced polyploidization far more frequently than previously suspected. Adams and Wendel (2005) and De Bodt et al. (2005) believe that angiosperms underwent two genome duplication events in the early stage of evolution (Fig. 7.1). Actually genome duplication seems to have been important for the evolution of the entire set of seed plants (Jiao et al. 2011). This indicates that de Vries's view of species formation by single mutational events is valid, though the extent of chromosomal variation may not be as high as in *Oenothera* species. Polyploid species are also abundant in ferns (Grant 1981; Wood et al. 2009). They are also known to exist in yeasts (Wolfe and Shields 1997; Kellis et al. 2004) and some species of insects, teleost fishes, and frogs (Lynch 2007, pp. 202–208).

In animals genome duplication occurs much less frequently than in plants, apparently because sex is

Species before genome duplication

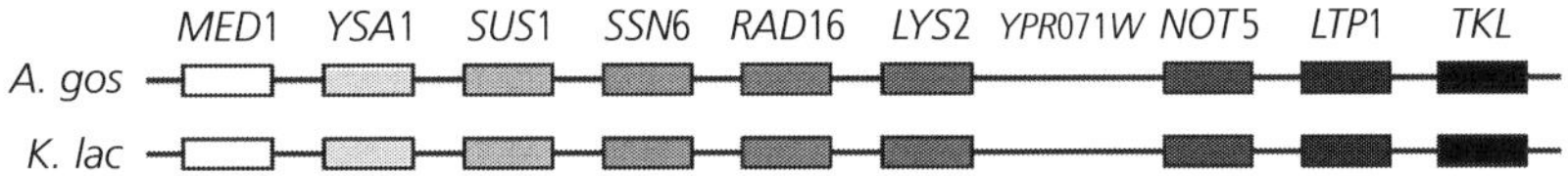

Species after genome duplication

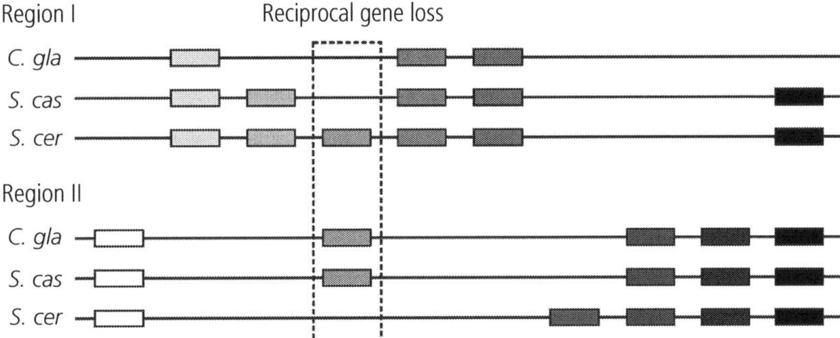

Fig. 7.2. Gene order relationships in a chromosomal region of *Saccharomyces cerevisiae SSN6* and its related species before and after genome duplication. Gene names are given at the top in italics. Reciprocal gene loss shown by a box supports the Oka model of speciation. Abbreviations are as follows: *A. gos, Ashbya gossypii*; *K. lac, Kluyveromyces lactis*; *C. gla, Candida glabrata*; *S. cas, S. castellii*; and *S. cer, S. cerevisiae*. Modified from Scannell et al. (2006). Reproduced with permission from Nature.

often determined by the XY or the ZW chromosomal system in animals and polyploidization would disturb this sex determination (Muller 1925). However, comparison of genome sizes of different groups of animals suggests that polyploidization has occurred quite frequently before chromosomal sex determination evolved (Nei 1969b). In fact, Ohno (1970, 1998) proposed that two rounds of genome duplication occurred in the early stage of vertebrate evolution.

Changes of Genomic Structures and Speciation

We have seen that genome duplication is an important mechanism of speciation in plants. Genome duplication occurs when autotetraploids are formed by duplication of the genome of an organism or when allotetraploids are formed by duplication of the genomes of a hybrid between two different species. In either case, the new polyploid species exhibits a sterility barrier from the parental species. Therefore, polyploidization establishes a new species as proposed by de Vries (1901–1903).

However, it was recently discovered that the number of genes in polyploid species does not necessarily increase in proportion to the number of genome duplications (Wendel 2000; Adams and Wendel 2005; Doyle et al. 2008). Some chromosomes or genes are often lost after polyploidization, and therefore the new species established may not have twice the number of genes of the parental species (Fig. 7.2). The loss of genes is usually species-specific or gene-family-specific (Rensing et al. 2008; Flagel and Wendel 2009). Under certain conditions, the number of genes in some gene families may decrease, yet this decrease appears to be beneficial for the plants (Flagel and Wendel 2009). At the same time, the number of gene copies in some gene families may increase. This process of increase and decrease of gene number may mimic the case of evolution by segmental gene duplication and deletion. If this is the case, de Vries's mutation theory, which encompasses any type of hereditary mutations including chromosomal rearrangements, would not be as unrealistic as generally thought. In other words, tetraploids, aneuploids, or trisomics, which de Vries identified as varieties or elemental species, may become new species. In fact, Scannell et al. (2006) showed that an ancestral species of the yeast

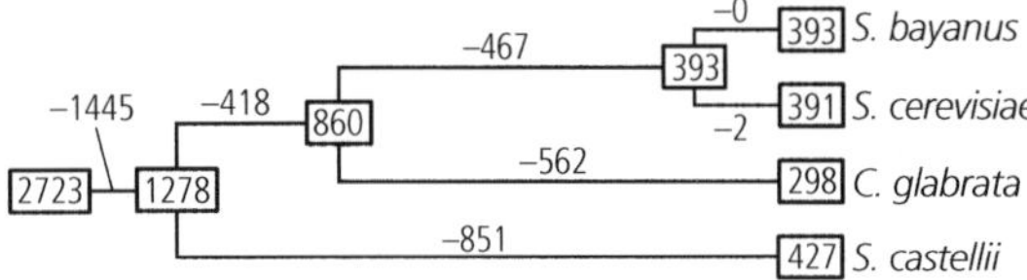

Fig. 7.3. Loss of duplicate genes after genome duplication in four species of yeasts. The numbers in squares represent the numbers of loci which were derived by genome duplication in the ancestral species and have been retained in the genome. Numbers (–) on branches indicate the numbers of loci, in which one of the duplicate genes was lost. 2723 duplicate loci were analyzed. Modified from Scannell et al. (2006). Reproduced with permission from Nature.

Saccharomyces cerevisiae apparently experienced polyploidization and then generated at least four well-established species with a reduced number of genes (Fig. 7.3).

In plants there are many allopolyploid species of which the genomes are known to come from specific extant species. One of the most interesting cases is the allopolyploid species, *Tragopogon miscellus* ($2n = 24$), which belongs to the aster family. The ancestral species of this plant are known to be *T. pratensis* ($2n = 12$) and *T. dubius* ($2n = 12$), and *T. miscellus* was produced only about 80 years ago. Because all three species are biennial plants, this corresponds to about 40 generations. However, *T. miscellus* is widely distributed in the Spokane area of the State of Washington. Chester et al. (2012) studied the chromosomes, karyotypes, and gene contents of the allotetraploid species and showed that extensive chromosomal variation was observed in all populations studied and one population was fixed for a particular karyotype: 76 percent of the individuals showed intergenomic translocation, and 69 percent were aneuploid for one or more chromosomes. Yet, the chromosomal number was mostly 24, indicating that even if chromosome number remains the same, the gene content of a new allopolyploid can vary extensively.

Chromosomal Rearrangements and Speciation

As mentioned above, de Vries did not know the chromosomal structure of *O. lamarckiana* and simply compiled various forms of morphological mutations. However, it is interesting to note that *O. lamarckiana* apparently had several sets of reciprocal translocations of chromosomes (Cleland 1923). It was soon realized that this type of plant generates gametes with balanced and unbalanced sets of chromosomes and only those with balanced sets are fertile (Dobzhansky 1951). It was also noted that individuals with different sets of balanced chromosomes will be isolated by reproductive barriers because the hybrids between them will be partially or completely sterile. Similar situations are known to occur when telomeric inversions or other chromosomal rearrangements are generated and recombination occurs (White 1969; Brown and O'Neill 2010).

Population geneticists such as Wright (1941) showed that the probability of fixation of these chromosomal rearrangements is so low that they would not be easily established in a population unless population size is very small (say less than 10). For this reason, the idea that new species are formed by chromosomal rearrangements was almost abandoned. In selfing plants like *O. lamarckiana*, however, the probability of fixation of new chromosomal rearrangements would not be very small because the effective population size is small. This suggests that some of the elementary species de Vries discovered in his experimental farm might have been reproductively isolated from others by this mechanism even if they were not tetraploids. It should also be noted that chromosomal rearrangements can be fixed even in a randomly mating population if it goes through bottlenecks multiple times.

In fact, recent studies suggest that this form of speciation is quite common in plants (Rieseberg 2001; Badaeva et al. 2007; Rieseberg and Willis 2007; Feldman and Levy 2012). Plant populations are sedentary and often reproduce asexually or by selfing. These reproductive systems enhance the chance of fixation of chromosomal rearrangements, and therefore speciation by this process should be reconsidered (Nei and Nozawa 2011). This type of speciation by chromosomal rearrangements is also known to occur in yeasts and mammals (Delneri et al. 2003; Brown and O'Neill 2010; Nei and Nozawa 2011). Of course, de Vries (1901–1903) did not have any idea about chromosomal variation, but his study of morphological mutations stimulated other workers to study the chromosomal mutations and their importance in speciation. Unfortunately, this type of speciation is still underappreciated in the current literature.

7.2. Evolution of Reproductive Isolation by Genic Mutation

According to the biological species concept (Dobzhansky 1937; Mayr 1963), a group of individuals is called a species when they are isolated from other groups of individuals by pre-mating or post-mating isolation mechanisms. It is therefore important to know how the reproductive barrier is generated at the genetic level. In the case of polyploidization, the reproductive barrier is instantly generated in self-fertilizing organisms because the hybrid of a new polyploid and its parental species is generally sterile as mentioned in Section 7.1. However, how does the reproductive barrier occur in the absence of chromosomal rearrangements? There are various genetic models that can explain the evolution of reproductive isolation. Here, I would like to discuss only the genetic models that have been studied empirically at the gene level. In practice, hybrid sterility or inviability is a complex character and is controlled by a large number of genes, and it is difficult to study the effects of all these genes simultaneously. Therefore, most experimentalists extract a small number of major genes and then study the mechanism of reproductive isolation at the genic or molecular level. This approach is certainly important, but we should not forget that it may lead to biased conclusions. Note also that the biological species concept is not always applicable to plants and fungi because these organisms often reproduce by selfing or asexual reproduction and populations are not well defined (Rieseberg and Willis 2007). Initial reproductive isolation is also generally achieved by pre-zygotic isolation rather than by post-zygotic isolation. For these reasons, speciation occurs more easily in plants and fungi than in animals.

During the last dozen years, many investigators have used the so-called Dobzhansky-Muller (DM) model (see the subsection after next) as a guideline for conducting experimental studies and interpreting their results. In practice, however, this is only

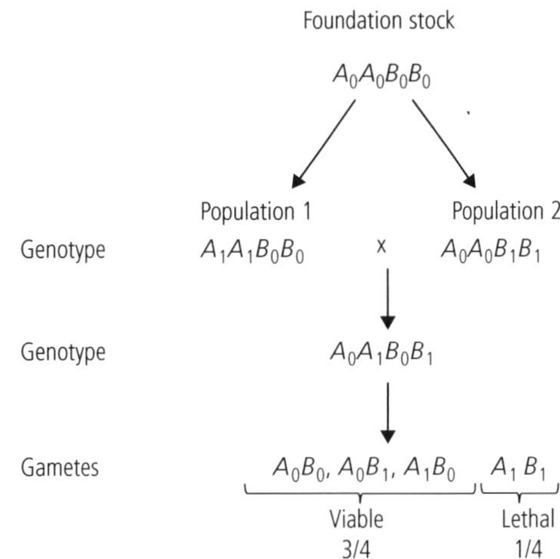

Fig. 7.4. Oka model of speciation by duplicate gene mutations. *A* and *B* are duplicate genes. A_0 and B_0 are the original normal alleles, and A_1 and B_1 are lethal mutations.

one of the many possible models for the evolution of reproductive isolation. Here I would like to discuss these models considering their molecular basis.

Oka Model of Speciation by Duplicate Gene Mutations

One of the simplest models of speciation is Oka's (1953, 1957, 1974) model by lethal mutations occurring in duplicate genes. Being apparently unaware of Oka's work, Werth and Windham (1991) proposed essentially the same model, which is better known in the United States. In this model the foundation stock is assumed to diverge into two geographically isolated populations (populations 1 and 2) and these populations evolve independently (Fig. 7.4). It is also assumed that the original foundation stock contains two duplicate genes (loci), A_0 and B_0, which have redundant functions, and that in population 1 allele A_0 mutates to a lethal allele, A_1, and in population 2 allele B_0 mutates to another lethal allele, B_1 (see Fig. 7.4). If these evolutionary events occur and populations 1 and 2 are crossed, the hybrid genotype will be $A_0A_1B_0B_1$. This genotype will produce gamete A_0B_0, A_1B_0, A_0B_1, and A_1B_1 each with a probability of 1/4 if the two loci are unlinked. Therefore, one quarter (A_1B_1) of them will be sterile.

Drosophila experiments have shown that the rate of lethal mutations per locus is about 10^{-5} per generation. Therefore, the probability of occurrence of hybrid sterility would not be very small. Note that the rate of fixation of a recessive lethal mutation in one of the two duplicate loci is nearly equal to the mutation rate when the effective sizes of local populations are relatively small (Nei and Roychoudhury 1973). These considerations suggest that reproductive isolation may occur in this way relatively easily.

The extent of gamete sterility obviously increases when there are many sets of duplicate loci. In fact, when there are n independent sets of duplicate loci that control the formation of sperm or eggs, the expected proportion of sterile gametes will be $1 - (3/4)^n$, which becomes 0.9 for $n = 8$ and 0.99 for $n = 16$. Therefore, this type of gamete sterility is likely to occur in the progeny of a newly generated polyploid, which contains a large number of duplicate genes. In recent years, however, it has been found that even non-polyploid organisms contain a large number of small-scale duplicate genes (copy number variation) in their genomes (e.g. Redon et al. 2006). Therefore, the Oka model is likely to apply to virtually all species. In practice, the relationship between the number of lethal genes and the extent of male sterility would not be as simple as mentioned above. Lynch and Force (2000) suggested that the functional divergence of duplicate genes may enhance the probability of occurrence of hybrid sterility.

In rice, *Oryza sativa*, there are two subspecies called *japonica* and *indica*, which diverged about 400 000 years ago. These subspecies have two duplicate genes *DPL1* and *DPL2*, which encode highly conserved, plant-specific small proteins and are highly expressed in the mature anther. Mizuta et al. (2010) showed that *japonica* carries functional ($DPL1^+$) and nonfunctional ($DPL2^-$) alleles at the *DPL1* and *DPL2* loci, respectively. By contrast, *indica* has nonfunctional ($DPL1^-$) and functional ($DPL2^+$) alleles at the two loci. The inactivation of allele $DPL1^-$ is caused by a transposon insertion in one of the exons of the gene, whereas the nonfunctionality of $DPL2^-$ is due to the A → G mutation at an intron splicing site. Alleles $DPL1^+$, $DPL1^-$, $DPL2^+$, and $DPL2^-$ correspond to alleles A_0, A_1, B_0, and B_1 in Fig. 7.4, respectively, and therefore the partial sterility of the hybrid observed between *japonica* and *indica* can be explained by the Oka model. A similar reproductive isolation caused by duplicate gene mutations has been observed between *O. sativa* and its related species *O. glumaepatula* (Yamagata et al. 2010). In this case the genes involved are the duplicate gene copies, *S27* and *S28*, encoding mitochondrial ribosomal protein L27. It was shown that the *S27* gene is absent in *O. glumaepatula* and the *S28* gene from *O. sativa* contains nonfunctional mutations. Therefore, a quarter of hybrid pollen does not have any functional gene, and this will cause pollen sterility. Another example of this type of reproductive isolation has been reported in *Arabidopsis* (Bikard et al. 2009; Nei and Nozawa 2011).

Actually, using classical genetic techniques, Oka (1953, 1974) had identified a number of hybrid sterility genes, which apparently occurred by duplicate gene mutations. In his time, however, no molecular techniques were available to study the evolutionary changes of genes, and therefore his conclusions have remained as conjectures. In this sense recent molecular studies have provided solid empirical evidence for his theory. Actually, Oka (1974) was aware of the possibility of ancient polyploidization of rice based on the cytogenetic study by Sakai (1935) and Nandi (1936).

At this point, it should be noted that A_1 and B_1 in Fig. 7.4 represent lethal mutations but they may also represent the loss of the duplicate genes A_0 and B_0, respectively, because they have the same effect as that of lethal mutations in generating reproductive isolation. In fact, the formation of new species in yeasts after the genome duplication in their ancestral species (Figs 7.2 and 7.3) can be explained by the Oka model (Scannell et al. 2006). It should also be noted that most authors who studied the duplicate gene mutation hypothesis mistakenly called it the DM model instead of the Oka model (e.g. Werth and Windham 1991; Lynch and Force 2000; Mizuta et al. 2010). In the Oka model lethal mutations or gene losses are the causal factors, and there is no need of interaction between A_1 and B_1. In the DM model (Fig. 7.5), however, A_1 and B_2 are functional genes and a special form of gene interaction between alleles A_1 and B_2 is assumed to exist, as will be discussed in the next subsection. In the DM

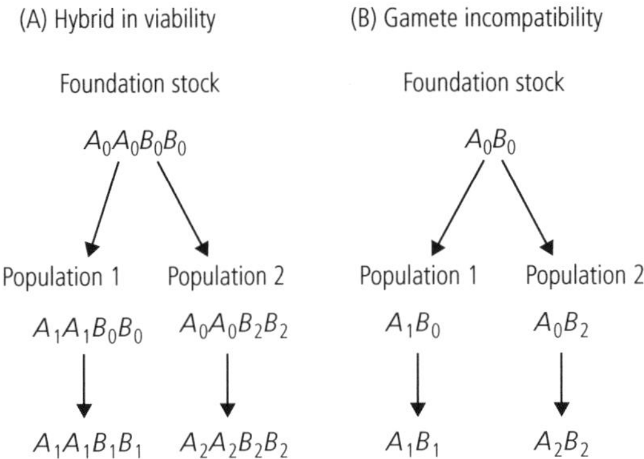

Fig. 7.5. Dobzhansky-Muller model of evolution of reproductive isolation. A_0, A_1, and A_2 represent alleles at the *A* locus, whereas B_0, B_1, and B_2 represent alleles at the *B* locus. (A) Diploid model. (B) Haploid (gamete) model.

model the fixation of alleles A_1 and B_2 by positive selection is also often assumed.

Some authors are not enthusiastic about the importance of the Oka model of speciation. Coyne and Orr (2004) stated that polyploidization does not occur so often in animal species and this minimizes the importance of this model. As mentioned above, however, recent genomic studies indicate that small-scale gene duplications are abundant and there is no reason to believe that the Oka model is less important in animals than in plants. Coyne and Orr stated that the ultimate fate of duplicate genes is to acquire new gene functions rather than nonfunctionality. Actually, this statement is incorrect. Duplicate genes become pseudogenes much more frequently than they gain new functions (Lynch and Force 2000; Nei and Rooney 2005). For these reasons the Oka model may play an important role in speciation in both plants and animals.

Dobzhansky-Muller (DM) Model of Evolution of Reproductive Isolation

In the Oka model of speciation, it is necessary to have duplicate genes. However, reproductive isolation may be developed without duplicate genes if there are two or more genes that interact with each other negatively when they are brought together in hybrids. One such model is the so-called Dobzhansky-Muller (DM) model (Dobzhansky 1937; Muller 1940, 1942). The essence of this model is presented in Fig. 7.5A. In this figure, two loci, *A* and *B*, are considered, and $A_0A_0B_0B_0$ represents the genotype for these loci in the foundation stock from which populations 1 and 2 were derived. If these two populations are geographically or ecologically isolated, it is possible that A_0 mutates to A_1 in population 1 and this mutant allele is fixed in the population by natural selection or genetic drift. Genotype $A_0A_0B_0B_0$ may then be replaced by $A_1A_1B_0B_0$ without loss of viability and fertility (Fig. 7.5A). Similarly, B_0 may mutate to B_2 in population 2 and the mutant allele may be fixed. However, if there is gene interaction such that any combination of mutant genes A_1 and B_2 in an individual results in inviability or sterility, the hybrid ($A_0A_1B_0B_2$) between the two populations will be inviable or sterile. In Fig. 7.5A, we have assumed that the foundation stock had genotype $A_0A_0B_0B_0$. Theoretically, however, it is possible to assume that the ancestral genotype is $A_1A_1B_1B_1$ and that this genotype remained unchanged in population 1 but it changed to $A_2A_2B_2B_2$ in population 2.

Orr (1996) argued that the first person who proposed the DM model is neither Dobzhansky nor Muller but William Bateson (1909), and that Bateson's model was identical with that of Dobzhansky and Muller. In my view, this argument is disputable. It is true that Bateson considered a two-locus model of complementary genes to explain hybrid sterility, but he never considered how such a system can evolve. By contrast, Dobzhansky and Muller spelled out the evolutionary process of hybrid sterility genes, albeit very crudely. In evolutionary biology it is important to understand the process of evolution. For this reason, I will refer to the model as the DM model in this book. However, Dobzhansky and Muller presented only a verbal argument and never explained why only A_1 is fixed in population 1 and B_2 is fixed in population 2. Theoretically, the $B_0 \rightarrow B_1$ mutation may also happen in population 1 and the $A_0 \rightarrow A_2$ mutation may occur in population 2 (Fig. 7.5A). How is then only A_1 fixed in population 1 and only B_2 fixed in population 2? Both Dobzhansky and Muller argued that allele A_1 may affect a secondary character through the pleiotropic effect and this effect may confer a selective advantage for A_1 over A_0 in population 1. Similarly, B_2 may have a selective advantage over B_0 in population 2 because of the pleiotropic effect.

Table 7.1. Fitnesses and frequencies of the four genotypes for the two incompatibility loci in the haploid model.

Alleles		A_0	A_1
B_0	Fitness	1	$1 + s_A$
	Frequency	$(1 - x)(1 - y)$	$x(1 - y)$
B_2	Fitness	$1 + s_B$	$1 - t$
	Frequency	$(1 - x)y$	xy

The first mathematical study of this problem was conducted by Nei (1976). Here let us present a summary of his results. For simplicity, we consider the haploid model given in Fig. 7.5B instead of the diploid model, because essentially the same result is obtained by both models. Note also that the haploid model is directly applicable to sperm or egg fertility. In the haploid model, four possible genotypes may be generated for the two alleles at each of loci A and B, and we assign the fitnesses for the four genotypes as given in Table 7.1. Here x and y represent the frequencies of alleles A_1 and B_2, respectively, whereas s_A and s_B are selective advantages conferred by pleiotropy for alleles A_1 and B_2, respectively, and t is the selective disadvantage of genotype A_1B_2, which becomes 1 when the interpopulational hybrids are completely sterile. Note that alleles A_0, A_1, B_0, and B_2 are all vitally important in this model. Here we have assumed no linkage disequilibrium for simplicity.

If we use this model, the amounts of changes (Δx and Δy) of allele frequencies x and y per generation are given by

$$\Delta x = x(1-x)[s_A - (s_A + s_B + t)y] / \overline{w} \quad (7.1)$$

$$\Delta y = y(1-y)[s_B - (s_A + s_B + t)x] / \overline{w} \quad (7.2)$$

where $\overline{w} = 1 + s_A x(1-y) + s_B(1-x)y - txy$ (Nei 1976).

Therefore, x increases if y is smaller than $\hat{y} = s_A / (s_A + s_B + t)$, while it decreases if y is greater than $\hat{y}$. Similarly, y increases if x is smaller than $\hat{x} = s_B / (s_A + s_B + t)$ but decreases if x is greater than $\hat{x}$. This means that if mutant allele A_1 occurs before B_2 and starts to increase in frequency allele A_1 tends to be fixed in the population, whereas mutant allele B_2 would be fixed if it occurs first and starts to increase before the occurrence of A_1. Therefore, selection is exclusive, and in any population either A_1 or B_2 may be fixed depending on the allele that starts to increase in frequency earlier than the other. Because fixation of A_1 or B_2 occurs at random, the probability that two populations show hybrid sterility or inviability is 1/2. However, if there are many loci controlling reproductive isolation, any pair of populations would eventually develop reproductive isolation.

One problem here is whether alleles A_1 and B_2 have selective advantage ($s_A > 0$ and $s_B > 0$) conferred by pleiotropy or not. Generally speaking, it is very difficult to identify any character affected by pleiotropic effects of speciation genes A_1 and B_2, and even if a character is identified, the selection coefficients s_A and s_B are unlikely to be large and stay constant for the entire process of fixation of alleles A_1 and B_2. However, even if s_A and s_B are 0, alleles A_0 and B_0 may be replaced with A_1 and B_2, respectively, by repeated mutation and genetic drift. In this case too, only A_1 or B_2 must be fixed in a population, and the average replacement time will be $1/v + 2N$ generations approximately, where v and N are the mutation rate and the effective population size, respectively (Nei 1976). Therefore, it will take a long time for alleles A_0 and B_0 to be replaced by A_1 and B_2, respectively. Even if A_1 and B_2 are selected with positive values of s_A and s_B, the replacement time will not be much shorter because it primarily depends on the mutation rate (Li and Nei 1977).

It should also be noted that the mutation rate v refers only to those mutations that generate strong deleterious effects when they are brought together in hybrid individuals. No one has measured the mutation rate for this type of mutation, but the rate must be very low because only special mutations would be able to produce such deleterious gene interaction in hybrids. We know that the DM model is currently very popular (e.g. Coyne and Orr 2004), but there are only a small number of experimental data sets that support the model in the strict sense (Nei and Nozawa 2011). Orr's (1995) paper is often cited as the theoretical justification of the model. Actually, he assumed the validity of the DM model from the beginning and simply studied the possibility of continuous accumulation of incompatibility genes without considering mutation rates. He assumed that reproductive isolation is developed by positive Darwinian selection caused by their pleiotropic effects. This is in contrast to the Oka model, where reproductive isolation is assumed to occur due to deleterious mutations in duplicate genes.

Let us now examine some recent experimental data that have been regarded as supporting the DM model. The first data set I consider is that of Presgraves et al.'s (2003) paper, in which the evolutionary change of a nuclear pore protein (nucleoporin), Nup96, has been studied in *D. melanogaster* and *D. simulans*. Nuclear pores are large protein complexes that cross the nuclear envelope and allow the transport of water soluble molecules such as RNAs, DNA polymerases, and carbohydrates between the nucleus and the cytoplasm. This nuclear pore is composed of a large molecular structure called the nuclear pore complex, which contains about 30 different protein components, each with multiple copies (Presgraves and Stephan 2007). One of the proteins is the nucleoporin Nup96, and Presgraves et al. (2003) showed that this protein is involved in causing hybrid male inviability between the two *Drosophila* species. This hybrid inviability occurred only when the *D. simulans Nup96* gene is associated with the *D. melanogaster* X chromosome. They therefore assumed that the hybrid inviability occurs when the *D. simulans Nup96* gene negatively interacts with one or more genes of the *D. melanogaster* X chromosome. Furthermore, McDonald and Kreitman's (1991) test of neutrality suggested that the *Nup96* gene evolved by positive selection after divergence of the two species. They then concluded that their observations support the DM model of speciation and the hybrid inviability is a consequence of adaptive evolution at the *Nup96* locus. A similar study was conducted by Tang and Presgraves (2009), who identified another nucleoporin gene, *Nup160*, involved in the hybrid male inviability between the two *Drosophila* species. This gene in *D. simulans* was inferred to interact negatively with the *D. melanogaster* X-chromosome genes as well as with the *D. simulans Nup96* gene.

However, there are a few problems with their conclusions. First, they have not really identified the *D. melanogaster* X-chromosome genes that are supposed to interact with the *D. simulans Nup96* or *Nup160*. This identification is critical, because otherwise we do not know how the interaction between the two genes leads to hybrid male inviability. Theoretically, the X-chromosome genes need not be protein-coding genes but merely the heterochromatin that is often involved in hybrid inviability (e.g. Ferree and Barbash 2009, see Section 7.3). Second, Presgraves and his colleagues obtained a signature of positive selection for the increase in frequency of *Nup96* and *Nup160* by using the McDonald-Kreitman (MK) test. However, the MK test depends on a number of simplifying assumptions and it may give erroneous conclusions when these assumptions are not satisfied (Nei et al. 2010). In fact, another test of neutrality based on synonymous (d_S) and nonsynonymous (d_N) nucleotide substitutions did not detect positive selection (Nei and Nozawa 2011). Rather the test suggested that the *Nup96* and *Nup160* genes are under purifying selection.

What is important here is not to know whether positive selection has occurred for the new alleles but rather to understand how these genes have generated hybrid inviability. Some may argue that positive selection is important because it would speed up the speciation process. In reality, there is no need for any organism to undergo rapid speciation. Reproductive isolation occurs merely as a consequence of mutational changes of the genes involved, and therefore it must be a passive process, as was suggested by Darwin.

However, there are a few data sets that apparently support the DM model. Long et al. (2008) discovered that a pair of closely linked loci *SaF* and *SaM* in rice contain different alleles in subspecies *japonica* (*SaF*$^-$ and *SaM*$^-$) and *indica* (*SaF*$^+$ and *SaM*$^+$) and the pollen of their hybrids is sterile. Gene *SaF* encodes an F-box protein involved in protein degradation, whereas *SaM* produces a small ubiquitin-like modifier E3 ligase-like protein. The protein encoded by *SaF* is 476 amino acids long, and there is only one amino acid difference between alleles *SaF*$^+$ and *SaF*$^-$. By contrast, *SaM*$^+$ and *SaM*$^-$ encode proteins with 257 amino acids and 217 amino acids, respectively, the latter being a truncated protein. Alleles *SaF*$^+$ and *SaM*$^+$ in *indica* are considered to be the ancestral genes, and *SaF*$^-$ and *SaM*$^-$ are regarded as mutants generated in the process of evolution of *japonica* (Fig. 7.6). The haplotype *SaF*$^-$; *SaM*$^+$ that is found in *indica* could be the ancestor of the haplotype *SaF*$^-$; *SaM*$^-$ in *japonica*. It hybridizes both with *indica* and *japonica* without any problem (Fig. 7.6). If

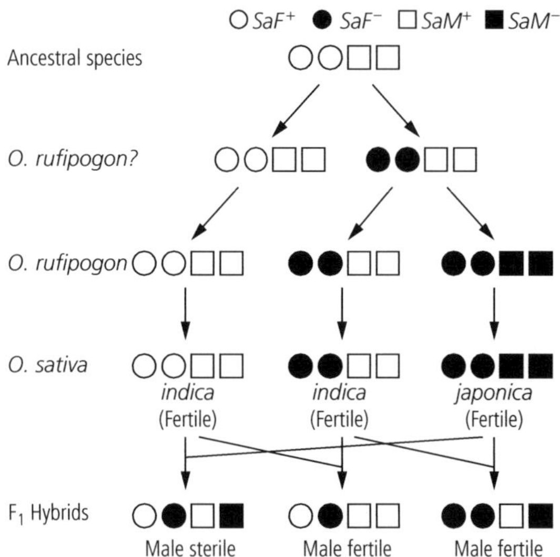

Fig. 7.6. Male sterility caused by different combinations of alleles at the *SaF* and *SaM* loci in rice (*Oryza*). Data from Long et al. (2008).

this is the case, this haplotype may represent an intermediate stage in the process of evolution of SaF^-; SaM^- in *japonica*. These evolutionary changes of SaF^+; SaM^+ to SaF^-; SaM^- are consistent with the DM model. However, the molecular basis of the gene interaction to generate the hybrid sterility is still unknown.

Another data set that purportedly supports the DM model is that of Chou et al. (2010), who studied a pair of genes causing the F_2 sterility between yeast species *Saccharomyces cerevisiae* and *S. paradoxus*. The genes studied are the nuclear-encoded mitochondrial RNA splicing gene (*MRS1*) and the mitochondria-encoded cytochrome oxidase 1 gene (*COX1*). In *S. paradoxus*, the introns of *COX1* are properly spliced out by its own *MRS1*. In *S. cerevisiae*, however, one (M1) of the introns has been lost, and the *MRS1* gene has lost its splicing function. The hybrid sterility between the two species is generated, because the MRS1 protein in *S. cerevisiae* cannot splice out the M1 intron of *COX1* from *S. paradoxus*. This functional change of *MRS1* is found only in *S. cerevisiae*, and this change is caused by three amino acid substitutions. This suggests that the original forms of the *COX1* and *MRS1* genes are those of *S. paradoxus* and both genes in *S. cerevisiae* are derived forms. This scheme of evolution of reproductive isolation is superficially in accord with the DM model. In this case, however, it is likely that the fertility of *S. cerevisiae* was once impaired when the M1 intron of the *COX1* gene was lost, but the fertility was later restored when the *MRS1* gene in this species lost its splicing function. If this is the case, this example does not represent the original DM model. A similar evolutionary change has been reported to explain the hybrid sterility caused by the *AEP2* and *OLI1* genes between *S. cerevisiae* and *S. bayanus* (Lee et al. 2008).

There are many other papers that have been regarded as support of the DM model (see Presgraves et al. 2003; Coyne and Orr 2004; Wu and Ting 2004). However, close examination of the papers indicates that the authors often misunderstood the concept of the model or that the demonstration is incomplete. Therefore, more careful studies are necessary into the genes reported in these papers (Nei and Nozawa 2011).

Multiallelic Complementary Genes Model

Nei et al. (1983) proposed an extended version of the DM model to explain species-specific gene compatibility and other reproductive isolation. As a concrete example, let us consider the evolutionary changes of sperm protein lysin and its egg receptor VERL in abalone species. In abalone, the egg is enclosed by a vitelline envelope, and sperm must penetrate this envelope to fertilize the egg (Shaw et al. 1995). The receptor VERL for lysin is a long acidic glycoprotein composed of 22 tandem repeats of 153 amino acids, and about 40 molecules of lysin bind to one molecule of VERL (Galindo et al. 2003). The interaction between lysin and VERL is species-specific, and therefore this pair of proteins apparently controls species-specific mating. Figure 7.7 shows a genetic model explaining the species-specificity between the lysin and VERL genes. Within a species (species 1 or 2), the lysin and VERL genes are compatible, so that mating occurs freely. However, if species 1 and 2 are hybridized, lysin and VERL are incompatible, and therefore the fertilization is blocked. This guarantees the species-specific mating when the two species are mixed.

However, it is not very simple to produce the gene for species 2 from that for species 1 or those for species 1 and 2 from their common ancestral genes by a single mutation at the lysin and VERL loci, because a mutation ($A_i \rightarrow A_k$) at the lysin locus makes the lysin gene incompatible with the wild-type allele (B_i) at the VERL locus. A mutation ($B_i \rightarrow B_k$) at the VERL locus also results in the incompatibility with the wild-type allele (A_i) at the lysin locus. Therefore, these mutations would not increase in frequency in the population. Of course, if mutations $A_i \rightarrow A_k$ and $B_i \rightarrow B_k$ occur simultaneously, lysin A_k and VERL B_k may become compatible. However, the chance that these mutants meet with each other in a large population would be very small.

For this reason, Nei et al. (1983) and Nei and Zhang (1998) proposed that the evolutionary change of allele A_i (or B_i) to A_k (or B_k) occurs through intermediate alleles and that closely related alleles have similar functions and therefore they are compatible. For example, A_i may mutate first to A_j and then to A_k, whereas B_i may mutate to B_j and then to B_k. If A_i is compatible with B_i and B_j but not with B_k and if B_i is compatible with A_i and A_j but not with A_k, then it is possible to generate the species-specific combination of alleles at the lysin and VERL loci in each species by means of mutation, selection, and genetic drift.

There are several other examples of ligand and receptor gene incompatibilities involved in fertilization or reproduction. For example, sea urchin protein bindin mediates the fertilization of eggs by sperm. The receptor of bindin is called EBR1, and its interaction with bindin is species-specific (Kamei and Glabe 2003). In mammals, a protein called ADAM2 (or fertilin β) plays a role of sperm ligand for the egg plasma membrane receptors, integrins

♂	♀	Sperm (Lysin)		Egg (VERL)	Fertility
Sp. 1	Sp. 1	Ai	×	Bi	High
Sp. 1	Sp. 2	Ai	×	Bk	Low
Sp. 2	Sp. 1	Ak	×	Bi	Low
Sp. 2	Sp. 2	Ak	×	Bk	High

Fig. 7.7. A model of species-specificity of gamete recognition between lysin and VERL in abalone. Modified from Nei and Zhang (1998).

(A)

Stepwise (reversible) mutation model

$A_{-3} \rightleftarrows A_{-2} \rightleftarrows A_{-1} \rightleftarrows A_0 \rightleftarrows A_1 \rightleftarrows A_2 \rightleftarrows A_3$

(B)

	A_{-3}	A_{-2}	A_{-1}	A_0	A_1	A_2	A_3
B_{-3} (A_{-3})	1	1	0	0	0	0	0
B_{-2} (A_{-2})	1	1	1	0	0	0	0
B_{-1} (A_{-1})	0	1	1	1	0	0	0
B_0 (A_0)	0	0	1	1	1	0	0
B_1 (A_1)	0	0	0	1	1	1	0
B_2 (A_2)	0	0	0	0	1	1	1
B_3 (A_3)	0	0	0	0	0	1	1

Fig. 7.8. Stepwise and infinite-allele mutation models for hybrid sterility (or inviability) genes. (A) In the stepwise mutation model forward and backward mutation may occur, whereas in the infinite-allele model no backward mutation is allowed. (B) The fertilities for various haplotypes for loci *A* and *B* (two-locus model) and genotypes for locus *A* (one-locus model) are given by 0 (infertile) or 1 (fertile). Distantly related haplotypes or genotypes are assumed to be infertile. In reality, of course, the fertility need not be 0 or 1 but may take an intermediate value, particularly for incompatible matings or haplotypes. Modified from Nei et al. (1983).

(Evans and Florman 2002; Desiderio et al. 2010), and the interactions between these proteins seem to be species-specific. The protein-protein interaction in various biochemical processes required for development and physiology is also often complementary. Similarly, the control of expression of protein-coding genes by *cis*-regulatory elements is complementary by nature.

Nei et al. (1983) developed several models of evolution of reproductive isolation by means of multiallelic compatibility genes. They considered both one-locus and two-locus models. Mutation was assumed to occur following either the stepwise or the infinite-allele model (Kimura 1983), and the fitness of a genotype was assumed to be either 1 or 0 depending on the mutation model and the genotype generated (Fig. 7.8). Pre-mating and post-mating isolations were also considered. Their conclusions are summarized in the following way. (1) The single-locus model generates speciation more quickly than the two-locus model. (2) The infinite-allele model generates speciation more quickly than the stepwise mutation model. (3) With the epistatic gene model used (Fig. 7.8), the evolution of reproductive isolation occurs more rapidly in small populations than in large populations. (4) Generally speaking, the time to the occurrence of speciation is very long and is roughly proportional to the inverse of the mutation rate.

However, these results are model-dependent, and I cannot apply the results to natural populations without qualifications. For example, the single-locus stepwise model, which will be considered in the last subsection of this Section, may be applicable only to certain characters such as flowering time in plants and developmental time in animals. At present, I also do not know which of the stepwise and infinite-allele models is more realistic than the other, though I believe that the latter model is generally more realistic because reproductive isolation is controlled by a large number of genes affecting different phenotypic characters. Our results imply that speciation occurs more rapidly through bottlenecks. This conclusion is in agreement with Mayr's (1963) theory of the founder principle, which has been criticized by many authors (e.g. Coyne and Orr 2004). However, Nei et al.'s (1983) study was done by using specific mathematical models of epistatic gene interaction, unlike Mayr's verbal argument without any genetic model. This would also mean that self-fertilizing organisms may develop reproductive isolation more easily than random mating populations.

In general, however, speciation occurs very slowly, and it takes millions to tens of millions of years for well-established species to be developed (Coyne and Orr 2004, pp. 419–421). This suggests that Nei et al.'s conclusion about speciation time may not be so unrealistic. It is hoped that some experimental studies will be conducted about the theoretical predictions presented here.

Single-Locus Speciation

Theoretically it is possible to develop reproductive isolation by single-locus mutations. Suppose that allele A_1 mutates to A_2 and the genotypes A_1A_1 and A_2A_2 are normal but the heterozygote A_1A_2 is lethal or semilethal. Then the individuals within the population fixed with either A_1 or A_2 are fertile but the hybrids between the two populations will be completely or partially inviable. The problem is how to get the population fixed with A_2 from the original population fixed with A_1. In outbreeding species the mutant allele A_2 causes such a deleterious effect in heterozygous condition that it is unlikely to be fixed in the population unless the population size is very small. For this reason, Dobzhansky (1937) and Muller (1942) rejected the idea of achieving reproductive isolation by single-locus mutations.

However, if multiallelic mutations occur at a locus and they are compatible with one another when they are closely related but mutant alleles become incompatible when they are distantly related, hybrid sterility or inviability may be generated at a single locus. Figure 7.8 shows one such example, where allele A_0 is compatible with allele A_{-1}, A_0, and A_1 but not with other alleles (ignore the alleles at locus *B*). Thus, genotype A_0A_0 may be compatible with A_0A_1 and A_1A_1 but not with A_2A_2 in mating ability or zygotic viability. Furthermore, A_1A_2 may be compatible with A_2A_2. We can then have genotype A_2A_2 from A_0A_0 passing through genotypes A_0A_1 and A_1A_2.

In rice, there is an example that supports single-locus speciation at the molecular level. The reproductive barrier between *indica* and *japonica* is controlled by many genes. One of them is the *S5* gene that encodes an aspartic protease determining embryonic sac fertility. The proteins encoded by this gene in *indica* (*S5i*) and *japonica* (*S5j*) are different at two amino acid sites (Chen et al. 2008). One of these differences (F273L) in *japonica* seems to be responsible for the sterility of the hybrid between *indica* and *japonica*. Interestingly, however, there is a special group of rice varieties that produce fertile hybrids both with *indica* and *japonica*. The *S5* gene (*S5n*) in these varieties encodes a protein with a deletion of a fragment with 115 amino acids, and this might have been an intermediate allele between *S5i* and *S5j*. This type of single-locus speciation is plausible particularly in self-fertilizing organisms like rice.

Reproductive isolation by single-locus mutations may also occur at loci controlling flowering time in plants. This would occur easily in self-fertilizing plants. At the molecular level, flowering time is controlled by a large number of loci, many of which are duplicate genes. Environmental factors such as photoperiodicity and vernalization also affect flowering time. In recent years many genes involved in the regulation of flowering time have been identified (Simpson et al. 1999; Boss et al. 2004; Pouteau et al. 2008). Although flowering time is controlled by many genes, a single mutation may change flowering time drastically and may produce reproductive isolation. One of the interesting cases is a mutation that occurred at the *flowering locus* (*FLC*), a repressor of flowering involved in the vernalization pathway. The genomes of *Brassica* species contain several *FLC* paralogous genes. Yuan et al. (2009) discovered that one of the *FLC* genes, *FLC1* in *Brassica rapa*, is polymorphic with respect to flowering time in nature and this polymorphism is caused by the mutation of a splicing site (G → A) in intron 6. This mutation was then shown to change flowering time substantially. Reproductive isolation due to heterogeneity in developmental time also occurs in insects (Tauber and Tauber 1977), though the molecular basis has not been studied.

7.3. Reproductive Isolation by Complex Genetic Systems

Segregation Distorters and Speciation

In diploid organisms a male heterozygote (*Aa*) for a locus produces two different types of sperm (*A* and *a*) with an equal frequency. However, there are genes that distort the Mendelian segregation ratio in their favor so that their frequency in sperm is much higher than 50 percent (sometimes nearly 100 percent). These genes are called segregation distorters (D^-). The segregation distortion occurs because the distorter gene destroys a high proportion of chromosomes carrying the opposite allele in the process of spermatogenesis (Hartl 1969; Wu and Hammer 1991; Kusano et al. 2003). The distorter gene is often located on the X chromosome, and therefore the sex ratio in the offspring is distorted (Presgraves 2008). Because these males produce more sperm with the X chromosome than sperm with the Y, there will be more female offspring than male offspring, and this distorted sex ratio is disadvantageous for the species. Furthermore, distorter genes themselves are often deleterious, but their frequencies may increase drastically by segregation distortion.

Interestingly, the expression of D^- genes is often suppressed by suppressor genes (S^-). Therefore, if a new distorter mutation (D^-) occurs in a population, its frequency would initially increase rapidly because of segregation distortion despite its deleterious effects. However, this increase in frequency will be stopped if a new suppressor mutation (S^-) arises and suppresses the deleterious effect of the D^- gene. The D^- and S^- genes may then be fixed in a species simultaneously. After fixation of these mutations, there will be no segregation distortion and no deleterious effects of the D^- gene (Wu et al. 1988; Frank 1991; Lyttle 1991; Tao et al. 2001; Phadnis and Orr 2009).

However, if this new species with mutant genes D^- and S^- is crossed with its sibling species having wild-type alleles D^+ and S^+, the effect of D^- may reappear in the F_1 hybrids if S^- is not dominant over S^+ or in the F_2 hybrids if D^- is recombined with S^+ and genotypes $D^-D^-S^+S^+$ or $D^-D^+S^+S^+$ are produced (Frank 1991; Hurst and Pomiankowski 1991; Tao et al. 2001). If these events reduce the fitness of hybrid individuals, it will constitute a new way of generating reproductive barriers between the two species. The genetic nature of suppressor genes is not well known. However, in the case of the *Segregation Distorter* (D^-) haplotype first reported by Sandler et al. (1959) in *D. melanogaster*, the *S* locus (Responder) consists of a large number of about 120 bp repeats. It was shown that the segregation distortion becomes stronger as the number of repeats increases and that when the suppressor locus contains a small number of repeats no segregation distortion occurs (Wu et al. 1988).

D^- genes are present on autosomal chromosomes as well as on the Y, but they seem to be less frequent than those of the X chromosome (Frank 1991; Jaenike 2001). This observation provides an explanation for Haldane's rule (Haldane 1922), which states that when two species or subspecies are intercrossed the heterogametic sex with XY or ZW chromosomes are sterile or inviable more often than the homogametic sex with XX or ZZ chromosomes (Frank 1991; Hurst and Pomiankowski 1991). Dozens of distorter genes have been reported in insects, mammals, and plants, though the molecular basis of the distortion is not well understood (Jaenike 2001). In *D. simulans*, at least three D^- loci have been identified and a unique S^- locus exists for each D^- locus (Presgraves 2008).

Heterochromatin-Associated Hybrid Incapacity

A number of investigators (e.g. Henikoff and Malik 2002; Brideau et al. 2006; Bayes and Malik 2009) have reported that repeat DNA elements in the heterochromatin regions of genomes are often associated with hybrid sterility or inviability. One of the interesting observations is the hybrid sterility caused by the *zygote hybrid rescue* (*Zhr*) locus in *Drosophila*. When *D. simulans* females are crossed with *D. melanogaster* males, hybrid females die in early embryogenesis. However, a mutant allele (Zhr^1) of *Zhr* is known to rescue the female viability (Sawamura et al. 1993). Ferree and Barbash (2009) showed that the wild-type allele *Zhr* contains a region of 359 nucleotide repeats in the *D. melanogaster* X chromosome that interact with some cytoplasmic factors of *D. simulans*. The number of 359 nucleotide repeats is small in *D. simulans*, so that the fertility within this species is high. In *D. melanogaster*, the number of the repeats is high, but this species also shows a high fertility apparently because the cytoplasmic factors are different from those of *D. simulans*. At present, however, the cytoplasmic elements have not been identified, and therefore the molecular basis of the interaction between the DNA and the cytoplasmic factors remains unclear. In this case, because the number of DNA repeats is species-specific and can change relatively rapidly due to concerted or birth-and-death evolution (Henikoff and Malik 2002), this system of hybrid inviability is apparently caused by mutation and random genetic drift. Although little is known about the evolutionary mechanism of cytoplasmic elements, it is possible that both DNA repeat elements and cytoplasmic factors coevolve as in the case of the two-locus multiallelic complementary genes model discussed in Section 7.2.

Another example is the *Odysseus* homeobox gene (*OdsH*), which causes hybrid male sterility when *D. mauritiana* females are crossed with *D. simulans* males (Ting et al. 1998; Sun et al. 2004). In this case the receptor for this transcription factor gene is not well defined. Recently, however, Bayes and Malik (2009) discovered that the *OdsH* protein produced from *D. mauritiana* localizes to the heterochromatic Y chromosome from *D. simulans* but the *OdsH* protein from *D. simulans* does not. They then proposed that the *OdsH* protein from *D. mauritiana* acts as a sterility factor in association with the Y chromosome heterochromatin of *D. simulans*. Again, however, the molecular mechanism of the interaction remains unclear. A similar mechanism appears to be operating with the *Prdm9* gene in mice (see Oliver et al. 2009a; Nei and Nozawa 2011).

7.4. Other Mechanisms of Evolution of Reproductive Isolation

There are many other speciation models which are based on relatively small numbers of observations or of which the theoretical basis is unclear (Nei and Nozawa 2011). For example, gene transposition may also cause hybrid incompatibility. Masly et al. (2006) showed that the *JYAlpha* gene encoding the alpha subunit of Na^+ and K^+ adenosine triphosphatase was transposed from chromosome 4 to chromosome 3R in *D. simulans* after the divergence from *D. melanogaster*. Consequently, when these two species are crossed, some of the F_2 individuals have no copy of *JYAlpha* in the genome and become sterile. This is a special case of classical reciprocal translocation of chromosomes discussed in Section 7.1. However, gene transposition or translocation may occur more frequently than chromosomal translocation, simply because there are more genes than chromosomes in the genome and transposons may mediate gene transfer. Several authors (Henikoff and Malik 2002; Brown and O'Neill 2010) have suggested that rapid concerted evolution of DNA repeat elements at the centromeric chromatin may generate speciation by distorting chromosomal segregation. The logic behind this argument is not very clear, but it is interesting to note that repeat elements are often involved in hybrid sterility and inviability. It should also be noted that epigenetic factors controlling photoperiodicity and vernalization in plants are apparently involved in speciation, though molecular study of these problems is still in its infancy.

In this chapter we have considered only the models of reproductive isolation for which molecular data have been obtained. If we consider other cases, there are many other models. Some of these models are supposed to be quite general, whereas others are applicable only for special cases. One of the former models is the so-called recombinational speciation model. In this model two species are supposed to interbreed and then their offspring strains receive special sets of chromosomes or genes to generate new species. This model of speciation is believed to apply to plants more often than to animals (Grant 1981; McCarthy et al. 1995; Rieseberg 1997). However, the applicability of this model to natural populations has been controversial.

One of the specific-case models is cytoplasmic incompatibility. This model was developed primarily by studies of *Wolbachia*, which are *Rickettsia*-like bacteria and inhabit the cells of many eukaryotes. *Wolbachia* are maternally inherited, but they may spread through to different host species horizontally (Chapter 6). For this reason, about 15–20 percent of insect species are infected with the bacteria. Infected host mothers pass the bacteria to most of their offspring, and paternal transmission is rare. When uninfected females are crossed with infected males, many offspring die at an embryonic stage. The reciprocal cross of infected females to uninfected males shows no lethality (see Coyne and Orr 2004, pp. 276–280). *Wolbachia* therefore cause unidirectional hybrid inviability. At present, however, the molecular mechanism of this hybrid sterility is not well understood.

If we consider these situations, it is clear that reproductive isolation is caused by many different factors and our knowledge about this subject is quite limited. In addition, we have not considered ecological factors such as sympatric speciation that will further complicate the problem of speciation. I suggest that the readers who are interested in these issues refer to Coyne and Orr's (2004) book "Speciation".

7.5. Speciation by Bottleneck Effects

In Chapter 3 we discussed the effects of population bottlenecks on genetic variability and genetic differentiation of populations, and concluded that Mayr's (1963) verbal arguments are generally acceptable but his conclusion about the bottleneck effect in speciation or the occurrence of genetic revolution is too vague because no specific genetic models are used. Actually, even at the present time it is difficult to find an appropriate mathematical model for studying the relationship between bottleneck size and speciation. In the study of hybrid sterility, however, we can study the bottleneck effect or the effect of population size mathematically because we have some crude genetic models of hybrid sterility, mentioned in Section 7.4.

Nei et al. (1983) have used several different models of multiallelic hybrid sterility or inviability, including the single-locus, two-locus, haploid, and diploid models. Figure 7.8 shows the one-locus and two-locus multiallelic-models. In this case mutation is assumed to occur stepwise in both the + and – direction (Fig. 7.8A). For simplicity, let us consider the one-locus case, ignoring the *B* locus. In this case, allele A_i mutates to allele A_{i+1} or A_{i-1} and these alleles have additive effects on some phenotypic character related to hybrid fertility or viability (e.g. flowering time in plants). We also consider the haploid model and assume that allele A_i is compatible with alleles A_{i-1}, A_i, and A_{i+1} but not with others. If this type of mutation and selection occur in a finite population, the allelic states may move to the + or – direction randomly in each generation. Therefore, if we consider two populations derived from the same ancestral population, one population may move to the + direction and the other to the – direction by chance and the difference in allelic states between the two populations may become sufficiently large to generate hybrid sterility. In fact, conducting computer simulations, Nei et al. (1983) showed that hybrid sterility between two descendant (sister) populations is eventually established in this way though the time required for its establishment depends on the mutation rate and population size.

Nei et al. (1983) also considered the infinite-allele model, in which the genic effect is no longer one-dimensional as in the case of the stepwise mutation model but mutation is assumed to occur in many (theoretically infinite) directions without backward mutation. They also assumed that allele A_i is compatible with allele A_i and with its immediate parental (A_{i-1}) or immediate descendant (A_{i+1}) allele, but not with other alleles. This infinite-allele model is much simpler but may be more realistic than the stepwise mutation model because fertility or viability is controlled by a large number of genes. Yet, the conclusions obtained by the two genetic models are qualitatively the same, though reproductive isola-

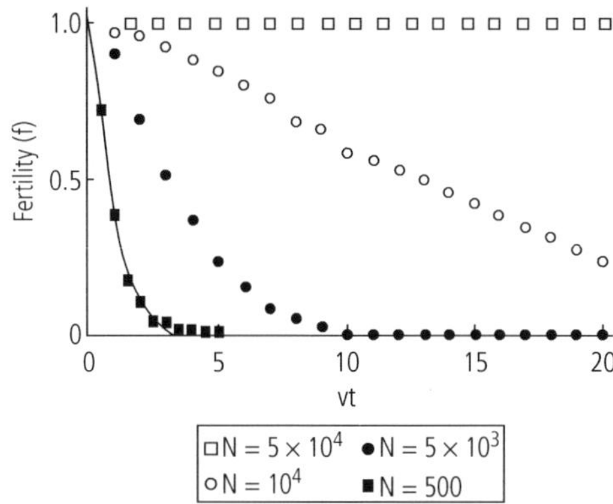

Fig. 7.9. Relationships between the average fertility over all replications (*f*) and evolutionary time (*vt*). The infinite-allele model with $v = 10^{-5}$ was used.

tion occurs much faster for the infinite-allele model than for the stepwise mutation model. Furthermore, Nei et al. studied the multiallelic model for two loci in relation to speciation, but the results were again qualitatively the same as those for single loci. I therefore present only the results for the single-locus infinite-allele model in the following.

Figure 7.9 shows the average fertility between two sister populations for different generations as measured by vt, where v and t are the mutation rate per locus per generation and the number of generations, respectively. Thus, when $v = 10^{-5}$, $vt = 5$ corresponds to 500 000 generations. In *Drosophila* there seem to be about 5 generations per year. If this is the case, 500 000 generations corresponds to 100 000 years. Similarly, $vt = 20$ corresponds to 400 000 years. The results given in Fig. 7.9 show that the establishment of interpopulational sterility (1 – fertility) is faster in small populations than in large populations. When N is 5×10^4, average fertility is close to 1 even when $vt = 20$. When $N = 500$, however, average fertility becomes almost 0 when vt is greater than 3, which may correspond to about 60 000 years. The reason why hybrid sterility does not evolve in large populations is that the incompatibility genes become polymorphic and therefore new mutations tend to be eliminated. By contrast, small populations tend to be monomorphic, so that new mutations may spread through the populations without the serious effect of incompatibility genes.

This result is consistent with actual observations from natural populations. Mayr (1970) examined the relationships between speciation and population size in many different groups of organisms and concluded that speciation occurs faster in small populations than in large populations. Carson (1971) also argued that many species of *Drosophila* in the Hawaiian islands emerged through small bottlenecks (see also Templeton 1980, 2008). Mayr and Carson regarded this observation as support for their theories of genetic revolution, but it is possible that the primary cause is random fixation of incompatibility genes such as those considered here.

In Mayr's or Carson's theory the occurrence of genetic revolution is triggered by the bottleneck effect. In this case, they considered epistatic gene interaction, but since their arguments were concerned with genetic coadaptation with overdominant selection, it is difficult to understand the meaning of their genetic revolution. Recent molecular studies show that such coadapted genetic loci are rare, and as we have seen, reproductive isolation is primarily caused by mutation, gene duplication, and gene interaction. In my view this explanation is much simpler for understanding speciation by the bottleneck effect than the genetic coadaptation theory.

7.6. Hybrid Sterility Generated by Passive Process of Phenotypic Evolution

Speciation is defined as a process in which one species diverges into two or more different species. Here different species mean different groups of individuals among which reproductive isolation has been established. In sexually reproducing organisms reproductive isolation is usually accomplished by generation of hybrid sterility or inviability. Therefore, the study of speciation is conducted by examining how hybrid sterility is generated. Many neo-Darwinians currently believe that species formation is aided by natural selection and therefore hybrid sterility is also generated by natural selection (Coyne and Orr 2004, p. 3). Certainly natural selection may occur when the genes involved in hybrid sterility differentiate between different allopatric populations. Furthermore, some authors have argued that hybrid sterility or inviability might be enhanced by natural selection because mixing of two incipient species by hybridization is disadvantageous in the formation of new species (see Dobzhansky 1951, pp. 206–211).

However, positive natural selection may not have anything to do with the generation of hybrid sterility because hybrid sterility is caused by mutations that have no deleterious effects within populations but have harmful effects when interspecific hybrids are produced. Furthermore, the idea of accelerated evolution of hybrid sterility is teleogical because there is no need for any organism to speed up reproductive isolation. Natural populations evolve without purpose as the genetic structure of a population changes as a consequence of mutation, selection, and genomic drift within populations.

As we have seen, there are various kinds of hybrid sterility genes, and they are always accumulating in the genome of any species without being noticed until hybridization occurs artificially or naturally. In other words, hybrid sterility or inviability is a mere consequence of the evolutionary change of interactive genetic systems within species. Some authors have suggested that hybrid sterility genes that are expressed in early stages of speciation would be more important in speeding up speciation than those that are expressed in later stages. This view is not acceptable because we know that any hybrid sterility genes are a mere consequence of mutations that disturb the gene interaction systems in interspecific hybrids. If a pair of species is kept isolated for a long evolutionary time, both genes expressed in the early and late stages of speciation are expected to contribute to sterility barriers.

We would also expect that hybrid sterility mutations would increase with evolutionary time and in the long run any pair of species would not be able to mate and produce offspring. For example, macaques and mice would never be able to produce any offspring, because they have accumulated so many mutations. By contrast, subspecies of mice are able to produce offspring because the genetic differences between them are small and the extent of disturbance of gene interaction systems would not be so large.

The above consideration suggests that hybrid sterility is a passive consequence of accumulation of mutations that would affect the gene interaction systems in development favorably within species but disturb the systems in interspecific hybrids. This view is the same as that of Nei et al.'s (1983) mathematical theory of evolution of reproductive isolation by incompatibility genes. In a broad sense it is also similar to Darwin's (1859) view mentioned at the beginning of this chapter, though he did not present any biological mechanism.

7.7. Summary

Recent genomic data have shown that polyploidization and chromosomal rearrangements play important roles for generating new species and even segmental gene duplication may lead to the formation of new species. We have seen that the elementary species *Oenothera gigas* discovered by de Vries (1901–1903) was actually a polyploid. Therefore, de Vries's assertion that new species may arise by mutational events has been vindicated. However, there are many different ways of generating reproductive isolation when genic mutations are considered. Many investigators have tried to understand speciation by means of natural selection. Natural selection may occur when the characters of different species diverge, but this can happen by genetic drift as well. What is really important in speciation is the development of hybrid sterility or inviability that would prevent gene admixture between species. Study of this problem is quite complicated, but since we now have molecular and genomic techniques to study the problem, more efforts should be given to understand the genetic basis of hybrid sterility and inviability. Theoretically, hybrid sterility or inviability is developed only when different populations are geographically or ecologically isolated.

At the present time, there is a tendency for many investigators to try to explain their results in terms of the Dobzhansky-Muller model even though their data do not really support the model. Actually, there are many different ways of generating hybrid sterility and inviability, and therefore we should try to understand the real molecular mechanism for each case without preconception. The classification of the mechanisms of evolution of hybrid sterility and inviability used in this chapter is somewhat arbitrary, and the distinction between different categories is not always clear-cut because the underlying molecular mechanisms are not well understood. In the future it would be important to differentiate different models of speciation at the molecular level.

Our understanding of the development of hybrid sterility is still poor partly because there are so many different mechanisms and partly because there are a large number of genes involved. It is certainly important to identify the gene sets that are involved in the early stage of speciation, but eventually we have to know the entire process of evolution of hybrid incompatibility genes. This process depends on the reproductive system of the organism, epigenetic effects, etc., and it would not be easy to understand the entire process at the same time. For this reason, we will first have to understand each step of generation of hybrid sterility separately.

As was mentioned at the beginning, the purpose of this chapter is to understand the roles of mutation and selection in speciation. I believe that mutation is essential for the evolution of reproductive isolation whether selection is involved or not. Dawkins (1987) criticized de Vries's mutation theory of evolution stating that there is no process of natural selection involved in his theory. Actually, de Vries was fully aware of the importance of natural selection in the establishment of new species. de Vries (1909, p. 212) stated *"The natural selection of newly arisen elementary species in the struggle for existence is an entirely different matter. They arise suddenly and without any obvious cause; they increase and multiply because the new characters are inherited. When this increase leads to a struggle for existence the weaker succumb and are eliminated."*

Our current knowledge of mutation is enormously rich compared with what was known at the time of de Vries. However, he was a visionary person and seems to have grasped the importance of both mutation and selection in the formation of new species, as is clear from this statement.

Some authors have contended that Charles Darwin did not really solve the "origin of species" because he did not study the speciation process. However, he examined the cause of hybrid sterility and inviability and concluded that these traits appear as a consequence of the evolution of other traits. This is consistent with the view that hybrid sterility is generated as a passive consequence of evolution of interactive gene systems in each species, as we have shown in this chapter. In this sense Darwin actually studied the origin of species.

CHAPTER 8

Adaptation and Evolution

8.1. Adaptation by Mutation

There are millions of species of organisms on earth, and they appear to be well adapted to their environments. Marine organisms including marine mammals have special physiological and morphological characters that are suitable for living in water and they cannot live on land. Bird species are equipped with wings and can fly in the air but cannot live under water for long. Cactus plants are adapted to dry environments by acquiring thick stems in which photosynthesis is carried out. Some species are adapted to very special environments. The teleost fish of the genus *Notothenia* live in the cold water below 0°C in the Antarctic Ocean. Some species of clams live deep at the bottom of the sea and obtain their energy from bacteria, which in turn live on the energy provided by underwater volcanoes. An exquisite case of adaptation is the so-called basket on the third pair of legs of the worker honeybee. The pollen collected from flowers is stored in this basket by the help of the other legs. This structure is unique and is not found in any other insects (Morgan 1903). Furthermore, it is present only in the worker bee and is absent in the queen and male bees. There is a virtually endless list of examples of this type of adaptation in nature, as was discussed by various authors including Darwin (1859), Morgan (1903), Stebbins (1950), Mayr (1963), and Arthur (2011).

The currently popular explanation of this type of adaptation is of course natural selection (Ridley 2003; Futuyma 2005; Stearns and Hoekstra 2005; Ayala 2007). According to neo-Darwinism, ample genetic variation is believed to exist in the population, so that adaptation to new environments is almost always possible. For this reason, some authors (e.g. Fisher 1930; Williams 1966; Dawkins 1976) have argued that adaptation occurs only by natural selection, and therefore any indication of adaptation is a reflection of the occurrence of natural selection in the past. Using this type of argument, evolutionary biologists have developed various theories of natural selection to explain the evolution of sex (Maynard Smith 1989), adaptation (Williams 1966), evolution of altruism (Hamilton 1964), etc. In reality, however, these theories are largely based on speculation, whether mathematical formulation is used or not, and there has been little empirical data to support the theories.

It should be noted that there are many examples in which the roles of natural selection are not very clear for explaining a particular relationship between organisms and their environments. Morgan (1903, p. 6) stated "*the polar bear is the only member of the bear family that is white, and while this can scarcely be said to protect it from enemies because it is improbable that it has anything to fear from the other animals in the ice-fields, yet it may be claimed the color is an adaptation to allow the animals to approach unseen its prey.*" Is this explanation valid? He continued "*in the tropics and temperate zones there are many greenish and yellowish birds whose colors harmonize with the green and yellow of the trees amongst which they live; but on the other hand we must not forget that in all climes there are numbers of birds brilliantly colored, and many of these do not appear to be protected in any special way. The tanagers, humming-birds, parrots, Chinese pheasants, birds of paradise, etc., are extremely conspicuous, and so far as we can see they must be much exposed on account of the color of their plumage.*" He then examined various other cases and concluded "*there can be little doubt that the color of the animal is a protection to it, but as has been hinted already, it is another question whether it acquired*

Mutation-Driven Evolution. First Edition. Masatoshi Nei.

these colors because of their usefulness." He suggested the possibility that some morphological characters are caused by neutral mutations or co-option.

A similar opinion was also expressed by Lewontin (1978). He first indicated the difficulty of defining adaptation in the real world. One definition would be that the environmental or ecological condition of an organism is optimal for the life of the organism. However, because there are so many environmental factors that can affect the life of an organism, it is difficult to identify the best niche for the organism. For example, birds can fly over many different kinds of trees, so that if a new species can eat tree leaves, it would have a considerable selective advantage. However, there seem to be few bird species that feed on leaves (excluding hoatzin and others). Therefore, even an empty niche may not be utilized. In reality, it is possible to find many such unoccupied niches for any organism.

Second, even if we consider only habitable niches, it is not easy to identify the optimal niche for any organism, because there are so many different niches in which the organism can live. There are many examples of species which proliferate when they are transferred from a particular environment (e.g. North America) to another (e.g. Asia) (Baker and Stebbins 1965). These examples might suggest that the original habitat was less appropriate for the species than the new one yet the original habitat obviously could support their survival without trouble. Third, it should be noted that natural selection does not necessarily optimize the adaptation of a species. A species may become extinct even when natural selection is operating (Chapter 2). This would happen when the climate for a species becomes unfavorable because of the arrival of an ice age or an asteroid hit on earth.

Furthermore, in Chapters 2 and 3 we have seen that natural selection is not as effective as previously thought and the evolutionary change of phenotypic characters are strongly affected by random factors. In Chapters 4 to 7 we have also seen that phenotypic evolution occurs by various forms of mutations such as nucleotide substitution, gene duplication, horizontal gene transfer, etc. Therefore, the notion that adaptation occurs solely by natural selection does not hold anymore. Adaptation is a difficult concept to define, because there is no objective way of measuring the adaptedness of an organism to its environment. It merely represents a human perception of the living status of an organism.

After developing the idea of natural selection, Darwin (1859) tried to explain the evolution of various characters by this theory, but he encountered a number of difficult cases, although at the end he managed to produce a reasonable explanation. In practice, however, the explanation would have been much simpler if he knew the occurrence of mutation. In the following I would like to discuss some of these problems. I will also consider some other examples that are directly related to our subject of the roles of mutation and selection in evolution.

8.2. Evolution of Some Specific Characters

Evolution of Eyes and Photoreceptors

The first example of Darwin's difficulties was the evolution of vertebrate eyes. He stated *"To suppose that the eye with all its inimitable contrivances for adjusting the focus to different distances, admitting different amounts of light, and for the correction of spherical and chromatic aberration, could have been formed by natural selection, seems, I freely confess, absurd in the highest degree"* (1872, pp. 143–146). However, he continued *"Reason tells me that if numerous gradations from a simple and imperfect eye to one complex and perfect can be shown to exist, each grade being useful to its possessor, as is certainly the case; if further, the eye ever varies and the variations be inherited, as is likewise certainly the case; and if such variations should be useful to any animal under changing condition of life, then the difficulty of believing that a perfect and complex eye could be formed by natural selection, though insuperable by our imagination, should not be considered as subversive of the theory."* Here Darwin apparently considered both new variations (mutations) and natural selection simultaneously, although he did not state it clearly.

Darwin's suggestion implies that all types of animal eyes originated only once in the past. Actually, there are many different types of animal eyes such as the camera eye, the compound eye, and the mirror eye, and some authors (e.g. Salvini-Plawen and Mayr 1961) proposed that eyes have evolved 40 to 60 times independently in different animal phyla. Interestingly, recent developmental studies suggest that Darwin was right and the animal eye originated only once. This result has been obtained by studies of the *Pax6* homeobox gene, a master control gene of eye development (Gehring and Ikeo 1999; Gehring 2005). This gene is highly conserved and is expressed in the eyes of all animal species so far examined. Of course, the eyes of simple animals such as jellyfish and planarians consist of only photoreceptor and pigment cells as Darwin postulated. These eyes are just for recognizing light and darkness in the environment. However, vertebrate and insect eyes are much more complex and require other organs such as eye caps, lens, retina and the brain to function properly (Gehring 2005). For this reason, it appears that about 2000 genes are involved in eye formation in *Drosophila*, and about 40 of them are transcription factors (Michaut et al. 2003). Apparently, these genes have been recruited into the morphogenetic pathways for a refined functional eye in more developed organisms.

Darwin suggested that the complex eye in vertebrates evolved gradually from a primitive invertebrate eye by natural selection. He did not mention the role of mutation, but without mutation a complex eye could not have evolved from a primitive one. Recent molecular studies have shown that this is indeed the case and gene duplication played an important role, generating complex developmental pathways of eye development. Furthermore, the evolutionary change of eyes would not have been gradual and smooth. Rather, the evolution was opportunistic and affected by random elements to a great extent (Gehring 2011). Note that the evolution of compound eyes has occurred only in insects and some crustaceans and the eye varies extensively among different phyla. This observation suggests that mutation played more important roles than natural selection. If all kinds of genetic variations existed in the initial population and natural selection was the major force of evolution as is often assumed in neo-Darwinism, so many different types of eyes would not have evolved. Although the discovery of *Pax6* as the master control gene of eye development is a great achievement, we still do not know why the octopus camera eye, for example, is different from the compound eye of Drosophila and how the difference emerged. To explain the difference, we will have to identify the genes responsible for the development of the eye in the two species and then compare their gene expression in the developmental processes. This is a daunting project, but we already have the molecular technology to do it.

Evolution of Caste Systems in Honeybees and some other Insects

The second example of Darwin's difficulties was the evolution of the caste system observed in many species of bees, wasps and ants in the insect order Hymenoptera. In the honeybee both queen and worker bees emerge from fertilized eggs and are of the female genotype. Whether a female individual becomes a queen or a worker bee is determined by the amount of royal jelly provided during embryogenesis. If the amount is high, a queen is produced, whereas if it is low, a worker bee is born. The queen has a larger body size than the worker bee and produces abundant offspring. By contrast, worker bees are effectively sterile and take care of the queen and her offspring. In this sense female workers perform an altruistic behavior. By contrast, all males are haploid emerging from unfertilized eggs. How did the caste system evolve then? Darwin admitted that there is a great difficulty in explaining the evolution of the caste system, yet he proposed that the queen and the worker bees were initially the same type of female but later the proportion of worker bees gradually increased by natural selection. He believed that the increase of sterile worker bees contributed to the success of the bee community in the same hive. In other words, he proposed that the differentiation of the queen and worker bees was generated by natural selection that occurred among different hives.

A more popular explanation of the evolution of the caste system or a phenomenon called eusociality was proposed by Haldane (1955) and Hamilton (1964). Here eusociality means the social organization in which adult members are divided into the reproductive and non-reproductive castes and the latter workers serve to raise the offspring produced by the former (queens). In particular, Hamilton presented a mathematical formula for the condition of the evolution of eusociality using Wright's (1921) coefficient of genetic relationship between two relatives. In his formula, the fitnesses of the donor individual (worker bee in this case) who sacrifices its reproductive potential and the beneficiary individual (queen) who receives a benefit from the donor were considered separately, and the following condition was derived.

$$R > c/b \tag{8.1}$$

where R is the coefficient of genetic relationship (proportion of genes identical by descent between two relatives), c is the cost in fitness to the donor (worker bee), and b is the benefit in fitness to the beneficiary (queen). Hamilton (1964) called b and c the inclusive fitnesses because these include the fitness effects of other individuals. In organisms like humans, where both males and females are diploid (diplodiploid sex determination), the R value between sisters or brothers is known to be ½. Therefore, if one brother sacrifices his life to save his three brothers in a flood, one may write $c = 1$ and $b = 3$. In this case $c/b = 1/3$, and $R = ½ > 1/3$. Thus the altruistic behavior of one brother to save his three brothers enhances the chance of their shared genes surviving. In the honeybee, however, males are haploid (haplodiploid sex determination), and therefore R becomes ¾ between sisters. This means that as long as the ratio of fitness cost of the worker bee to the fitness benefit to the queen is lower than ¾, altruistic behavior may evolve. In other words, altruistic behavior or eusociality may be developed more easily in honeybees with haplodiploid sex determination than in organisms with diplodiploid sex determination. This theory is often called the kin selection theory (Maynard Smith 1964).

During the last four decades, Hamilton's equation has been used as a cornerstone of sociobiology. The simplicity of the theoretical argument and the extensive amount of empirical data available from hymenopteran species appeared to support Hamilton's rule convincingly. According to Nowak et al. (2010), by the year 1990 almost all species that were known to have evolved eusociality belonged to Hymenoptera, which have the haplodiploid system of sex determination. This observation enhanced the credibility of Hamilton's theory.

However, Equation (8.1) was derived under various simplifying assumptions, and it is unclear how eusociality has evolved from nonsociality when eusociality is controlled by a few genes. The R value refers to the expected proportion of polymorphic genes shared by the genomes of two relatives and does not specifically refer to the genes that are responsible for the evolution of eusociality. In reality, it is likely that eusociality is developed by a set of genes involved in several developmental pathways that interact with one another. Furthermore, the gene involved in the highest hierarchy of the developmental pathways must be expressed epigenetically, because the initial differentiation between the queen and the worker bees is determined by nutrition (West-Eberhard 2003). Therefore, to understand the evolution of eusociality, it is important to identify the master-control gene or genes that trigger the development of different castes (see below). If we consider this developmental process of eusociality, it is difficult to justify Hamilton's theory. Actually, Hamilton's theory is incapable of explaining the very initial step of evolution of eusociality.

Recently, Nowak et al. (2010, 2011) criticized Hamilton's theory from both theoretical and empirical points of view. They studied the allele frequency change for a "eusociality gene," considering such factors as selection coefficient, mutation rate, population size, and others, but their results did not support Hamilton's theory. Hamilton's theory is based on the neo-Darwinian theory of panselectionism, and no consideration is made of new mutations and genetic drift. In reality, mutation cannot be ignored when we consider the evolution of new genetic systems such as eusociality. Genetic drift also seems to be important in bees, because the effective population size has been estimated to be only about 1000 (Yokoyama and Nei 1979). In fact, Graur (1985) showed that the average heterozygosity or gene diversity for protein loci is much lower in hymenopteran species than in other insect species, suggesting the importance of genetic drift in these species.

Nowak et al. also presented several sets of empirical data that do not support the association of eusociality with haplodiploidy. A well-known example is termites, which contain thousands of species with eusociality, but they depend on diplodiploid sex determination (Crozier and Pamilo 1996, p. 5). Nowak et al. also indicated that vast numbers of living species, spread across the major taxonomic groups, use either haplodiploid sex determination or clonal reproduction, the latter yielding the highest possible degree of genetic relatedness ($R = 1$). However, among the organisms with $R = 1$, only one taxonomic group, gall-making aphids, is known to have achieved eusociality. Nowak et al.'s criticism of Hamilton's theory generated hostile responses from a large section of the sociobiology community including five papers simultaneously published in Nature (e.g. Abbot et al. 2011). However, note that the mathematical formulation of natural selection is always difficult even in the case of two Mendelian alleles at a locus (Chapter 2).

In my view, this problem should be studied by using the technique of developmental biology rather than the mathematical method. In honeybees, sex is determined by multiple alleles at the complementary sex-determining (*csd*) locus (Beye 2004). When the genotype at this locus is heterozygous, it will be a female, but a haploid genotype produced from an unfertilized egg becomes a male. In practice, diploid homozygotes (males) are also produced with a low frequency, but they are eaten up by worker bees. Therefore, all queen and worker bees are genetically diploid and females. However, there is a second switch gene that initiates the developmental pathway of male and female phenotypes. This gene is called the *feminizer (fem)*, and it has been shown to be homologous to the *transformer (tra)* gene in Mediterranean fly, housefly, and *Drosophila* (Gempe and Beye 2010). Furthermore, Kamakura (2011) showed that the switch between queen and worker bees is determined primarily by the protein Royalactin contained in royal jelly. Therefore, the eusociality of honeybees is controlled by at least three genes, *csd*, *fem*, and *royalactin*. However, the genes involved in the formation of ovary and testis are not well understood.

Interestingly, Kamakura (2011) also showed that application of Royalactin to *Drosophila melanogaster* generates a queen-like phenotype even in this species though the *Drosophila* lineage diverged from the honeybee lineage about 300 million years ago. This suggests that the developmental pathway for a queen-like phenotype existed in the common ancestor of honeybees and *Drosophila* but that the signal protein for activating the developmental pathway evolved only in the hymenopteran species.

Actually, the signal protein may not always be Royalactin. In many species of the ant genus *Pheidole* there are three different castes: queens, soldiers, and minor workers. These three castes are determined by two juvenile hormone (JH) mediated switches in response to environmental cues. At the first switch the queen and the two other castes are divided, and at the second switch the soldier and the worker castes are separated (Fig. 8.1A). Queens have four wings, whereas soldiers and minor workers do not have any. Soldiers are bigger than minor workers and have one set of vestigial wing discs, whereas minor workers have none.

Rajakumar et al. (2012) showed that the formation of wings of queens is generated by the expression of the *spalt (sal)* gene in the wing hinges and pouches (Fig. 8.1), whereas in soldiers *sal* is expressed only in the hinges of vestigial discs and minor workers show no gene expression. Interestingly, some species have a subclass called supersoldiers, which are bigger than regular soldiers and have two sets of vestigial wing discs, where the gene *sal* is expressed. It is known that this type of supersoldier occasionally appears as a mutant in the regular three-caste system. Rajakumar et al. conducted experiments to study whether supersoldiers can be generated in a three-caste species when methoprene (a JH analog) is applied to larvae. Their results showed that this application indeed induces supersoldiers.

This observation suggests that all species of *Pheidole* have the potential for producing a subclass of supersoldiers. They constructed a phylogenetic tree for the eleven species examined and inferred the evolutionary changes of the occurrence of supersoldiers (Fig. 8.1B). Their conclusion was that the ancestral species of this group of ants probably had the *sal* gene but this gene was silenced in most descendant lineages. However, the genes for the developmental pathway to produce supersoldiers have been retained in the genome. Therefore, it appears that supersoldiers in some descendant species (e.g. *P. obtusospinosa*) were recovered by restoring the *sal* gene expression. They tested this hypothesis by applying methoprene to several non-supersoldier species and showed that supersoldiers are indeed inducible.

This study suggests that apparently independent evolution of a caste system can be due to mutations of regulatory genes controlling the upper hierarchy of the developmental pathway. If this is the case, various forms of caste systems existing in Hymenoptera could be due to different signal proteins generated by mutations. Kamakura (2011) showed that the first switch gene in honeybees is *royalactin*, but different switch genes may be used in different groups of hymenopteran species. Furthermore, different numbers of switch genes may be involved in different groups. Ethologists seem to believe that different caste systems must have evolved by natural

(A)

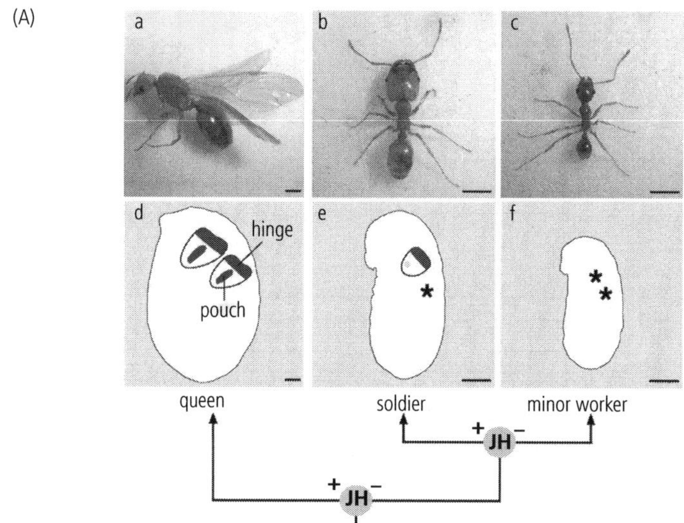

(B)

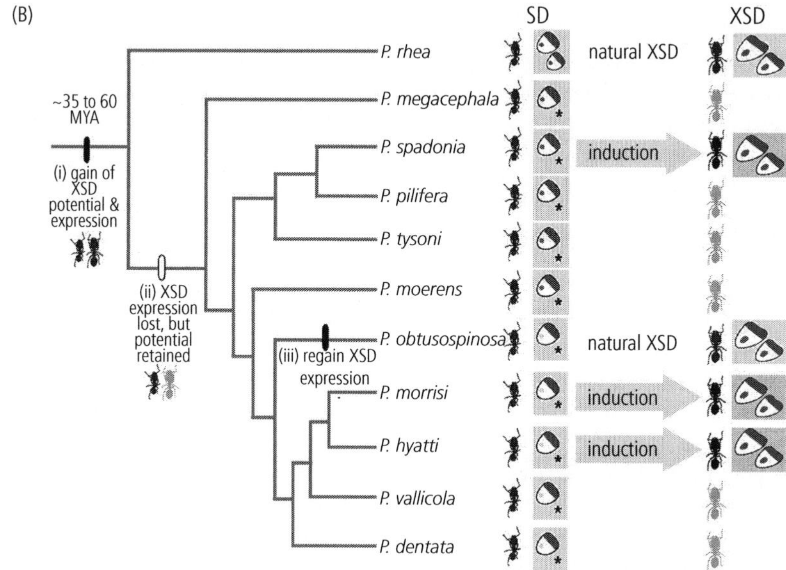

Fig. 8.1. A. Wing polymorphism in *Pheidole morrisi*: the ability of a single genome to produce (a) winged queens and wingless (b) soldiers and (c) minor workers. Caste determination occurs at two JH-mediated switch points in response to environmental cues. (d) Wing discs in queen larvae showing conserved hinge and pouch expression of *sal*. (e) Vestigial wing discs in soldier larvae showing a soldier-specific pattern of *sal* expression, where it is conserved in the hinge but downregulated in the pouch. Asterisks represent the absence of visible wing discs and *sal* expression in (e) soldier and (f) minor worker larvae. B. Evolutionary history of ancestral developmental potential and phenotypic expression of supersoldiers (XSDs). MYA, million years ago. Purple represents the pattern of *sal* expression; asterisks indicate the absence of vestigial wing discs and *sal* expression. Green arrows and boxes represent the induction of XSD potential. From Rajakumar et al. (2012). Reproduced with permission from the American Association for the Advancement of Science. See also Plate 4.

selection. However, if we consider that there are many different types of sex determination in both insects and vertebrates and some changes among them are apparently due to mutation and genomic drift (see Section 8.4), we have to entertain the possibility of non-selective changes of caste systems (Nei 2012).

The importance of studying the evolution of eusociality using developmental methods has also been indicated by Wilson (2008) and Nowak et al. (2010), though it is not at the molecular level. These authors proposed a scenario of the evolution of eusociality, considering change in four steps. According to this scenario the first step is the formation of groups within a freely mixing population. Different groups of individuals may be formed when nest sites or food resources show a localized distribution. When these groups are formed, the individuals within a group tend to be genetically related and form a kin group. The second step of evolution is the generation of other traits that make the change to eusociality more likely to occur. These pre-adaptive traits arise in the same way as a defensible nest of the solitary ancestors evolved with no anticipation of a potential future role in the evolution of sociality. These are products of co-option, in which species split and spread into different niches. The third step of evolution is the origin of eusocial alleles by mutation. In insect species with pre-adapted characters this may occur by a single mutation, and the mutation need not prescribe a novel behavior. It need simply cancel the old function (e.g. development of reproductive organs in the prospective worker bees). The only requirement for crossing the threshold to eusociality is that a female and her adult offspring do not disperse to start new nests but instead remain in the old nest. If environmental selection pressures are strong at this stage, the group may start cooperative interactions and initiate eusociality. The fourth and final step is the inter-colony selection that enhances the integrity of colonial life or the genetic organization of a hive by mutation and selection. In this case they support group selection advocated by Charles Darwin.

This scenario is highly speculative and no consideration of current molecular biology is taken into account. However, this hypothesis is now testable by identifying the genes at the molecular level. Because genomic sequences are obtainable relatively easily, experimental tests of the hypothesis are not impossible.

Evolution of Asymmetric Morphology in Flatfish, Snails, and other Organisms

Another example of an unusual adaptation is the asymmetric morphology of the flatfish. All species of flatfish belong to the *Pleuronectiformes*, one of about 40 different orders of teleosts, that includes the familiar plaice, turbot and flounder. These fish lie on one side and do not swim in a vertical position as do other fish. In some species of the flatfish the body lies on one side and in other species it lies on the other side. Some species include both right-sided and left-sided individuals (Arthur 2011). In adaption to this peculiar way of life, they show striking morphological changes in the developmental process. The eye that would be on the underside of the body shifts so that it appears on the upper side. Therefore, both eyes lie on the upper side of the body. This form of asymmetric morphology has been observed in no other vertebrates. As a result of the shifting of the eye, the bones of the skull have also become profoundly modified. Furthermore, the young fish that hatch from the egg swim at first upright, but after they have had a free living for some time they turn to one side and sink to the bottom of the sea. If the underside eye did not move to the upper side, it would be no use to the fish and the condition might be disadvantageous. Why then did the transformation of the eye evolve by natural selection? Darwin (1872) inclined to explain the evolution of flatfish by Lamarckism, while Morgan (1903) explained it by mutation.

However, this problem probably cannot be resolved without studying the regulation of gene expression in the developmental process. It is obvious that some regulatory genes or elements in the developmental process control the sinking of a symmetrical fish to the bottom of the sea and the start of a new life of flat living on a sandy area. The eye in the underside of the fish then moves to the upper side. This movement must be controlled by another set of regulatory genes or elements. At the same time the other anatomical changes such as the different coloring of the upper and bottom sides occur. Unfortunately, no molecular study seems to have been done on this interesting problem. However, we can argue that because there are polymorphic species with respect to the left-right asymmetry, the initial step of generation of this asymmetry must be controlled by a small number of genes. Note also that the asymmetric morphology evolved only once because all species of flatfish belong to the order *Pleuronectiformes*.

There are many other animals that display characters with left-right asymmetry. For example, the mammalian gut is arranged in the counter-clockwise direction in the abdomen. This asymmetry of mid-gut looping is determined by the left-right asymmetry in the cellular architecture that connects the primitive gut tube to the body wall (Davis et al. 2008). The molecular pathway that leads to this asymmetry uses the signaling protein Nodal, a member of the transforming growth factor-β superfamily, which is expressed in the left lateral mesoderm (Hamada et al. 2002).

A clear case of asymmetric body structure is also observed in various species of snails, where the shell structure is either clockwise (right-handed) or counter-clockwise (left-handed). This asymmetry is often species-specific, but there are species in which both types exist as polymorphism. In the case of snails, the left- and right-handedness is due to the expression of the *nodal* gene in the very early stage of development, when the embryo is composed of only four cells. It is known that the asymmetry can be changed by manipulating the arrangement of the cells artificially at this stage (Kuroda et al. 2009). It is also known that if *nodal* and its target gene (*Pitx*) are expressed in the right-side of the embryo, the individual (or species) becomes right-handed, and if they are expressed in the left-hand side of the embryo, the individual becomes left-handed (Grande and Patel 2009). However, the molecular mechanism for determining this asymmetry is still unknown, though breeding experiments with the polymorphic species *Lymnaea peregra* have shown that the asymmetry is determined by a single locus with a pair of alleles (Boycott and Diver 1923; Sturtevant 1923). It would be interesting to identify and clone the gene responsible for the asymmetry by using polymorphic species.

8.3. Regressive Evolution and Pseudogenes

Universality of Vestigial Characters

Many species of both prokaryotes and eukaryotes show so-called vestigial or regressive characters. Well-known examples are blind eyes and depigmented skin colors of cave-dwelling animals. These characters are universal in many different groups of organisms (e.g. insects, fish, and amphibians) and different geographical areas (e.g. United States, Mexico, and Europe). Darwin (1859) briefly discussed the inheritance of acquired characters under the heading of "Effects of Use and Disuse" and stated that "*as it is difficult to imagine that eyes, though useless, could be any way injurious to animals living in darkness, I attribute their loss wholly to disuse.*" Yet, he noted that when degeneration of an organ occurs due to disuse natural selection often causes other phenotypic changes such as an enlargement of the antennae of insects as a compensation for blindness.

Loss of organs or morphological structures is quite common when they are not used. In his book *Descent of Man*, Darwin (1871) enumerated various vestigial characters in humans, including the decreased amount of hairs compared with other mammals, the decay of posterior molar teeth, and the appendix attached to the intestine. He again expressed the view that disuse of ancestral characters is responsible for the development of these vestigial characters. After Darwin's argument, regressive evolution has become a controversial subject in evolutionary biology, and some authors have proposed that regressive evolution is caused by positive Darwinian selection. They stated that in the dark condition within a cave there is no need to have eyes and therefore natural selection operates to reduce them to save scarce nutrition for developing new characters. By contrast, Morgan (1903) ascribed the reduction of the eyes almost exclusively to destructive mutations, which are neutral in the cave condition.

Molecular Basis of Regressive Evolution

In recent years a number of authors investigated the molecular basis of regressive evolution. In these studies regressive evolution is defined as the reduc-

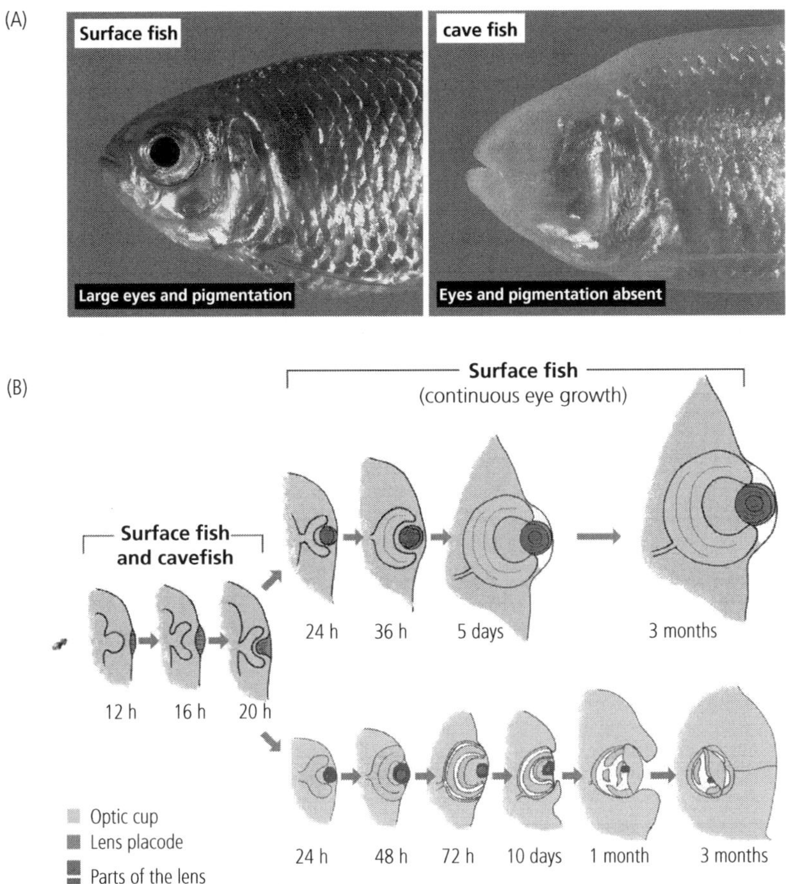

Fig. 8.2. A. *Astyanax mexicanus* surface fish and cave fish. B. A diagram showing the events of *Astyanax* eye development and degeneration. Left: Early events of eye primordium formation are the same in surface fish and cave fish until approximately 1 day after fertilization. Top: In surface fish, the eye differentiates and the eye parts grow in concert with increased body growth. Bottom: In cave fish, the eye primordium grows for a while, then arrests, degenerates, and is internalized by overgrowth of the body. Reproduced with permission from Annual Review of Genetics. See also Plate 5.

tion or loss of useless characters. One of the most intensive studies has been done with the cavefish *Astyanax mexicanus* from Northern Mexico. This fish entered into deep Mexican caves in the Pleistocene (probably about 500 000 years ago; Chakraborty and Nei 1974) from the nearby surface river.

There are several isolated fish populations in the cave, and they seem to have experienced regressive evolution almost independently. They can hybridize with one another as well as with the surface river population and produce fertile hybrids. Many cave populations consist of albinos, and these individuals often lack eyes (Fig. 8.2). In the early stage of embryogenesis both cavefish and surface fish develop the eye, but the eye of cavefish gradually decays, and in the adult stage the eye is nonfunctional and buried deep under the skin (Figs 8.2A and 8.2B). The decay of the eye is caused by apoptosis (cell deaths) of the lens, and there seem to be at least 12 genes that have contributed independently to the eye degeneration (Jeffery 2009). The actual molecular basis of eye degeneration is not yet very clear.

The molecular basis of pigment loss is better understood. There are at least two genes controlling the depigmentation of the cavefish. One is the albinism gene, *oca2*, mutation of which is known to cause albinism in humans and mice. This gene encodes an enzyme involved in the pathway of melanin formation. In humans, *oca2* is a very large gene, consisting of 24 exons in a DNA region of 345 kb (Oetting et al. 2005). The DNA sequences from different cavefish populations have mutations or deletions in different exons and all cause albinism. This indicates that the albinism mutations occurred independently in different cave populations. Another gene involved in depigmentation is the *Mc1r* gene that encodes a receptor for the melanogenesis regulating ligand MSHa (Rees 2003). In *Astyanax* this gene is responsible for producing brown-color pigments. In the cavefish this gene also shows several different mutations or deletions in its coding or regulatory regions (Jeffery 2009).

Therefore, it seems that regressive evolution is primarily caused by destructive mutations, as argued by Morgan (1903). Certainly, there is no need to consider the possibility of Lamarckian inheritance. However, one can still argue that the loss of the eye is beneficial to cavefish, because the eye is unnecessary in the dark condition and therefore eye loss would save energy that can be used for other purposes (Protas et al. 2007). One can also argue that the genes involved in eye development are pleiotropic and therefore give selective advantages to other characters. For example, the *hedgehog* gene is known to enhance the development of taste buds (receptors for taste) and the enlargement of the forebrain at the expense of eye development (Yamamoto et al. 2004). Therefore, eye loss may have had beneficial effects on the development of other characters.

Actually, this type of argument in favor of natural selection has been presented by many authors in the last 150 years. Here it is important to distinguish between the evolutionary events that would have occurred immediately after the entrance of a group of surface fish into the cave and the events that might have occurred after the fish population settled down in the cave. In this case the number of individuals that entered into the cave for the first time must have been very small and the population size should have remained small even after the cave population was established. According to Avise and Selander (1972), the number of individuals living in the Pachon cave of Mexico was 200–500 in 1971. This estimate suggests that the effective size of this cave population has been no more than 200 for the entire evolutionary process. If this is the case for all the cave populations, many mutations must have been fixed by genetic drift even if they are not neutral. In surface populations most mutations that generate albinos or eye reduction would be deleterious, but in cave populations they are expected to behave as nearly neutral alleles. In this case the probability of fixation of new mutations by the t-th generation is given by

$$P(1,t) = 1-(4Nv+1)e^{-vt} \tag{8.2}$$

approximately, where N and v are the effective population size and the mutation rate per locus per generation, respectively (Crow and Kimura 1970, p. 395). In this case the mutation rate could be of the order of 10^{-5}, because any destructive mutation can be included. Yet, $4Nv$ must be practically 0, because N is very small. Previously, Chakraborty and Nei (1974) estimated that the Pachon cave population, which is highly isolated from other cave populations, diverged from the surface population about 500 000 years ago and the generation time of this fish seems to be about 5 years (P. Sadoglu, personal communication). Therefore, $vt = 10^{-5} \times 5 \times 10^5/5 = 1$, and the probability of fixation of destructive mutations would become $1 - e^{-vt} = 1 - e^{-1} = 0.63$ approximately because $4Nv$ is practically 0. This suggests that many destructive but neutral mutations have been fixed in the population.

How then could some seemingly advantageous mutations (e.g. genes for taste buds) have been fixed? One possibility is that these mutations are advantageous in the cave population and they have been kept by purifying selection once they are fixed in the population. If this happens in the regulatory region of the genes, they may have affected the expression of many copies of protein-coding genes. Another possibility is that such characters as taste buds are controlled by multigene families and the family sizes of these gene families have expanded by recent gene duplication.

The above argument suggests that because genetic drift is so important in small populations neutral evolution is likely to occur more frequently than adaptive evolution in the cave condition. This conclusion is in accord with Darwin's disuse hypothesis.

Of course, our knowledge of the molecular basis of regressive evolution in caves is still poor, and more investigation is necessary. For example, if Darwin's hypothesis is right, the cave fish are expected to have more pseudogenes than the surface fish if they entered in the caves about 500 000 years ago. This hypothesis can therefore be tested more accurately by examining their genomes. By contrast, if positive selection is more important than disuse evolution, it may have been aided by regulatory gene mutations, expansion of multigene families or evolution of epigenetic control of gene expression.

If we consider long-term evolution, regression or inactivation of morphological characters seems to have occurred quite often in both animals and plants, and it is often associated with adaptive evolution of other characters. The largest animals that have ever lived on earth are whales. Especially, blue whales grow as big as 35 m long and can have a weight of 170 metric tonnes. Whales are believed to have evolved from a sister species of hippopotamus about 54 million years ago (Nikaido et al. 1999), and they have adapted to ocean life very well. The body shape is fish-like, the modified forelimbs or fins are paddle-shaped, and the tail has two flukes that propel the whale by vertical movement. Toothed whales and dolphins have the echolocation system to communicate among different individuals and measure the distance from other objects. However, like other mammals, they are warm-blooded, breath air, nurse their young with milk, and have body hair. At the same time, they have vestigial organs such as the invisible rear limbs and the reduced olfactory bulb in toothed whales. In toothed whales more than 70 percent of olfactory receptor (OR) genes are pseudogenes (McGowen et al. 2008). This indicates that destructive mutations are the primary source of degeneration of the olfactory system. As often argued, this loss of OR genes may have happened because olfaction is no longer needed after development of the echolocation system. However, the echolocation system too should have originally evolved by mutation.

This example shows that it is not always easy to separate the effects of mutation and selection, but it is clear that mutation is the primary cause of evolution.

Parasitic Organisms and their Genomic Changes

Most animals and plants carry some parasites in their body. Parasites include viruses, bacteria, protists, fungi, and multi-cellular worms. These parasites infect their hosts and live on the nutrition provided by the host, and therefore their digestive system is reduced. For example, the tapeworm lives in the digestive tract of vertebrates and grows several inches long. Their mouth attaches to the host intestine and absorbs nutrients generated by the host. Therefore, the tapeworm's own digestive system has almost disappeared. Similarly, the parasitic snail *Entoscolax ludwigii* has a digestive system reduced to a sucking tube. The remaining digestive tract has completely degenerated (Morgan 1903, p. 353). There are many other examples of parasitic organisms in which certain characters are degenerate. Unfortunately, no molecular studies seem to have been done about the evolution of degeneration of these digestive systems.

In recent years, however, the genomic sequences of many parasitic bacteria have been studied, and these studies have shown that the genome size and the number of genes of these parasitic bacteria are considerably smaller than those of free-living bacteria (Table 4.2). For example, the number of genes of the bacteria *Buchnera* species is about 500, compared with 5200 in the free-living *E. coli*. *Buchnera* is parasitic to the insect aphid, but actually they have developed a symbiotic relationship so that one cannot live without the other. This symbiosis is believed to have occurred about 200 million years ago (Moran et al. 1993), and during this symbiotic period the size of the *Buchnera* genome has decreased dramatically because many unused genes have been lost (Moran and Degnan 2006). *Buchnera* is a γ-proteobacterium and closely related to *E. coli*. A recent comparison of the *Buchnera* and *E. coli* genomes has indicated that almost all genes in *Buchnera* are present in *E. coli* and no new genes have been generated in *Buchnera* (Itoh et al. 2002). Furthermore, comparison of the genome sequences of a few different *Buchnera* species suggested that the gene loss occurred in the early stage of symbiosis and in recent years the number of genes has been more or less stabilized. In fact, using a recently evolved symbiotic bacterium, *Serratia symbiotica*,

Burke and Moran (2011) showed that massive loss of genes occurred in the early stage of symbiosis.

This situation is somewhat similar to the evolution of mitochondrial genes in animals. The mitochondrion is believed to be an endosymbiont derived from a bacterial species that started symbiosis with the host organism about 1.5 billion years ago (Javaux et al. 2001). At the present time, most animal mitochondria contain 37 protein- and RNA-encoding genes, which are involved in generating adenosine triphosphate (ATP) and regulatory cellular metabolism. The mitochondrial genes are highly conserved in animals, though nucleotide substitution occurs continuously within the genes.

Despite the high degree of conservation of the same set of genes in *Buchnera* and animal mitochondria, the rate of amino acid substitution in protein-coding genes is known to be about two times higher in these endosymbionts than in related free-living bacteria (Moran 1996; Lynch 1997; Clark et al. 1999; Itoh et al. 2002). This curious observation has often been explained by Muller's ratchet effect (See Chapter 3), because these symbionts reproduce asexually and their effective size is very small (Moran 1996; Lynch 1997). However, Muller's ratchet is concerned with the accumulation of slightly deleterious mutations, and it cannot continue for hundreds of millions of years. If the ratchet effect had operated continuously, the population would have become extinct much earlier (see Fig. 4.1B).

In reality, *Buchnera* and mitochondrial genes have survived for a long time, though the rate of amino acid substitution is much higher than that of free-living bacteria. This suggests that most amino acid substitutions are not deleterious but more or less neutral. Why, then, is the rate of amino acid substitutions so high? The answer seems to be a higher mutation rate in endosymbionts than in free-living bacteria, because the endosymbionts lack one class of DNA repair enzymes (Itoh et al. 2002). If the mutation rate is high, the rate of amino acid substitution would become high, other things being equal. One way to confirm this hypothesis is to show that the rate of synonymous substitution is higher in endosymbionts than in free-living bacteria. Unfortunately, this test cannot be done because synonymous substitutions have become saturated in these bacteria. Theoretically, advantageous mutations are expected to enhance the rate of amino acid substitution. Fares et al. (2001) reported that certain amino acid substitutions in the heat-shock protein BroEL of *Buchnera* are apparently caused by positive Darwinian selection. However, the gene for this protein shows a low d_N/d_S ratio like other genes (Wernegreen and Moran 1999). Therefore, the effect of positive selection on the overall rate of amino acid substitution appears to be negligible.

8.4. Evolution of Sex-Determination Mechanisms

In many species of animals and plants sex is determined by the X and Y or the Z and W chromosomes (genetic sex-determination, GSD). In the XY system, the male has one X and one Y (or no Y), whereas the female has two X chromosomes (XX). This system is used by most mammals, many amphibians, and many insect species. In birds, snakes, lepidopteran insects, and some other animals, the female is heteromorphic (ZW) and the male is homomorphic (ZZ). In general, the Y and W chromosome carry many inert genes and a small number of sex-determining or sex-related genes, whereas the X or Z chromosomes have a large number of functional genes. The X and Y or the Z and W chromosomes are believed to have originated from homologous pairs of autosomal chromosomes (Muller 1914, 1932).

However, what determines sex are not the sex chromosomes themselves but the genes that are responsible for the development of male and female phenotypes, and the molecular mechanism of formation of male or female phenotypes is still largely unknown. Note also that sex in many species of reptiles and fish is determined by environmental factors, particularly by the temperature in the incubation period. This sex-determination system is called the environment-dependent sex-determination (ESD) or the temperature-dependent sex-determination (TSD) system (see Chapter 6). It should also be noted that some orders or families of reptiles and amphibians include species with different sex determination systems (XY and ZW and ESD). Furthermore, some amphibian species are polymorphic with respect to the XY and the ZW systems.

Note also that insects use quite different genetic mechanisms. Therefore, it is not a simple matter to explain all different mechanisms in animal species. In the last two decades, however, sophisticated molecular and genomic techniques have been introduced for studying this problem, and our knowledge has advanced substantially. Furthermore, many investigators are now studying various groups of organisms that are outside the model organisms such as humans, mice, chicken, and *Drosophila*, and these studies have generated many new findings. In the following I would like to present some of the important findings in vertebrates and invertebrates separately.

Sex Determination in Vertebrates

One of the most important findings in recent years is the identification of the gene controlling testis formation of mammals, called the *Sry* (sex-determining region-Y) gene (Koopman et al. 1990; Sinclair et al. 1990). This gene is located on the Y chromosome and encodes a transcription factor that activates the developmental pathway for the formation of the male phenotype (Fig. 8.3). Females have no Y chromosome and no *Sry* gene. They then choose the developmental pathway for ovary formation as a default option. This finding stimulated other workers to study the molecular biology of sex determination in various organisms.

In all vertebrates a similar gonad differentiation pathway is used irrespective of the GSD and TSD systems. The undifferentiated (bipotential) gonad is formed from a ridge of cells on the embryonic kidney, and this gonad later differentiates into the testis in males and an ovary in females. There are many genes involved in the developmental pathways of the testis and the ovary, but the details of the pathways are still unclear. Gene *Sox9* is upregulated in the testis in all vertebrates, and there are several genes regulating the expression of *Sox9* in mammals. The *Sry* gene is one member of the *Sox* gene family, but it has been poorly conserved except for the DNA-binding domain called the HMG box. The closest relative of *Sry* is *Sox3*, which is located on the X chromosome. This suggests that *Sry* was derived from *Sox3*. In chickens, in which the ZW system is used, sex appears to be determined by the two copies of *Dmrt1*, one in each of the two Z chromosomes. The *Dmrt1* upregulates the *Sox9* gene without the *Sry*, and this leads to testis formation (Graves 2008; Smith et al. 2009). Ovary formation is initiated as a default option when a single copy of *Dmrt1* is present in the individual as in the case of the ZW female.

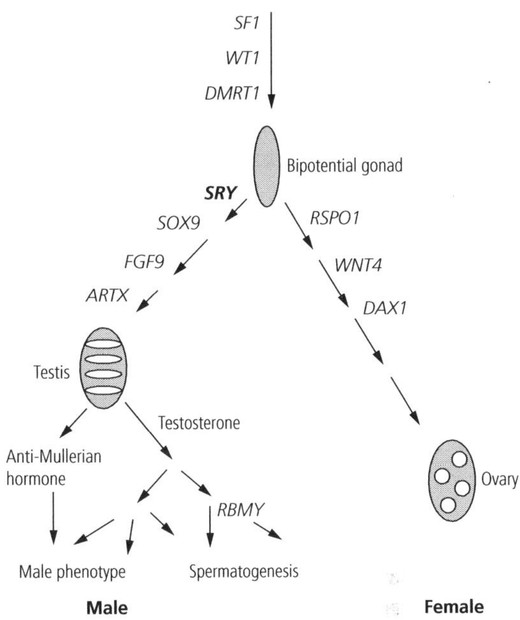

Fig. 8.3. Some of the genes involved in the mammalian gonad differentiation pathway. A bipotential gonad differentiates into a testis (left) under the influence of *SRY*, and testis hormones do the rest. If no testis forms, genes such as *RSPO1* promote the differentiation of the bipotential gonad into an ovary. Adapted from Graves (2008).

These two examples indicate the importance of *Dmrt1* in sex determination in vertebrates. Note that *Dmrt1* plays an important role in mammals as well (Fig. 8.3). Furthermore, there is at least one lizard species with a Z chromosome that contains the same set of genes as that of the chicken Z (Kawai et al. 2009). This suggests that *Dmrt1* is an ancient sex-determination gene. In fact, the egg-laying mammal platypus, which diverged from the other mammalian species about 200 million years ago, is known to have a sex chromosome complex that is unrelated to the mammalian XY but shares genes with the chicken ZW system including *Dmrt1* (Graves 2008). Note that a gene homologous to *Dmrt1* is known to be involved in the sex determination of fruitflies and nematodes as well (Raymond et al. 1998).

By contrast, the *Sry* gene appears to have originated relatively recently, and it has a sex-determining role only in mammals. In this sense, the gene is less conserved than the *Dmrt1* gene. It is interesting to note that the *Sry* gene can be lost from the genome and its function can be replaced by other genes. This has happened in two species of voles (Just et al. 1995) and two species of spiny rats (Sutou et al. 2001). In these species no Y chromosome and no *Sry* genes are found. In the case of spiny rats, the two species, *Tokudaia osimensis* and *T. tokunoshimensis*, inhabit two separate tiny islands, Amami Oshima and Tokunoshima, near Okinawa, Japan, respectively. The chromosome number of *T. osimensis* is $2n = 25$, and that of *T. tokunoshimensis* is $2n = 45$ for both males and females. In both species no Y chromosome is observed, so that the chromosome system is XO for both males and females. These two species have a small population size and are closely related to another species on Okinawa island, *T. muenninki*, which has 44 chromosomes with the standard XX/XY or XY system (Kuroiwa et al. 2010).

Because the two species living on the tiny islands do not have any *Sry* gene, Kuroiwa et al. (2011) studied the copy number and location of ten genes (*Artx, Cbx2, Dmrt1, Fgf9, NroB1, Nr5A1, Rspo1, Sox9, Wnt4, Wnt1*) that might be involved in the initiation of sex determination in the species. They then found that there are multiple copies of *Cbx2* genes in the *Tokudaia* species and that there are two or more copies of *Cbx2* genes in males than in females in both Amami and Tokunoshima spiny rats. Because *Cbx2* is known to repress ovarian development in humans and mice, they concluded that a larger number of *Cbx2* genes in males is probably responsible for testis development. However, this hypothesis has not been confirmed by cloning and characterizing the *Cbx2* gene.

Although the molecular mechanism of sex determination in these species is still unclear, it is interesting to note that the XO/XO genetic system can evolve from the XX/XY system within a short period of time (about two million years). This evolutionary change apparently occurred by fixation of new mutations and this fixation was facilitated by genetic drift because the population sizes of the two island species are very small. (The two species are listed as endangered species in Japan.) However, once the new sex-determination system evolved, it has worked well so that the new species continue to survive. This indicates that even such an important character as sex determination can change by mutation and genetic drift relatively rapidly. In this case the role of positive Darwinian selection is considered to be negligible, because there is no need to change the original genetic determination system (XX/XY system) which has been used for hundreds of millions of years.

Actually, the sex-determination system is known to have changed quite often in the evolutionary process of reptiles and fish. In these organisms, there are both the genetic sex-determination (GSD) and the environmental or temperature sex-determination systems (ESD or TSD), and the GSD includes both the XY and the ZW systems. Therefore, the same order or family of organisms may include different sex-determination systems. Figure 8.4 shows the distribution of different sex-determination systems (ZW, XY, and TSD) in mammals, birds, and reptiles. At present, it appears that mammalian species are essentially of the XY type and birds are of the ZW type, but reptiles show many different mechanisms of sex determination. For example, some species of the order of turtles have TSD, whereas others have either the XY or the ZW system. These systems have apparently evolved after they diverged from the bird and crocodile lineages. Similarly, species of gecko have various forms of sex determination. These data suggest that the sex-determination system can change during a relatively short period of evolutionary time.

In amphibians many different orders and families include both XY and ZW systems, and it is apparent that the XY ↔ ZW change has occurred many times (Hillis and Green 1990; Sarre et al. 2011). In the extreme case of the Japanese wrinkled frog (*Rana rugosa*) both XY and ZW systems are observed in different geographical areas. In addition, there are two more different homomorphic systems of sex determination observed in other geographical areas (Miura et al. 1998). Cytogenetic and molecular studies showed that the four different karyotypes of sex chromosomes were formed by repeated occurrence of chromosomal inversion and translocation but the sex-determining genes have remained the same. Therefore, the morphological changes of chromo-

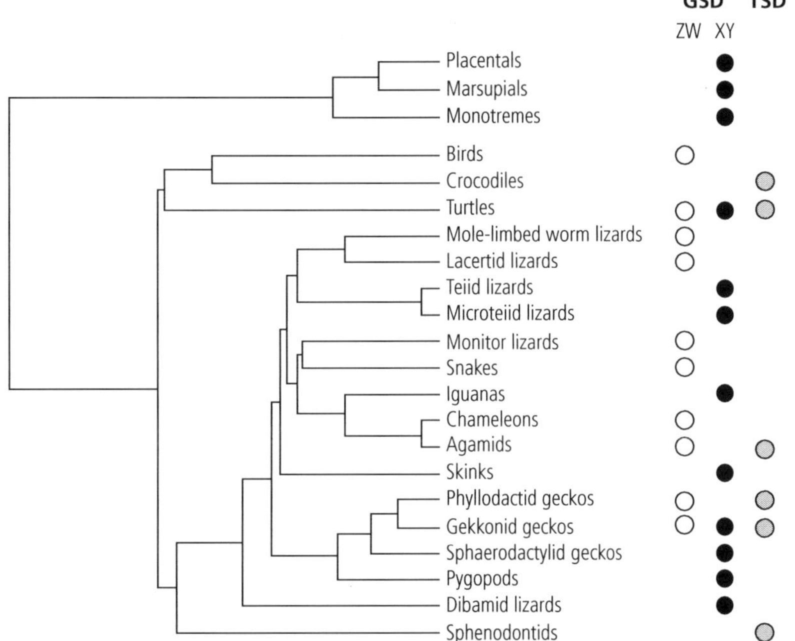

Fig. 8.4. Distribution of male and female heterogamety and temperature-dependent sex determination (TSD) among mammals, birds, and reptiles. Adapted from Sarre et al. (2011).

somes are superficial for the sex determination of this species. This finding suggests that the XY ↔ ZW change observed in many species of amphibians and reptiles is caused by similar chromosome inversion and translocation and it does not necessarily mean a significant change of sex-determining genes. It also suggests that the effective population size of some reptile and amphibian species is relatively small and therefore new chromosomal rearrangements can be fixed in the population rather easily. In the case of Japanese wrinkled frogs, populations with different chromosomal types are separated by mountainous areas, and this suggests that the population size was reduced when the frogs crossed the mountains and new chromosome types were generated at this time by bottleneck effects.

In reptiles and amphibians, the molecular basis of sex determination is not well understood. In the clawed frog *Xenopus laevis* with the ZW system, the *Dm-w* gene located on the W chromosome appears to be a sex (ovary)-determining gene (Yoshimoto et al. 2008). This gene has a high degree of sequence similarity with the *Dmrt1* gene. A similar study was done with the Japanese wrinkled frog, but no definitive conclusion has been obtained (Miura et al. 2012).

Sex Determination in Invertebrates

Because fruitflies have been used as a model organism for the last 100 years, a wealth of information is available about the chromosomal and molecular basis of sex determination in this organism. The first step of sex determination in fruitflies is the activation of the *Sex lethal* (*Sxl*) gene in females. This activation is initiated when two X chromosomes are present in an individual. The *Sxl* gene then produces a female-specific SXL protein through a special exon-splicing mechanism (Gempe and Beye 2010; Verhulst et al. 2010b). This protein is produced only in females and activates the *transformer* (*tra*) gene, which encodes a female-specific TRA protein (Fig. 8.5). This protein then stimulates the activity of the *doublesex* (*dsx*) gene. Activation of *dsx* produces a female-specific DSX^F protein, and

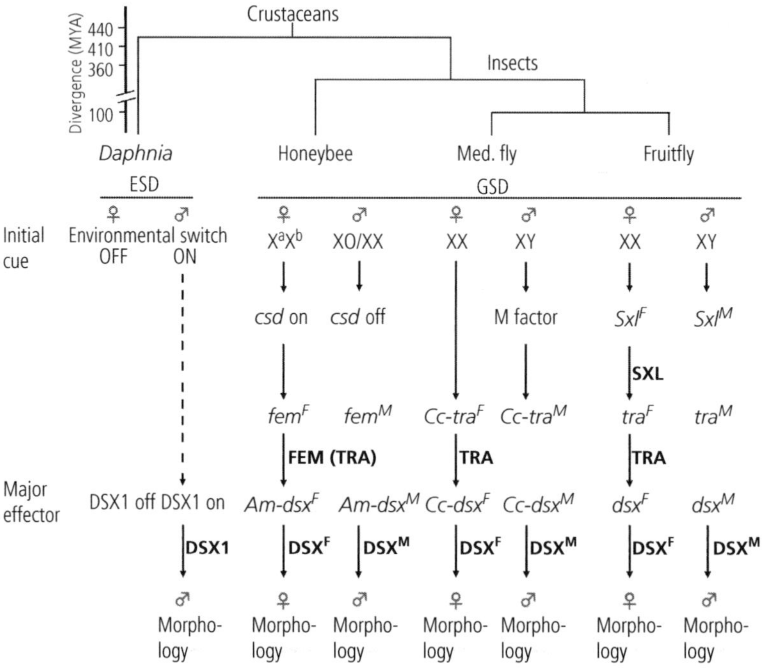

Fig. 8.5. Simplified view of sex-determining pathways in the crustacean *Daphnia* and insects. An ESD pathway in *Daphnia* is compared with GSD pathways in insect model species, honeybee, Mediterranean fruitfly (Med. Fly), and fruitfly. *Csd*, complementary sex determiner; *fem, feminizer*; *sxl, sex lethal; dsx, doublesex*; MYA, million years ago. Modified from Kato et al. (2011).

this protein initiates the formation of the ovary and female phenotype. In XY males the SXL^M protein produced is nonfunctional because of the special exon-splicing mechanism. Similarly, the transcript of tra^M is nonfunctional. However, in the absence of the TRA protein a male-specific DSX protein is produced, and this protein initiates the formation of the testis and male phenotype (Fig 8.5).

Essentially the same mechanism of sex-determining cascade is known to work in all the insect species so far examined (e.g. fruitfly, housefly, Mediterranean fly, and honeybee). The gene working at the bottom of the hierarchy is the *dsx* gene, which initiates the formation of female or male phenotype. This gene contains a DNA-binding domain (DM domain), which is known to be highly conserved and shared by all insect species so far studied, and by the crustacean *Daphnia magna* which has an environmental sex determination (ESD). The *tra* gene that activates *dsx* is also shared by all insect species. In honeybees this gene is called the *feminizer* (*fem*), but it is evolutionarily orthologous to *tra*. The function of this gene is less conserved, and in daphnia its function remains unclear (Kato et al. 2011).

The first step of sex determination in fruitflies is played by *Sxl* as mentioned above, whereas in honeybees the *csd* gene plays the role. The *csd* gene apparently has been derived from the *tra* gene and has no relationship with *Sxl*. The M factor in Mediterranean flies is encoded by an unidentified gene. Therefore, all of the three insect species presented in Fig. 8.5 use different signal proteins for the initiation of sex determination. A different initial signal protein is also used in Lepidoptera, where the ZW system of GSD is used (Verhulst et al. 2010b). Similarly, in the hymenopteran wasp *Nasonia*, another initial signal protein rather than *csd* is used (Verhulst et al. 2010a). Note also that in daphnia environmental factors initiate the activation of the *dsx* gene (Fig. 8.5). Therefore, the initial signal of sex determination varies extensively even within insects and crustaceans. This is in contrast to the *tra* and *dsx* genes, which have been conserved for a long evolutionary time. The *tra* gene controls the second deepest hierarchy of sex determination and appears to have been conserved in all insect species. However, the *dsx* gene working at the deepest level of the hierarchy is more conserved. Actually, *dsx* is homologous to the *Dmrt1* gene in vertebrates and the *mab-3* gene in *C. elegans* (Raymond et al. 1998). In fact, the gene name *Dmrt1* is the abbreviation of the doublesex and mab-3 related transcription factor 1 and is shared by virtually all animal species.

This observation suggests that the deepest level of the hierarchy for sex determination evolved early in animal evolution, and the currently observed bewildering variety of sex-determination systems have then been invented by recruiting and adding new signal molecules to the deepest level of the hierarchy. Wilkins (1995) called this view the moving-up-of-the-hierarchy hypothesis. Recent studies are giving increasing support for this hypothesis. If this hypothesis is correct, it is possible to explain the evolutionary change in sex-determination mechanisms relatively easily by various new mutations. In this case selection may not be so important because the final consequence of changes of the initial signal molecule may not be serious unless the *Dmrt1* system is affected. It is also possible that some parthenogenetic or asexual organisms retain the *Dmrt1* gene and resume sexual reproduction later when proper mutations occur.

8.5. Degeneration of the Y (W) Chromosome

Y Degeneration and Dosage Compensation

In Chapter 3 we have already seen that the chance of nonfunctional mutations accumulating in the non-recombination portion of the Y chromosome is quite high and that the inertness of the Y chromosome can be explained by the accumulation of these mutations alone. However, Charlesworth (1978) was critical of this view and proposed that inactivation of a majority of Y-chromosome genes has occurred by Muller's ratchet effect as formulated by Haigh (1978). He considered the ratchet effect theory in conjunction with the evolution of X-chromosome dosage compensation that equalizes the expression levels of X-linked genes in both XX females and XY males (Lucchesi et al. 2005; Mank 2009). He stated that because Y-chromosome degeneration is almost always associated with the dosage compensation of X chromosomes we must consider both evolutionary processes simultaneously as a compound theory.

However, recent studies have shown that the accumulation of lethal mutations and the development of dosage compensation occur independently, because nonfunctional mutations accumulate rapidly at sheltered genetic loci but the molecular mechanism of dosage compensation varies from organism to organism (Lucchesi et al. 2005; Meyer 2005; Payer and Lee 2008; Mank 2009). For example, dosage compensation in placental mammals is achieved by random inactivation of one of the two X chromosomes in the XX females so that the expression level of X-linked genes becomes the same for both females and males (Fig. 8.6). In *Drosophila*, however, the equal level of expression of X-linked genes in males and females is achieved by doubling the transcription level of male X-linked genes. For this reason, Charlesworth (1991) stated that "*the operation of Muller's ratchet on a proto-Y in heteromorphic sex, and reducing transcription from the Y while increasing transcription from the X in the heteromorphic sex would cause a Drosophila type of dosage compensation to evolve in concert with the inactivation of the Y... A mammalian type of dosage compensation could have evolved as a response to this kind of situation because it creates a selective advantage to reducing X activity in the homomorphic sex, thereby restoring the balance between autosomal and X chromosomal gene products. The endpoint of this would be total inactivation of the X in the homogametic sex, as is now observed in eutherian mammals.*"

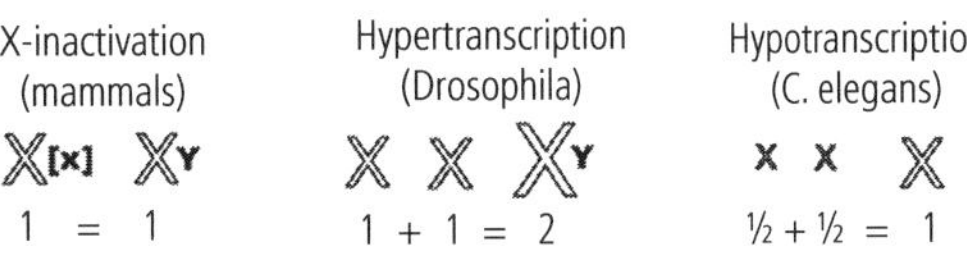

Fig. 8.6. Dosage compensation in mammals, *Drosophila*, and *C. elegans*. Some investigators believe that the gene expression balance between autosomal and X-chromosomal genes and the expression level of mammalian genes is doubled after one of the two chromosomes in females are inactivated (Mank 2009). Modified from Lucchesi et al. (2005).

There are several weaknesses in this argument. (1) Charlesworth has not provided why the genes in the Y are devoid of recombination. In *Drosophila melanogaster* and silkworm this is achieved by the lack of recombination in the heteromorphic sex and therefore even autosomal chromosomes do not show recombination. However, how was this accomplished? This seems to be due to an enzymatic control, but the enzyme has not been identified. In most other organisms, a reduced level of recombination occurs only between the X (Z) and the Y (W) chromosomes. Without knowing this recombination-reducing mechanism, it is difficult to argue the ratchet effect. (2) The ratchet effect is effective only when the selection coefficient (s) is small and finally causes the deleterious alleles to reach the equilibrium value (v/s) (see Chapter 3). Therefore, this does not necessarily degenerate the Y chromosome. (3) Experimental data suggest that the X-chromosome dosage compensation does not occur for each gene separately as expected from Charlesworth's theory. Rather Y-chromosome inactivation and X-chromosome dosage compensation occur independently, as discussed below

Molecular Basis of X-Chromosome Dosage Compensation

The mechanism of X-chromosome dosage compensation varies extensively with organism, and details of the mechanism are controversial. As mentioned above, the dosage compensation in mammals occurs by inactivating one of the two X chromosomes in the female, whereas in *Drosophila* it is accomplished by doubling the transcription level of the X chromosome of the XY males. By contrast, the dosage compensation in the nematode *C. elegans* is achieved by reducing the transcription level of each of the two X chromosomes in the female (hermaphrodite) (Fig. 8.6). Birds and moths with the ZW sex-determination system do not show a global sex chromosome dosage compensation (Mank 2009).

Furthermore, recent molecular studies indicate that dosage compensation occurs by an epigenetic control of transcription of X-linked genes by chromosome-specific localization of activities that acetylate and deacetylate histones, methylate DNA, condense chromosomes or otherwise modify chromatin structure (Park and Kuroda 2001; Lucchesi et al. 2005; Meyer 2005; Payer and Lee 2008; Mank 2009). In placental mammals the X chromosome contains the X inactivation center (XIC), and in this center the X inactive-specific transcript (Xist RNA) is transcribed and coats one of the two X chromosomes in the XX female strictly in the *cis* form. The X chromosome coated is then inactivated. The choice of which chromosome is to be inactivated is made in early embryogenesis and occurs at random in placental mammals. In marsupials, however, the paternal X chromosome is inactivated preferentially. Inactivation of the X chromosome is accomplished by spreading Xist RNA over the entire region of the chromosome, which induces DNA methylation and histone hypoacetylation and leads to transcription of no X-linked genes. Some investigators (Ohno 1967; Nguyen and Disteche 2006) believe that transcription of the single active X chromosome is then upregulated in both sexes to equalize the transcription level with that of autosomal genes, but this view is controversial (Lin et al. 2012). These observations suggest that dosage compensation evolved after the occurrence of inactivation of Y-chromosome genes.

In *Drosophila* the hypertranscription of male X-linked genes occurs for dosage compensation, and the molecular mechanism for this hypertranscription is more complicated than that for the X inactivation in mammals. The hypertranscription is achieved by the action of a large protein-RNA complex containing at least two RNAs and five polypeptides, collectively called the Male-Specific Lethal (MSL) complex. MSL complexes bind to one of about 150 chromatin entry sites (CES), from which the complexes spread to other regions of the male X chromosome (Alekseyenko et al. 2008). MSL complexes also bind to acetylated histones and enhance the transcription of male X-linked genes.

In *C. elegans* the X chromosome of XX hermaphrodites is downregulated by half to equalize the expression levels of males and females (Fig. 8.6). This dosage compensation is achieved by special dosage compensation complexes that are targeted at the X chromosomes of one sex to regulate transcription levels by changing chromosome structure. This regulation is controlled by the dosage compensation complex (DCC), a complex of at least ten proteins that acts as a molecular switch. This DCC is recruited to X chromosomes by two hermaphrodite-specific genes (*sdc*-2 and *sdc*-3) that jointly control sex determination and dosage compensation. The SDC-2 protein is crucial for the X chromosomes of hermaphrodites to have dosage compensation. The DCC complex appears to have many target sites at the X chromosome for reducing transcription level by half. Incidentally, the molecular structure of the DCC is similar to that of the mitotic/meiotic condensin complex and 13S condensin complex so that it is believed that the DCC is constructed by modifying these complexes (Meyer 2010).

In the above discussion we considered three examples of X-chromosome dosage compensation to show how the mechanism varies with organism. In other organisms, the molecular mechanism has not been studied well, but we know that there are many different forms of sex determination and dosage compensation even if we exclude the cases of environment-dependent sex determination. For example, birds and snakes do not seem to have clear-cut dosage compensation. These observations suggest that the degeneration of the Y chromosome occurs almost independently of dosage compensation and that the latter mechanism evolved probably after the occurrence of Y-chromosome degeneration. The reason is that dosage compensation is for correcting the imbalance of gene expression between males and females and between X-linked genes and autosomal genes after the occurrence of Y-chromosome degeneration. Theoretically, of course, dosage compensation may evolve before the completion of Y-chromosome degeneration. However, this is a secondary problem, and we had better not worry too much. It is probably more important to clarify the molecular mechanism of X-chromosome dosage compensation for various organisms. These problems should be attacked by studying the developmental process of testis and ovary formation pathways (Meyer 2005; Graves 2008; Payer and Lee 2008).

It should also be noted that in the above view both Y-chromosome degeneration and X-chromosome dosage compensation have occurred by the mutational process and natural selection probably has occurred primarily for eliminating unfit mutations. If this is the case, mutation has played a more important role than natural selection in the evolution of sex chromosomes.

Evolution by Sexually Antagonistic Mutations

In the above models of the evolution of the Y (or W) chromosome, degeneration of the Y is believed to have occurred by accumulation of deleterious mutations. However, Rice (1984, 1996) proposed the hypothesis that degeneration of the Y chromosome occurs by natural selection of sexually antagonistic mutations and Muller's ratchet effect. Sexually antagonistic mutations refer to the mutations that confer a selective advantage to the XY male but a disadvantage to the XX female. This idea was originally presented by Fisher (1931), when he used Winge's (1927) observation of Y-linked polymorphism of male ornamental color patterns in the guppy fish for supporting his theory of evolution of dominance. Rice then argued that if the Y-linked genes are advantageous in males and disadvantageous in females, the Y chromosome will have a selective advantage and may be fixed in the population. He also assumed that if many sexually antagonistic genes of this type accumulate on the Y, the recombination values in the Y chromosome would be reduced, and the Y chromosome will then become a clonal entity similar to an asexual organism. If this happens, deleterious mutations may accumulate continuously on the Y chromosome by Muller's ratchet effect (see Section 3.5), and if this process continues, the Y chromosome will eventually be lost. Therefore, in this hypothesis the final destination of the XY chromosome evolution is the formation of individuals with the XO chromosome. In recent years, Rice's theory has become popular, partly because some experimentalists such as Graves (2008) proposed on the basis of empirical data that the Y chromosome in mammalian species is destined to become smaller and smaller and be lost eventually.

In my view there are several problems in this hypothesis. First, it is quite unreasonable to assume that the mutations controlling ornamental characters in males are advantageous in males but deleterious in females. If the existence of different color patterns in males represents a case of sexual selection as is often assumed, selection should occur among males rather than between males and females. Actually, because males and females collaborate with each other to produce offspring, it is difficult to justify Rice's assumption. Second, as discussed in Section 3.5, Muller's ratchet is effective only in a limited condition, so it is unlikely to work for the Y in the presence of deleterious mutations on the X chromosome. Furthermore, if the Y is initially subject to a strong positive selection, why should the Y chromosome accumulate deleterious mutations later? Third, the male secondary sex features are generally controlled by sex-limited genes, so it is unclear how important the Y-linked antagonistic genes are for the Y-chromosome degeneration. Fourth, when the male sex is determined by two doses of DMRT1 in the two Z chromosomes as in the case of birds (Section 8.4), why does the W chromosome accumulate deleterious mutations? Some more detailed study seems to be necessary for Rice's theory.

8.6. Evolution of Behavioral Characters

Selfish Gene Theory of Evolution

In Section 8.2 of this chapter, we discussed Hamilton's (1964) kin selection theory in relation to the evolution of altruism or eusociality. Actually, nearly at the same time Williams (1966) had proposed another theory of evolution of complex characters. This theory asserts that adaptation occurs exclusively by natural selection and natural selection operates by preservation of beneficial alleles or allelic combinations in the population for the next generation rather than by choosing well-adapted phenotypes. This is against the generally accepted view that natural selection acts on phenotypes rather than on genotypes or genes. Williams argued that his gene-centric theory explains the evolution of any complex characters including genetic systems such as mitosis and meiosis as well as behavioral characters.

This gene-centric view of natural selection has then been popularized by Dawkins (1976), who wrote a book called *The Selfish Gene* by using many arguments with nicely crafted metaphors. In this book he proposed that genes are the fundamental unit of selection and preserve information about the effects of natural selection for future generations. By contrast, phenotypes or individuals are regarded as vehicles that carry the genes and are directly subject to natural selection, but because they are destroyed every generation, they are only for temporary use in evolution. Phenotypes are reconstructed by genes every generation, and they may display new characters generated by natural selection. Yet, they are merely survival machines.

This theory is particularly popular in the area of ethology, the study of animal behaviors and their evolution. Ethology is an active area of evolutionary biology, and a large number of researchers are working on the subject. One of the important problems in ethology is to explain the evolution of seemingly purposeful characters such as instinctive mating in animals, parental care of children, seasonal migration of birds, and altruistic behavior of worker honeybees. I have no intention to cover the recent progress in this field, but I would like to explain the essence of the selfish gene theory and comment on it.

According to Williams (1966), the random forces of evolution (e.g. genetic drift) were important only when life originated on earth, and the adaptation of organisms in the subsequent evolutionary process has occurred exclusively by natural selection. He argued that natural selection certainly occurs on phenotypes but the selection can be described ultimately by the average fitness of two alleles at each genetic locus. Here he used the concept of the gene pool in population genetics theory, in which a collection of alleles at a locus for all individuals in the population are considered. Suppose that this gene pool contains two alleles, A_1 and A_2 at a locus with the frequencies of x and $1–x$ at a generation, respectively. If allele A_1 is advantageous over A_2, the frequency of x increases in the next generation, and the amount of increase is given by Equation 2.5 in Chapter 2. In this equation, the crucial part is $(w_1 - w_2)$, and if the allelic fitness w_1 is greater than w_2, the frequency of A_1 increases whether A_1 is dominant, semidominant, or recessive in relation to A_2.

In reality, the allelic fitnesses w_1 and w_2 would be affected by interaction with many other gene loci. Williams then argued that if we redefine w_1 and w_2 as the allelic fitnesses averaged over all loci, the following equation, similar to Equation 2.5, may be derived.

$$\Delta x = xy(\overline{w}_1 - \overline{w}_2)/\hat{w} \tag{8.3}$$

where Δx is the amount of change of allele frequency x per generation at locus A, and $\overline{w}_1$ and $\overline{w}_2$ are the average fitness of alleles A_1 and A_2, respectively, taking into account all the effects from other loci, and $\hat{w}$ is the average fitness for the entire population.

Theoretically speaking, Equation (8.3) may not be correct, because natural selection for multiple loci is quite complicated and the allele frequency change should be given by Equation (2.12a) or its extension for many loci. In other words, in the presence of gene interaction and linkage disequilibria the allele frequency change for a locus can be positive or negative (see Fig. 2.3). Yet, Williams assumed that Δx is always positive irrespective of the number of loci and the type of selection involved. Δx in (8.3) may become negative even when the environmental condition changes with generation (Fig. 2.2), but Williams did not consider the possibility. Note also that new genetic variations are always generated by new mutations, new gene duplications, new epigenetic systems, etc. Therefore, natural selection at the genomic level is very complicated and his argument will no longer be applicable.

Furthermore, Williams did not consider any specific genetic model or mathematical model, arguing that it is not necessary as long as Δx is positive. This would be acceptable if the evolution of a phenotypic character can occur without new mutations, gene interaction, and genetic drift. His view is clearly dependent on the classical neo-Darwinian theory. If we consider various evolutionary factors discussed in Chapters 4 to 6, Williams's argument is no longer acceptable.

Actually, the selfish gene or gene-centric theory of evolution has been criticized by a number of authors in the past. Gould (1980, Chapter 8; 2002, p. 638–641) argued that the concept of a gene as a unit of selection is incorrect because natural selection occurs when a certain group of individuals produce more offspring than the other group and it is a matter of birth and death of individuals. Actually, both Williams (1966) and Dawkins (1976) admit that natural selection occurs on phenotypes or individuals but argue that because allele frequency changes describe the process of evolution in population genetics the unit of selection should be the gene. Unfortunately, Gould's criticism was also based on intuitive arguments rather than population genetics theory, so that Dawkins (1982) could make another intuitive argument to defend his selfish gene theory. Mayr (1997) has made a more incisive comment, indicating that current population genetics theory is based on phenotypic (or genotypic) fitnesses taking into account all types of natural selection for all types of characters and the units of selection are clearly phenotypes rather than genes.

In real populations, the evolution of complex characters is controlled by many nonselective factors such as mutation, gene duplication, gene regulatory systems, and genetic drift. Recently, Nowak et al. (2010, supplementary information) studied the allele frequency changes for eusociality genes under various conditions and concluded that the allele frequency does not necessarily increase when mutation and genetic drift are taken into account. Williams's negative view on group selection has also been reexamined by a number of authors (Borrello 2005; Wilson and Wilson 2007), and these authors reached the conclusion that both individual and group selection can be justified theoretically depending on the condition.

Furthermore, the natural selection considered in the selfish gene theory is intuitive and can cover many different types of selection without any specific model. Williams and Dawkins argued that this simplicity is an advantage of the selfish gene theory, but it is also a deficiency of the theory. Because they do not give any specific model of selection, the theory is not falsifiable. Therefore, it is not a scientific theory according to Karl Popper. Evolution by omnipotent natural selection is similar to creationism, in which natural selection is replaced by God. In practice, mutation and genetic drift also play an important role in the evolution of complex characters. This indicates that their intuitive analysis of natural selection is not biologically meaningful. To understand the effect of natural selection, we will have to have a specific genetic model of evolution and compare the theoretical prediction with empirical data.

Molecular Studies of Behavioral Genes

Although many theoreticians are still interested in mathematical studies (e.g. Frank 1998; Charlesworth and Charlesworth 2010; Bourke 2011), there has been

Table 8.1. Examples of studies of social behaviors at the molecular level

Behavior	Organism	Gene	Molecular function
Rover versus sitter phenotype	Fruitfly	*foraging (for)*	Protein kinase G
Roamer versus dweller phenotype	Nematode	*egl-4 (egg-laying defective 4)*	Protein kinase G
Division of labour: onset age of foraging	Honeybee	*foraging (For)*	Protein kinase G
Division of labour: foraging related?	Honeybee	*Period (Per)*	Transcription cofactor
Foraging specialization: nectar versus pollen	Honeybee	*Protein kinase C*	Protein kinase C
Queen caste	Honeybee	*Royalactin*	EGFR
Pheromone communication	Mouse	*V1R, V2R* (vomeronasal receptor, families 1 and 2)	G-protein receptors
Male courtship; timing of mating	Fruitfly	*Period (Per)*	Transcription cofactor
Female receptiveness (lordosis)	Rodent	Estrogen responsive genes	Various functions
Refractoriness to mate, ovipositioning, decreased longevity	Fruitfly	Genes for seminal proteins	Various functions
Monogamy, parental care	Rodent	*V1aR, OTR*	Vasopressin and oxytocin receptors
Maternal care	Rat	*GR*	Glucocorticoid receptor
Territorial males	*Haplochromis burtoni*	*GnRH1 (gonadotropin releasing hormone 1)*	Gonadotropin-releasing hormone
Dominant males	*Procambarus clarkii*	*5HTR1, -2 (serotonin receptor type 1 and 2)*	Serotonin receptors
Aggression	Honeybee	*Maoa* (monoamine oxidase A)	Monoamine oxidase
Aggression	Macaque	*5HTT* (serotonin transporter)	Serotonin transporter

Compiled from various sources of literatures including Robinson et al. (2005). EGFR: Epidermal growth factor receptor.

substantial progress in the study of the genetic and molecular basis of evolution of behavioral characters. Seymour Benzer (1967) pioneered this field by generating many mutations and studying their molecular basis. However, since artificially generated mutations are mostly deleterious, recent investigators are trying to use polymorphic alleles in natural populations to understand the evolution of behavioral characters.

There are a large number of such examples (Robinson et al. 2005 and others), and some of them are presented in Table 8.1. Here I would like to discuss a few interesting cases. One of the earliest examples is the *period* (*per*) gene that controls circadian rhythm (24-hour cycle of body clocks) in fruitflies (Sawyer et al. 1997). Molecular analysis of the *per* gene showed that the per protein contains a Thr-Gly repeat region and there are two major alleles with respect to the number of repeats: 17 and 20 Thr-Gly repeats. The 17-repeat allele is known to be more abundant in southern Europe, whereas the 20-repeat allele is more abundant in northern Europe. How the two different alleles affect circadian rhythm is not well understood, because there are several other genes that interact with the *per* gene. It should be noted that the *per* locus affects mating behavior in addition to circadian rhythm (see Table 8.1).

Another interesting example from fruitflies is the *foraging* (*for*) gene that controls the range of foraging for food in the larval stage. There are two polymorphic phenotypes: rover and sitter. The allele for the rover is known to be dominant over the sitter, and the rover phenotype exhibits long-range foraging for food, whereas the sitter searches for food over small ranges. The *for* gene encodes a cyclic GMP (cGMP)-dependent protein kinase (PKG), and PKG enzyme activities and mRNA levels are higher in rovers than in sitters. Therefore, the difference between rovers and sitters appears to be caused by one or more regulatory mutations. However, the real mechanism for generating the rover and sitter behavior is still unknown (Sokolowski 1998).

In the nematode *Caenorhabditis elegans*, some wild-type strains are solitary foragers, moving across their food (*E. coli*) and feeding alone, whereas others are social foragers that aggregate together on the food while they feed. One of the genes that control foraging traits is the *PKG* (protein kinase G) gene, which is responsible for variation in food-dependent locomotion. Allelic variation at this locus affects the proportion of time which the worm spends for roaming or dwelling. This is caused by the effects on sensory neurons that are involved in locomotion and olfaction (Robinson et al. 2005). Mutations that decrease PKG signaling lead to an increase in roaming, suggesting that their behavior is PKG-dependent. PKG-dependent foraging is also observed in honeybees. Worker honeybees stay in the hive when they are young and start roving outside in search of food when they become older. The onset age of foraging is socially regulated by the need of the colony. For example, precocious foraging occurs when young bees sense a lack of foragers by the aid of pheromones. It is known that the *For* gene, the orthologous gene of the *for* in fruitflies, is involved in the regulation of onset age for foraging in honeybees. Levels of *For* mRNA in the brain are higher in foragers than in worker bees in the hive (Robinson et al. 2005). Foraging is therefore socially regulated in honeybees. These observations suggest that the foraging behavior is controlled by the same or similar regulatory genes in the two distantly related species of insect and nematode and that the foraging systems are highly conserved. However, the regulatory systems of gene expression in honeybees seem to have evolved relatively rapidly.

Behavioral characters are often controlled by multigene families. For example, the genes that synthesize sex pheromones for moth mating are members of a multigene family of acyl-CoA desaturase genes. In moths, mating is initiated by emission of female moth sex pheromones that attract males of the same species. The pheromones are composed of complex blends of long-chain fatty acid hydrocarbons, and to synthesize the complex blends, three key enzymatic reactions are necessary. The enzymes involved in their reactions are called desaturases, and the genes encoding these enzymes are known to compose a gene family. This gene family is known to be subject to a birth-and-death process of evolution, which generates species-specific sets of genes. For this reason, different species produce different pheromone blends, which are responsible for species-specific mating (Roelofs and Rooney 2003; Rooney 2009).

In general, the genetic basis of behavioral characters is very complex and is controlled by many different genes that interact with one another as well as with environmental factors. In animals, mating behavior, parental care, and aggressive behavior are important for survival, and they are controlled by many factors. In the case of aggressive behavior of fruitflies, Edwards et al. (2006) conducted a genome-wide analysis of gene expression and identified 34 mutations controlling aggressive behavior. At the present time, the study of behavioral genes has just begun, and our understanding of the evolution of behavioral characters is very limited. However, because genome-wide analysis of gene expression can now be done, it is expected that we will soon have better understanding of this important subject (Chandrasekaran et al. 2011; Hunt et al. 2011).

8.7. Summary

There are a large number of examples in which an organism is exquisitely adapted to a particular niche or lifestyle. Many evolutionists have stated that this type of adaptation occurs only by natural selection. However, there are examples in which the adaptive significance of mutant characters is not easily recognizable, yet their frequency in the population is quite high. Furthermore, recent molecular studies suggest that most innovative characters in phenotypic evolution are actually generated by mutations and the role of natural selection is for shifting allele frequencies.

It is well known that Charles Darwin had difficulties in explaining the evolution of complex characters such as the vertebrate eye in terms of natural selection. Yet, he proposed that these complex characters evolved gradually by natural selection from the primitive characters in lower organisms under the assumption that natural populations contain almost any kind of genetic variation. Actually, his explanation would have been much simpler if he had assumed that new variations are generated by mutations. Unfortunately, in Darwin's time no clear notion of mutation existed. Some of Darwin's questions can now be explained more straightforwardly by mutation at the molecular level. Evolution of the caste systems in hymenopterans can also be explained by mutation.

Darwin also had difficulty in explaining vestigial characters observed in humans and other animals, and in this case he often used the idea of evolution by use and disuse, a form of Lamarckism. At the present time, this can easily be explained by accumulation of nonfunctional mutations when the characters are no longer used. Parasitic organisms are also known to accumulate nonfunctional mutations when the original organs or genes are no longer needed. Rates of amino acid substitution in parasitic bacteria are unusually high but these high rates are apparently due to small effective population sizes and the lack of DNA repair enzymes.

In mammals the sex is determined by the XX/XY system, whereas in birds it is controlled by the ZZ/ZW system. In amphibians and reptiles the XX/XY and the ZZ/ZW systems are interchangeable, and in some reptile species the sex is determined by environmental factors, but the molecular basis of the evolutionary changes in these systems is not well understood. In all vertebrate species, however, the signal protein that triggers the formation of male or female phenotype appears to be the same. Similarly, the signal protein that determines the male and female phenotypes is the same for all invertebrate species and is apparently homologous with the vertebrate protein, though the initial sex-determining signal protein varies extensively among different groups of species. The evolutionary differentiation of the X and Y (or Z and W) chromosome is initiated by the reduction of recombination value for sex-determination related genes in the Y chromosome and the inactivation of other Y-linked genes. Dosage compensation of X-linked genes is then established in some groups of organisms.

At the present time, behavioral biologists are largely panselectionists and pay little attention to mutation, genetic drift, and gene co-option. However, molecular data about behavioral variation are supportive of the occurence of nonselective as well as selective forces of evolution. Behavioral biologists are currently formulating complex mathematical theories to understand the evolution of behavioral characters. However, this subject should be studied at the molecular level before any meaningful theoretical study can be made.

CHAPTER 9

Mutation and Selection in Evolution

9.1. Distinct Processes of Mutation and Selection

In the study of evolution the following question is often asked: what is the relative importance of mutation and natural selection? In my view this is not an appropriate question, because the roles of mutation and selection are qualitatively different. In its simplest form, a mutation refers to any change of DNA molecules. The genotype generated by this mutation may form an innovative phenotype compared with the pre-existing genotypes. The genotype will then have a higher fitness than the pre-existing ones. The selective advantage of the new genotype is expected to be high when the extent of the phenotypic innovation is high. Therefore, both the extent of phenotypic innovation and selective advantage are dependent on the type of mutation. It is then obvious that mutation and selection are not independent of each other but that selection is initiated by the occurrence of advantageous mutation. This means that if we identify the mechanism of selective advantage of the new mutant allele at the molecular level, we are able to understand how evolution would proceed. This was one of the reasons why I proposed that mutation is the driving force of evolution and natural selection is of secondary importance (Nei 1983, 1987, 2007). This view is different from Morgan's (1925, 1932) mutation-selection theory, in which mutation and selection were thought to be independent events. At his time, the molecular basis of a mutation was not known, and therefore it was difficult to know why a particular mutation is advantageous over the pre-existing alleles. He was also concerned with only genic mutations.

In the era of neo-Darwinism, it was already known that mutation is the ultimate source of genetic variation, but it was difficult to identify the molecular change of a gene caused by mutation. Therefore, neo-Darwinians believed that there are various kinds of mutations in the population and previously deleterious mutations may become advantageous when the environmental condition changes. For this reason, an environmental change was often considered to be the primary cause of initiation of natural selection. However, this idea was generated mainly by the example of the melanic and light forms of *Biston betularia*, and the actual number of cases supporting this idea was rather small. Mendelian geneticists certainly observed that most new mutations are deleterious, and this observation also contributed to the idea of the importance of environmental changes for initiating allelic substitution. Here again, however, there were not many examples of fitness changes due to environmental changes. Therefore, the idea of environmental changes initiating allelic substitution was largely conceptual rather than factual.

Another argument which was often used in the neo-Darwinian era is that natural selection enhances the chance of recombination of advantageous mutations into single individuals so that the recombinant genotypes may have selective advantage (see Chapter 3). This is certainly a legitimate argument, but note that recombination is a mutational event and therefore the above argument about the importance of mutation applies to this group of genotypes as well.

During the Darwinian or neo-Darwinian era, mutation was treated essentially as a black box, and therefore it was natural to emphasize the importance of natural selection. In the last few decades, however, our knowledge of the molecular mechanisms of mutation has increased so much that the mutational process is no longer a black box. We can study

Mutation-Driven Evolution. First Edition. Masatoshi Nei.

mutational changes of DNA sequences by comparing closely related individuals or closely related species (Lynch et al. 2008; Ossowski et al. 2010; Keightley 2012). We can also study how phenotypic changes occur by mutation at the molecular level and why the newly produced organism is advantageous over the existing ones. Of course, the formation of innovative characters is often accomplished by a complex genomic change, and it is not always easy to solve this question. We have indicated that the molecular basis of altruistic behavior of worker honeybees is still largely unknown though the initial signal protein has been identified. Even the molecular basis of left-right handedness in flatfish and gastropod shells is unclear. In reality the molecular mechanisms of formation of complex characters such as human stature and flowering time of plants are quite complicated. However, we already have scientific methodologies to study these problems and explain the evolution of complex characters at the molecular level, as was discussed in Chapters 6 and 8.

By contrast, natural selection is a more complex process. Theoretically, it is a process of allele or gamete frequency changes in the population, and this process is often studied by using the mathematical methods presented in Chapter 2. In reality, however, it is very difficult to prove the occurrence of selection and estimate the selection coefficients in natural populations, as was repeatedly emphasized by Richard Lewontin (see Chapter 2). At the present time, the only gene for which the action of natural selection in the wild has been established unequivocally is the sickle cell anemia gene (*S*). The homozygote (*SS*) for this gene is effectively lethal, but the heterozygote (*AS*) is known to have a higher fitness for a biochemical reason than that of the wild-type homozygote (*AA*) in the malaria-ridden area of Africa (Allison 1954; Motulsky 1964). For this reason, the sickle cell anemia gene is maintained in high frequencies in some areas of Africa and India. There are many other genes in which selection was detected statistically or inferred by inter-specific comparison of amino acid sequences (see Chapters 2–6). Yet, they represent a small proportion of genes contained in the genomes of many organisms. This is mainly due to the difficulty of estimating selection coefficients because the fitnesses of organisms are controlled by a large number of genes and many environmental factors, and the differences in mean fitness between different genotypes are often very small.

What should we do then with natural selection? I certainly do not recommend pursuing studies aimed at the estimation of small magnitudes of selection coefficients, which would vary from generation to generation. Actually, what is important in the study of evolution is to know how different species (or individuals) with different phenotypes have evolved. For example, the human and the chimpanzee are the closest relatives among primate species, but the phenotypic difference is enormous, at least in our eyes. This difference is certainly caused by fixation of different mutations in the two species, whether the fixation occurred by natural selection or random genetic factors. Furthermore, the phenotypic difference can be studied at the molecular level at least theoretically. For this purpose, it does not matter whether the fixation of mutant alleles is caused by natural selection or random factors. If we take this approach, we can avoid the question of natural selection, of which the study has proved to be so difficult.

In practice, of course, it would not be easy to clarify the molecular basis of the phenotypic difference between the human and the chimpanzee. There are a large number of genes involved, and these genes interact with one another in a complicated way. However, this approach can be used for closely related populations or species. In the case of marine and freshwater sticklebacks, we have a reasonably good understanding of the molecular basis of the presence and absence of hind-fins (Chapter 6).

Some evolutionists still cling to panselectionism and state that natural selection is the only process by which species adapt to their environment. In the genomic era, this interpretation is clearly incorrect because without mutation no adaptation can occur and adaptation may occur by non-selective genetic processes, as discussed in the previous chapters. We now know that all mutations occur by some kind of DNA changes and these changes are identifiable at the molecular level at least theoretically. Natural selection is a complicated process because of environmental effects and gene interaction, but its essential role is to save advantageous mutations and eliminate deleterious mutations as mentioned repeatedly.

The importance of mutation in evolution is obvious if we compare the imaginary world in which no mutation occurs and the world in which no selection occurs. In the mutation-less world, evolution cannot occur because no new variation is generated. In this mutation-less world, even life cannot originate (see Section 9.4). In the selection-less world, however, evolution can happen because new genetic variants can be created, and if we consider that only individuals which have a harmonious genomic structure can survive, evolution continues to occur by constraint-breaking mutations (see Section 9.4) without struggle for existence. Of course, this does not mean that natural selection is unimportant in the real world. Although it is difficult to measure the extent of natural selection in the world, many examples of exquisite adaptation of organisms discussed in the preceding chapter appears to have evolved by the aid of natural selection.

9.2. Random Factors and Gene Co-Option in Evolution

As was discussed in Chapter 2, the theory of random genetic drift was established in the 1930s. However, most evolutionists appear to believe that adaptive evolution occurs only by natural selection without drift. This is particularly so with the evolution of complex characters such as altruism and parental care of offspring. However, recent molecular studies of phenotypic evolution do not support this view and indicate that phenotypic evolution is affected by various random factors such as mutation, gene duplication, recombination, chromosomal rearrangement, gene transposition, genetic drift, etc. These random factors are expected to generate neutral components of phenotypic evolution. We have also seen that the initial stage of evolution by gene co-option is influenced by random factors.

An interesting class of neutral gene co-option is the gene sharing first discovered by Piatigorsky et al. (1988). Gene sharing refers to the case where the same protein shows two different functions. For example, the protein argininosuccinate lyase is known to function as an enzyme in many different tissues of vertebrates, but it is also used as the lens structural protein δ-crystalline that is water-soluble and maintains lens transparency. Another protein, known as the small heat-shock protein, is also used for the lens structural protein as well as the protein that regulates the folding and unfolding of other proteins in vertebrates. Similarly, other crystallines (e.g. β- and γ-crystallines) are known to be bifunctional. These bifunctional proteins are not duplicate genes, but they are produced by changes in gene expression regulation. Therefore, there are no amino acid differences between the proteins expressed in different tissues.

In recent years many such examples have been discovered (Piatigorsky 2007), and these genes are often called moonlighting genes, because they have a primary and a secondary job (function) (Jeffery 2003). For example, cytochrome c has a primary function in energy metabolism and a secondary function of apoptosis, whereas enolase in fungi has a primary function as a glycolytic enzyme and a secondary function of mitochondrial tRNA import.

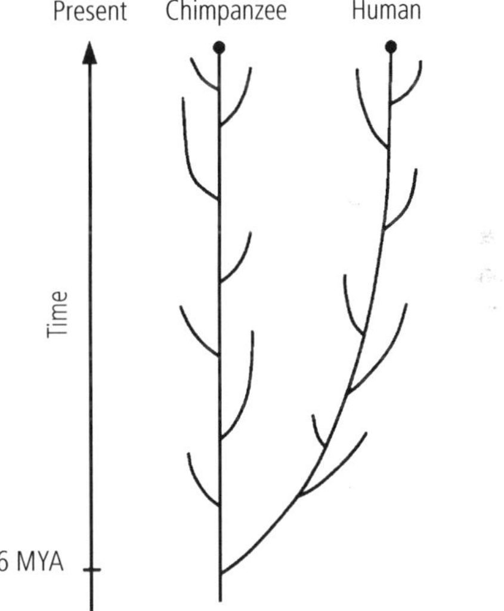

Fig. 9.1. Schematic representation of human and chimpanzee evolution. The branches derived from the human and chimpanzee lineages are the species or subspecies that have become extinct in the past. The retrospective view of evolution is depicted by the smooth lines aiming at the current morphologies of the two species. It is assumed that the morphology of chimpanzees is similar to that of the common ancestor of the two species, whereas the human morphology has changed substantially. In the prospective view, the future evolution is unpredictable, and therefore the evolutionary process might have deviated considerably from the smooth lines.

In Chapter 8 we described an evolutionary scenario of the honeybee caste system proposed by E.O. Wilson and his colleagues. This scenario suggests that some steps of the evolution of eusociality may have been achieved by co-option (recruitment) of redundant genes or gene regulatory systems. It would be interesting if some steps of evolution of a complex character such as eusociality could be explained by essentially a neutral process.

In real populations, any phenotypic character is controlled by a large number of genes and environmental factors, and most natural populations contain a large amount of quantitative genetic variation, of which the inheritance is subject to genetic and environmental random errors. In the past, evolutionists have believed that the most important random factor is genetic drift, but in the previous chapters we have seen that there are many other random factors that affect evolution.

9.3. Retrospective and Prospective Studies of Evolution

Every biologist is aware that any evolutionary theory should be free of teleology if the theory is to be scientific. In fact, this is the approach Charles Darwin took in his study of evolution, and for this reason he was so successful. However, human minds appear to be easily susceptible to teleological thinking. It is well known that the great biologist J.B.S. Haldane in the twentieth century once remarked "*Teleology is like a mistress to a biologist; he cannot live without her but he is unwilling to be seen with her in public.*" In evolutionary biology it is not uncommon to assume that the humans have evolved in a special way because they are so different even from their close relative, the chimpanzee. Therefore, we are tempted to believe that the genes controlling phenotypic characters have been subjected to positive selection more often in the human lineage than in the chimpanzee (e.g. Sabeti et al. 2006).

Lewontin (1978) stated "*The modern view of adaptation is that the external world sets certain problems that organisms need to solve and that evolution by means of natural selection is the mechanism of creating these solutions. Adaptation is the process of evolutionary change by which the organism provides a better and better solution to the problem.*" Although Lewontin is a critical evolutionist, I find some elements of teleology in this statement. Organisms evolve passively as a consequence of mutation, selection, and genomic drift (random changes of genetic materials) and never try to solve any problem actively. Any organism does not evolve to achieve any purpose by its own will.

If we study human evolution retrospectively by using the knowledge of the current morphological and physiological characters of humans and chimpanzees, we can always make a sensible story of evolution, although the story may be somewhat teleological. One such story would be that the population leading to the human lineage moved to a new habitat (grassland), whereas the chimpanzee lineage stayed in the original habitat (forest) of the human-chimpanzee common ancestor (Fig. 9.1). For this reason, many new adaptive mutations may have been fixed in the human lineage and these mutations led to the evolution of current *H. sapiens*. However, there is no reason to believe that a greater number of adaptive mutations are fixed in the human lineage than in the chimpanzee. The chimpanzee lineage also may have experienced positive selection to adapt to its own habitat, of which the climate and ecological conditions surely changed over geological time. In fact, this view is supported by microarray studies of gene expression levels between humans and chimpanzees (Khaitovich et al. 2006).

However, it would be very difficult to predict future evolution, if the evolutionary change occurs without purpose. This happens because we do not know what kinds of mutations occur in the future and how the environmental or ecological conditions will change. Note also that a species usually splits into many evolutionary lineages but many of them usually become extinct. In the evolution of the human lineage, it is known that this happened many times and only one species, *Homo sapiens,* has survived (Fig. 9.1). The cause of extinction of many sub-lineages is not known, but the main reason must be the random change of genetic material (genomic drift) and extrinsic environmental changes. If the *H. sapiens* lineage had become extinct and a lineage of Asian *Homo erectus* had survived, this world would have been very different. This indicates that evolution is opportunistic at the species level too. Let us imagine that we can go back to the time when the two populations leading to humans and chimpanzees diverged, about 6 MYA, and we are asked whether one can predict which of the populations would produce humans later. Most evolutionists would say "it is impossible." This would be the case even if the two populations are genetically differentiated considerably. In other words, although we know that mutation, selection, genetic drift, etc. are responsible for evolution, we cannot predict future evolution. Evolution occurs without purpose, and therefore it is intrinsically unpredictable.

The environment in which proto-humans lived should have affected the direction of human evolution, but the direction also must have been influenced by the type of mutations that occurred. It is often said that mutation is a random factor and cannot control evolutionary direction and that only natural selection can decide the evolutionary direction. This view is clearly based on the idea that any population contains all kinds of mutations and the evolutionary direction is determined only by natural selection. In practice, most mutations are deleterious or neutral and only a small proportion of mutations seem to be responsible for generating innovative characters. If this is the case, mutation must be important in determining evolutionary direction. Note also that most organisms are able to live in suboptimal environments.

9.4. Genomic Constraints and Constraint-Breaking Evolution

Progressive Evolution

If evolution is affected by so many random factors, why should we see that the evolutionary change of organisms appears to be so orderly and progressive when long-term evolution is considered? For example, the proportion of brain size in the body is higher in humans than in any other primate species, and it appears that the proportion has increased from New World monkeys to Old World monkeys and then to apes and humans. A similar progressive evolution of body size is observed among whale species after they entered oceans. Some neo-Darwinians like Dawkins (1997) believe that evolution occurs progressively and it is caused by positive Darwinian selection. These authors often write as though natural selection has the power of determining the future direction of evolution whereas mutation merely provides raw material for evolution. However, we have seen in the previous chapters that at the molecular level the major force of evolution has been mutation and that there is little empirical evidence for natural selection to have determined the future direction of evolution. For example, the adaptive radiation of the Antarctic notothenioid fish mentioned in Section 6.5 would not have been possible unless the complex proteins called antifreeze glycoproteins (AFGFs) were produced by mutation. Natural selection was certainly important in preserving beneficial proteins, but this was of secondary importance in evolution. Comparing simple organisms with complex organisms, Dawkins also states that the evolution of eyes has been progressive and this has occurred by natural selection. However, we have already seen that human eyes must have evolved by mutation, selection, and genetic drift but the driving force has been mutation.

A more proper way of explaining the orderly evolution of organisms is to note that the evolutionary changes of genetic materials and phenotypic characters occur almost always in a conservative fashion, but the conservative characters are occasionally subject to innovative changes and these changes result in improved phenotypic characters. To discuss this form of evolution, which may be called conservation-breaking or constraint-breaking evolution, I would like to present a brief explanation of the origin of life.

Origin of Life Without Purpose

The most fundamental property of life is metabolism and reproduction. At the present time, the genetic information for controlling metabolism and reproduction is carried by DNA in most organisms. DNA encodes RNAs and proteins and these molecules are used for carrying out metabolism and DNA replication. DNA itself is not capable of playing the roles of enzymes. By contrast, RNAs are known to work both as enzymes and self-replicators. For this reason it is now believed that in the origin of life the RNA world preceded the current DNA world (Woese 1983; Gilbert 1986).

The RNA world hypothesis proposes that life originated by using RNA molecules and this RNA world was later transformed to the current world of life based on DNAs, RNAs, and proteins. RNAs are capable of storing genetic information, like DNAs, and catalyzing biochemical processes, like enzymes. Therefore, it may have supported pre-cellular or early postcellular life before solid cellular organisms appeared. In fact, although the formation of RNA molecules in natural conditions is a complex problem, it is now generally believed that an RNA world evolved in the very early stage of life (Joyce 2002; Orgel 2004; Mortiz 2010).

It is unclear how large the first self-replicating aggregate of RNA molecules was, but if we do not accept teleology, the aggregate of RNA molecules must have been a product of a random combination of nucleotides and subsequent mutational changes of nucleotide sequences. It is therefore likely that it took hundreds of millions of years for the first successful living cell to be generated. Once a primitive living cell was produced, the efficiency of metabolism and self-replication must have improved by further mutations and natural selection. For this reason, many investigators (e.g. Wilson and Szostak 1999; Joyce 2004) have conducted experiments to understand how a particular RNA sequence spreads through the population when a large number of randomly generated sequences are provided. These investigators emphasized the importance of natural selection for generating an efficient set of RNA molecules.

However, it should be noted that an efficient RNA molecule must have been generated by addition of new nucleotides and rearrangements of different nucleotides (mutations) that are compatible with the existing sequences. In this case most new mutations were probably incompatible with the existing set of RNA molecules or a primitive genome and were therefore eliminated by purifying selection. Only occasionally, the function of the primitive genome was improved by rare beneficial mutations. Therefore, life originated without purpose as a consequence of random changes of nucleotide sequences of RNA molecules and elimination of harmful mutations. However, this does not mean that the first life was generated instantly as an appropriate combination of nucleotides. Rather it must have been a historical consequence of many mutational events that generated a specific set of compatible RNA molecules capable of enzymatic activity and self-replication. Once this set of molecules was generated, the efficiency of enzymatic function and self-replication must have improved by further rare beneficial mutations and natural selection. In this view, life originated as a consequence of a conservative evolutionary process that generated a very rare but efficient set of RNA molecules. It should be noted that the transition of the RNA world to the DNA world should also have occurred by similar constraint-breaking mutations, because natural selection does not create new variations.

Constraint-Breaking Evolution

In the above, I stated that evolution occurs without purpose. However, this does not mean that evolution occurs with random mutations only. Fixation of mutations occurs with a specific genetic background which has been developed in the long history of the species. If an organism is adapted to a particular environment, it should have a specific genetic background. Therefore, only mutations that are not incompatible with the genetic background are incorporated into the genome. It also depends on the environmental condition in which the species is located. If there is a vacant niche to which a new mutant variant is adapted, it would occupy the niche and flourish there. In other words, most genes evolve under functional constraints, so that the d_N/d_S ratio is much lower than 1, as we have seen in Chapter 4 (Fig. 4.4). This result indicates that evolution occurs by rare beneficial mutations rather than by abundant positive Darwinian selection whether we consider molecular or phenotypic characters. This same principle apparently operated even in the origin of life, as mentioned above. I would like to call this conservation-breaking or constraint-breaking evolution. In my view, this is the general principle that applies for the evolution of all organisms as well as for the origin of life.

In this view of evolution, life originated by rare mutations with a very low probability, and subsequent evolution also occurred by rare mutations without purpose, the evolutionary direction being determined by the nature of the mutation, the genetic background previously established, and the environmental condition. If this view is correct, one would expect that different evolutionary lineages diverge continuously as evolution continues, because new mutations, genetic backgrounds, and environmental condition are almost always different. This is indeed the case with the evolution of all organisms in all kingdoms. In the past, evolutionists have had a tendency to believe that the differentiation of different phyla, classes, orders, families, etc. has occurred primarily by ecological differences. I am skeptical of this view and believe that the primary factor of differentiation has been mutation, though ecological conditions should have been important for screening different mutations. If evolution occurs in this fashion and if we compare different groups of organisms retrospectively, we would expect that at least some groups (e.g. classes) of organisms should appear to have evolved in an orderly and progressive fashion even if the real mechanism has been conservative and constraint-breaking.

Genome conservation and constraint-breaking evolution are also capable of explaining the gradual increase of hybrid sterility or viability as incipient species diverge. As discussed in Chapter 6, an organism is a product of expression of conservative genes, conservative gene interaction, and conservative maintenance of genetic systems. Yet, the genomic structure of an individual gradually changes with occasional advantageous or neutral mutations, and consequently the genomic diversity among different populations would increase with time. The intrapopulational genomic variation must be determined by the mating ability of individuals within the population. However, once different populations are ecologically isolated, they would start to accumulate different mutations that are compatible for development and reproduction within populations but are incompatible between different populations. Therefore, if the ecological or geographical isolation continues for a long time, hybrid sterility or inviability will be generated. Thus, we can explain both phenotypic divergence and hybrid sterility by the same mechanism of genome conservation and constraint-breaking evolution.

As mentioned in Section 8.6, Williams (1966) stated that the random forces of evolution operated only when life originated and the subsequent evolution occurred exclusively by natural selection. My view of evolution is different, and the entire process of evolution from the origin of life to the emergence of humans can be explained by the same principle of genome conservation and constraint-breaking evolution. We can also explain the evolution of all kinds of genetic systems such as single-cell organisms, multicellular organisms, sexual reproduction, and hybrid sterility by the same principle. In this view of evolution there is no need of including teleological considerations.

9.5. Genetic Variation within Species

Although genomic evolution is conservative, we are aware that a randomly mating species contains a large amount of phenotypic variation. Any individual from any species is phenotypically different from any other individual. If we consider that phenotypic variation is caused primarily by amino acid differences, this observation indicates that the amino acid variation within a species is very large. However, this large phenotypic variation appears to be largely neutral or nearly neutral for two reasons.

First, a majority of amino acid substitutions in a protein is apparently neutral as mentioned in Chapter 4. In the case of hemoglobins and color vision genes, only about five percent of amino acid substitutions have been estimated to be important for functional changes. If the proteins controlling morphological characters follow this pattern and the major gene effect hypothesis (see Chapter 4) is applicable, it would be safe to assume that a large proportion of phenotypic variation is more or less neutral if we exclude the environmental effects. Second, if the majority of phenotypic variation is controlled by strong selection, the number of offspring born to a mating pair of male and female is expected to vary substantially. In practice, however, the number of offspring living to adulthood appears to follow the Poisson distribution in human populations when no birth control is practiced (Imaizumi et al. 1970). Furthermore, the correlation between the fertilities (number of offspring) of the parent and the offspring generations was very low in the same no-birth-control population. These observations suggest that phenotypic variation is generated largely by nonselective forces (Nei 1987, pp. 422–423).

This conclusion means that although there is an enormous amount of genetic variation within a species the fitness is nearly the same for most individuals except for those affected by deleterious mutations. In human populations some people certainly can produce a large number of offspring, but it is questionable that the fertility of an individual is inherited to the next generation.

In Chapter 2, I mentioned that at the time of neo-Darwinism there was an intense controversy over the classical and balanced theories of maintenance of genetic variability within populations. If the majority of genetic variation is due to neutral or nearly neutral mutations, the controversy is no longer very important. However, if we consider only the genetic variation maintained by selection, one may conclude that a majority of variation is due to mutation-selection balance, because most genes are subject to purifying selection as we have seen in the previous chapters. Molecular studies have revealed that there are some genes that are subject to overdominant selection (e.g. MHC and disease-resistant genes) and show a high degree of polymorphism, though overdominant loci themselves are not permanent. MHC studies have shown that these loci are often subject to birth-and-death evolution and therefore the genetic loci are subject to evolutionary changes. Yet, it is interesting to see that the hotly debated controversy over the maintenance of genetic variation in the neo-Darwinian era has finally been solved by the molecular approach and that the final conclusion is partially favorable and partially unfavorable for both schools of thought.

Another major issue that has been controversial in the past is the extent and pattern of genetic components in behavioral characters. I have argued that even in the evolution of the caste system in hymenopterans mutation has played important roles. Once a caste system evolves in an evolutionary lineage, it is expected to stay in the lineage for a long evolutionary time. However, the initial signaling pathway appears to change from lineage to lineage or occasionally from time to time in the same lineage, apparently because of mutation or genomic changes. At the present time, detailed aspects of the change remain unclear, but since we now have advanced molecular technology, we should be able to solve this complex problem in the near future.

The sociobiology in human populations is much more complicated than that in hymenopterans, because human development is affected by culture and education as well as by genetic and environmental factors. In fact, this is one of the most controversial issues in human biology, and it is part of the age-old Nature-Nurture controversy in human society (Wilson 1975, 1978; Gould and Lewontin 1979; Dawkins 1982; Lewontin et al. 1984). This controversy is expected to continue for many more centuries. However, it is now the time to study this problem at the molecular level. Fortunately, some progress in this direction is already being made (e.g. Bakermans-Kranenburg and van IJzendoorn 2006; Rutter 2006).

9.6. Niche-Filling Evolution

In Chapter 1, I introduced the concept of niche-filling evolution. Here I would like to elaborate the significance of this concept in relation to the theory of mutation-driven evolution. In Chapter 3 we indicated that Fisher's fundamental theorem of natural selection has several conceptual problems when long-term evolution is considered. One of them is that the theorem was developed under the assumption of a constant environment and whenever the environment changes, the biological meaning of the theorem becomes unclear. That is, when the environment changes, the fitnesses of different genotypes also change, and in this case one cannot predict the increase of mean fitness of the population. This is particularly so when geological upheavals occur or when a species colonizes a new environment. The second major problem is that Fisher's theorem has nothing to do with the absolute fitness of the population and the population may become extinct even if a sufficient amount of genetic variation exists within the population (Chapter 3).

These problems would disappear if we use the concept of niche-filling evolution. In this concept, evolution is regarded as a process in which a new species survives by occupying a new niche rather than by a process in which the previous species is replaced by the new one due to the struggle for existence. The former process, i.e. niche-filling evolution, may occur whenever there is a vacant niche and a new variant (mutant) form can occupy the niche. Of course, the new variant form entering the new niche may not be well adapted to the niche, and in this case subsequent mutations and natural selection may be necessary for the adaptation.

One example of this type of evolution is the adaptation of the Mexican fish *Astyanax mexicanus* to the cave condition. As mentioned earlier, the population size of this fish is so small that most mutations identified must have occurred after the fish entered the cave. In the early stage of evolution the fish apparently lost unnecessary characters such as pigmentation and the eye, but in later stages they gained new characters such as taste buds to adapt to the cave condition. These changes in phenotypic characters were generated by mutations whether they were subjected to positive selection or not. This indicates that the adaptation of a population to a given environment may occur primarily by mutation.

The Darwinian and neo-Darwinian theory of evolution usually assumes that when a new species is generated the old species (or population) is replaced by a new one because of the higher competitive ability in the struggle for existence. In my view, however, many new species may be generated when there are different ecological niches. Of course, the definition of a niche is a difficult problem as emphasized by Lewontin (1978), but here I consider a niche rather loosely and call any isolated habitat a niche.

It is well known that if different populations inhabit different niches they often lead to different species. At present, the cave fish populations of *A. mexicanus* still hybridize with the surface fish as mentioned earlier. However, if the different groups of fish live as separate populations for a long time, they are expected to become different species. Actually there are many examples of rapid speciation when different populations occupy different niches. For example, the ancestor of whales was a sister species of hippopotamus, but after they entered into the sea about 50 MYA, they found many different niches or environments and have become one of the most divergent group of species in terms of morphology and life style. The well-known mammalian radiation may also have been caused by the opening of new niches for an ancestral species of mammals around 100 MYA.

Of course, the importance of different niches in speciation is well known (Darwin 1859). However, the theory of niche-filling evolution is somewhat different from the standard theory of evolution by natural selection. Obviously, Fisher's theorem or Wright's shifting-balance theory of evolution has nothing to do with this niche-filling evolution. Yet, it is likely that the niche-filling evolution is quite common. As indicated by Lewontin (1978), there are many vacant niches which have never been exploited by organisms. This is clearly caused by the historical restriction of genetic constitution or genomic structure. Therefore, it is impossible for organisms to fill every conceivable ecological niche. As mentioned above, genetic or phenotypic evolution is intrinsically conservative. This conservative evolution would be broken up only by rare mutations or rare combinations of mutations that would create innovative characters. Evolution occurs as a passive consequence of genomic and environmental changes.

CHAPTER 10

General Summary and Conclusions

As mentioned in the preface, the purpose of this book is to present a theory of mutation-driven evolution and show that this theory can explain organismal evolution in a more logical way than the theory of selection-driven evolution. For this argument, I have considered various aspects of evolution covering works conducted during the last 150 years. Because the subject matters discussed were so wide-ranging, I first would like to summarize my discussions and then present the conclusion in a comprehensive manner.

As is well known, the essence of Darwin's theory of evolution is natural selection. He was aware of the importance of the generation of phenotypic variation on which natural selection operates and assumed that the variation was generated by correlation of growth, use and disuse, and direct action of the physical conditions of life (Darwin 1859, p. 466). We now know that these factors are not important for generating mutations, but we should appreciate his insight in understanding the basic process of evolution. The origin of genetic variation on which natural selection operates came to be known only after Thomas Morgan and his followers clarified the nature and frequency of occurrence of genic and chromosomal mutations in the 1910s to the 1930s.

Another problem in Darwin's theory of evolution was his acceptance of pangenesis and blending inheritance. This problem was indicated by Fleeming Jenkin as early as 1867, as was discussed in Chapter 1. These deficiencies in his theory were removed only after Mendelian inheritance was rediscovered in 1900. Despite these problems, Darwin proposed a basic model of evolution in terms of heritable phenotypic variation and natural selection. The only problem was that apparently because he was not sure about the cause of new heritable variations he emphasized the importance of natural selection excessively. Nevertheless, Darwin's theory of evolution was the start of the mechanistic explanation of evolution, and in this sense it was revolutionary.

In the early twentieth century a new evolutionary theory based on particulate inheritance and spontaneous mutation was proposed. Initially, Hugo de Vries (1901–1903, 1909, 1910) proposed the mutation theory of evolution, in which new species were assumed to be generated by single mutations. However, this theory was thought to be unrealistic and was soon abandoned in the 1920s. Interestingly, this theory has been revived recently and is believed to play important roles in the evolution of plants and fungi. Meanwhile, extensive studies of genic mutations with *Drosophila* and maize showed that many mutations occur at Mendelian genetic loci and that most mutations are deleterious but a small proportion of them are advantageous. This observation led Thomas Morgan to propose the mutation-selection theory of evolution or mutationism, in which genic mutation generates new phenotypic characters and natural selection saves beneficial mutations and eliminates deleterious ones.

This simple theory of evolution by genic mutation was quite popular in the 1920s and 1930s, and it is still maintained by many biologists. Generally speaking, however, the theory declined gradually as neo-Darwinism became dominant in the 1940s and 1950s. Neo-Darwinians criticized mutationism by stating that most new mutations are deleterious and therefore evolution is unlikely to occur with these mutations. They then proposed that evolution occurs when previously disadvantageous mutations become advantageous because of environmental changes. One example, which was often used to

Mutation-Driven Evolution. First Edition. Masatoshi Nei.

illustrate this idea, is the increase in frequency of the dark-color phenotype of the peppered moth *Biston betularia* that occurred in the industrial area of England in the nineteenth century.

However, there are several problems with the frequency change of the dark-color phenotype of *B. betularia*. First, the frequency change is affected by long-range migration, and the dark-color phenotype tends to move to the dark geographical area. Therefore, the allele frequency change does not necessarily represent the effect of natural selection. Second, the real mechanism of natural selection at the dark-color locus is still unclear and controversial. Third, the molecular basis of the phenotypic difference between the dark-color and light-color phenotypes remains unclear. It is possible that the difference is caused by several mutations or even a gene complex locus. This is true with many phenotypic polymorphisms studied in the era of mutationism and neo-Darwinism.

Nevertheless, neo-Darwinians introduced a significant change in the study of evolution. That is, they introduced mathematical methods for studying the genetic change of populations. Because the evolutionary change of a population is a complicated and slow process, it is very difficult to visualize a long-term change intuitively. Therefore, the mathematical theories introduced by population geneticists R. A. Fisher (1930), Sewall Wright (1932), and J.B.S. Haldane (1932) were very useful for understanding the long-term evolution of populations or species. In this era both deterministic and stochastic theories of allele frequency changes within populations were developed, and some of the theories are still used in the study of molecular evolution.

The paradigm of neo-Darwinism was panselectionism represented by R. A. Fisher (1930). Mathematical theories were also developed for dealing with the evolution of sex (Maynard Smith 1978), altruism (Hamilton 1964), and complex characters (Williams 1966), etc., and these theories were based on the idea that evolution occurs almost exclusively by natural selection and population size is infinitely large. The theories were developed for solidifying intuitive arguments about evolutionary changes rather than for testing any evolutionary hypothesis.

In the twentieth century, a large number of investigators attempted to obtain empirical evidence that natural selection actually occurs in the wild population. They found abundant evidence that natural populations contain many deleterious mutations that are kept at low frequency. This finding clearly showed that disadvantageous alleles are eliminated by purifying selection and they are maintained in the population by the balance between mutation and selection with a nearly constant frequency. However, it was much harder to prove that any advantageous mutations increase in frequency and replace previously abundant alleles in the population. There were several reasons for this difficulty. First, human life is too short to observe the long-term change of allele frequencies. Second, because the fitnesses of genotypes depend on environmental factors that vary every generation, it was difficult to estimate the selection coefficients for any genotypes. Third, natural selection occurs as a process of differential rates of birth and death of individuals, and therefore it is impossible to isolate and estimate the effect of natural selection due to a single locus or interaction of different loci. Fourth, the selective advantage of an allele usually depends on climatic and geological conditions, but because these conditions vary with time, the direction of natural selection may change during evolution particularly when big geological upheavals such as the Cretaceous mass extinction occur.

For the above reasons, natural selection is often inferred from the adaptability of organisms to their environments. In this case it was difficult to measure the selection coefficient for any allelic substitution, but it is possible to infer the selective advantage of one allele over the other(s). Nevertheless, there is always some danger of making an erroneous inference, because it is easy to draw up any evolutionary story retrospectively. In this case it is important to consider the possibility of non-selective evolution as well as selective evolution.

With respect to complex characters such as intelligence, altruism, and aggression, it is difficult to identify the genes involved and their interactions. Therefore, investigators such as Williams (1966) and Dawkins (1976) have assumed that the evolution of these characters occurs solely by natural selection with no effects of random factors such as mutation, gene duplication, genetic drift, epigenetics, etc. They then produced various evolutionary scenarios by intuitive arguments. Because these scenarios are not based on solid population genetics theory or molecular biology, they are not testable scientifically. In practice, there is plenty of evidence that even the evolution of such complex characters as sex determination is affected considerably by genetic drift, genomic drift, and epigenetics.

In the era of neo-Darwinism, evolution of a population was studied by using polymorphic alleles controlling phenotypic characters, but the molecular basis of allelic differences was not known. At the molecular level, the allelic difference could be due to a single nucleotide substitution, deletion/insertion, or haplotype differences in complex loci such as the Rh phenotypes. In other words, the mutational process of evolution was a black box in the era of neo-Darwinism. Therefore, the investigators could not identify homologous genes between different species, and evolutionary studies were confined mainly to genetic variation within species.

Partly for this reason, one of the most important issues of evolution at the time of neo-Darwinism was to understand the mechanism of maintenance of genetic variation within species. There were two main different theories for explaining the maintenance of genetic variation as spelled out by Dobzhansky (1955), i.e. the classical theory and the balance theory. The classical theory was essentially the view maintained by Thomas Morgan and Hermann Muller and asserted that most genetic variations within species are caused by the balance between weakly deleterious mutations and purifying selection. By contrast, the balance theory assumed that most genetic variations are maintained by overdominant selection or heterozygote advantage. This view was advocated by Dobzhansky (1951), Ford (1964, 1975), Lewontin (1974), and others. At this time, however, it was difficult to determine whether a locus is heterozygous or homozygous for many genes, and therefore the arguments were always controversial. Only after the electrophoretic technique for detecting protein polymorphisms was introduced in the 1960s could we study the problem empirically. When protein polymorphism data became available for a large number of loci from various species, the average mutation rate for protein loci could also be estimated from the rate of amino acid substitution. These data were then used for testing the neutral theory of molecular evolution, taking into account the population size (Kimura 1968b).

The results obtained were quite interesting, because the observed level of heterozygosity was much lower than the expected value under the overdominance theory. Actually, various statistical studies of protein polymorphism showed that the pattern of protein polymorphism is consistent with the neutral theory. For these reasons, the balance theory of genetic polymorphism was gradually rejected by many evolutionists. This rejection coincided with the general acceptance of the neutral theory of molecular evolution proposed by the study of long-term evolution of protein molecules (Kimura 1983; Nei 1983).

As mentioned earlier, the theoretical foundation of neo-Darwinism was laid out mainly by Fisher (1930), Wright (1931, 1932), and Haldane (1932), and in this work the mathematical approach played important roles. Similar mathematical approaches were used in the subsequent development of population genetics (e.g. Malecot 1948; Wright 1968; Crow and Kimura 1970). The mathematical principles developed were then used by empirical evolutionists (e.g. Dobzhansky 1937, 1951; Simpson 1944, 1953; Stebbins 1950; Mayr 1963). However, the theory was used mainly for conceptual purposes, and it was rarely used for analyzing real data concerning the evolutionary change of any particular character or a specific evolutionary lineage. For this reason, Mayr (1963) questioned the utility of the mathematical theory for evolutionary studies. He criticized that the theory cannot explain the evolution of any specific morphological character, of which the developmental formation occurs through complicated biochemical processes. To understand the evolution of phenotypic characters, we have to know the molecular basis of development of these characters. For this reason, we had to wait for about 30 years before we could answer Mayr's question.

Study of the evolutionary change of protein and DNA molecules was initiated by a small number of protein chemists in the early 1960s (e.g. Ingram 1961; Zuckerkandl and Pauling 1962; Margoliash 1963). They first compared amino acid sequences of such proteins as hemoglobins and cytochrome c from a diverse group of organisms. They then found that the number of amino acid substitutions between two species increases roughly in proportion to the time since divergence of the two species. Here they discovered the molecular clock for the first time. They also found that amino acid changes occur more often in the functionally less important parts of the proteins than in the more important parts. Later studies showed that these conclusions apply to many other proteins, and this finding led Motoo Kimura, Jack King, and Tom Jukes to propose the neutral theory of molecular evolution. This theory claimed that natural selection plays minor roles in the evolutionary change of proteins and therefore the rate of amino acid substitution is roughly proportional to the mutation rate.

This theory was initially thought to be antagonistic to neo-Darwinism, because in the latter theory almost every genetic change is assumed to be due to natural selection and depend on environmental change. For this reason, many neo-Darwinians criticized the neutral theory. However, as more data on neutral evolution accumulated, they became less critical of the theory. Actually, many neo-Darwinians now seem to accept the neutral theory of molecular evolution but still criticize it stating that neutral changes of molecules do not affect any phenotypic evolution and therefore it is not interesting. However, recent studies suggest that neutral changes of genes or genomes often affect phenotypes and that phenotypic evolution is not always caused by natural selection.

Molecular study of evolution has also made it possible to separate the mutational change of a gene and the process of natural selection for the first time. This separation initiated a new era of the study of evolution at the DNA level. In the early days of the study of molecular evolution, the mutational change of a gene was shown to be an amino acid change in the protein encoded by the gene. It was then clear that this mutation would spread through the population if the new protein generated is selectively advantageous over the pre-existing proteins. Fixation of the mutant protein may also occur by random genetic processes if it is functionally equivalent to the pre-existing alleles. Here, natural selection is regarded as a passive process of evolution that saves advantageous mutations and eliminates deleterious ones.

Further studies of molecular evolution have shown that a large number of mutations occur every generation in a population, but many of them are eliminated by purifying selection and genetic drift. It has also been shown that a large proportion of mutations that are fixed in the population are more or less neutral and a small proportion of them are advantageous. In this view, because the majority of mutations fixed in the populations are more or less neutral, the evolution of proteins may be regarded as a neutral process. This is the essence of the neutral theory of molecular evolution.

In practice, however, the extent of purifying selection varies considerably with the type of protein. Some proteins like histone and ubiquitin are highly conserved because of strong functional constraints, whereas olfactory receptors and MHC molecules are less conserved and show rather high rates of evolutionary changes. Occasionally, the functional constraints of proteins are relaxed or enhanced in different evolutionary lineages, and in this case the rate of amino acid substitution may change. One example of the enhancement of the rate has been observed in the insulin protein in guinea pigs and related species. In this case the rate of amino acid substitution has increased several times compared with that in other mammalian species. By contrast, the rate of amino acid substitution in histones has decreased substantially from protists to vertebrates. In these proteins, therefore, the molecular clock does not work.

Note that violation of the molecular clock is generated by a temporal change of mutation rate as well. In parasitic bacteria, such as *Buchnera* species, the rate of amino acid substitution is often higher than that in non-parasitic bacteria. It appears that many parasitic bacteria have lost DNA-repair enzymes, and for this reason the mutation rate has increased. In these cases, the molecular clock often fails when individual genes are considered. One of the most extreme examples showing varying rates of evolution is the mitochondrial genes in species belonging to the plant genus *Silene,* where the rate of synonymous nucleotide substitution varies enormously among different species apparently because of the loss of RNA editing in some species. However, the examples of extreme variation in the rate of amino acid or nucleotide substitution are rare, and the average rates for thousands of genes are roughly constant per year.

Study of molecular evolution has been accelerated after a set of new techniques of DNA sequencing was introduced in 1977 (Maxam and Gilbert 1977; Sanger et al. 1977). DNA sequences are much more informative than protein sequences for the study of evolution because they provide data on the non-coding regions as well as the coding regions of DNA sequences, and the non-coding regions often include genetic elements controlling the transcription of DNAs and translation of messenger RNAs. It is also possible to study the rates of synonymous and non-synonymous nucleotide substitutions, which are useful for studying the effects of natural selection, because synonymous substitutions do not change amino acids and therefore the rate of synonymous substitution (r_S) may be regarded as an estimate of the rate of neutral nucleotide substitution. By contrast, the rate of non-synonymous nucleotide substitution that causes amino acid substitution (r_N) may be related to the rate of nucleotide substitution due to positive or negative Darwinian selection.

In fact, the ratio of $w = r_N/r_S$ (or d_N/d_S in Chapter 4) for a locus is now often used as a measure of the extent of natural selection, $w > 1$, $w = 1$, and $w < 1$ representing positive, neutral, and negative selection, respectively. It has been shown that w is less than 1 for most genes and therefore almost every gene or protein is subject to purifying selection. In other words, for a protein to function properly, a given tertiary structure is required, and any mutation that distorts the structure is eliminated. For this reason, the rate of non-synonymous substitution (r_N) is generally lower than the rate of synonymous substitution. That is, $w < 1$ for most genes, and this indicates that almost all genes are evolving under functional constraints. The extent of functional constraints is usually high for transcription factors, because they interact with DNA and many other proteins.

This observation is consistent with the classical theory of maintenance of genetic variation, which was controversial in the middle twentieth century. If most genes are under functional constraints, many mutations would be deleterious, and therefore genetic variation is maintained primarily by the mutation-selection balance. Occasionally, advantageous mutations would generate transient polymorphism, but this is not expected to generate a large amount of genetic variation. Study of molecular evolution has shown that a large proportion of amino acid substitutions are more or less neutral and therefore generate a substantial amount of neutral polymorphism in any population. Most of this polymorphism was apparently unobservable when population genetics studies were conducted by using morphological characters. Only when the electrophoretic technique was introduced could we see the variation due to amino acid changes in protein. Statistical studies of protein variation detected by electrophoresis suggested that a large portion of protein variation is caused by neutral mutation. Therefore, we can see that polymorphism data are consistent with data on long-term evolution of proteins. Recent DNA sequence data also show that DNA polymorphism is largely neutral.

However, it should be noted that neutral evolution or polymorphism is mainly caused by amino acid or nucleotide substitution in functionally unimportant sites of protein or gene sequences. Every gene has a certain proportion of functionally important nucleotide sites where nucleotide substitution occurs at a much lower rate than the neutral rate because the mutations occurring at these sites are eliminated by purifying selection. In other words, almost all genes evolve under functional constraints, and neutral evolution is observed only when unconstrained nucleotide sites are considered.

There are a small number of genes that show extensive intraspecific polymorphism. Examples are MHC and other immune system genes in vertebrates and the self-incompatibility alleles in some plants. Particularly, the polymorphism of DNA sequences at the MHC loci is known to be notoriously high. It is not uncommon that a local population contains more than 20 different alleles. In principle, this observation is in conformity with Dobzhansky's (1955) balance theory of maintenance of genetic variation. Therefore, we can say that both classical and balance theories are correct depending on the genetic loci considered. Of course, a large number of alleles maintained in MHC loci or the self-incompatibility loci in plants do not create a large amount of genetic load, because in these loci almost all individuals are heterozygous and there is not much selection caused by the differences in fitness between different heterozygotes.

Note also that a high degree of heterozygosity in these loci is actually caused by a relatively small proportion of nucleotide sites which interact with highly variable foreign antigens (in the case of MHC genes) or other ligands (e.g. self-incompatibility genes) and other parts of the genes are structurally or functionally constrained and do not show a high rate of evolutionary change. Therefore, even highly polymorphic genes actually evolve under functional constraints. This finding indicates that virtually all genes evolve under functional constraints, and this occurs because any gene interacts with many other genes. Occasionally, however, the partners of a gene may change and in this case the structural and functional requirements also may change.

The significance of gene duplication in evolution has been known since the early twentieth century, but its importance as the mutational source of genetic variation has been realized only after genomic sequences became available about a decade ago. We now know that the numbers of protein-coding genes as well as the non-coding regions of DNA in the genome have increased dramatically from simple organisms (e.g. bacteria) to complex organisms (e.g. mammals). Some parts of the non-coding regions of DNA are known to encode RNA molecules that are essential for the regulation of protein synthesis, and therefore it is important to have an increased number of non-coding regions of DNA as well as the protein-coding genes in complex organisms. Generally speaking, genome duplication or polyploidization is the most important mechanism of increasing the amount of DNA content, but tandem or segmental gene duplication also plays important roles in increasing gene numbers.

However, the number of genes and non-coding regions of DNA may also decrease under certain circumstances. This decrease occurs particularly after genome duplication or polyploidization. This happens partly because of the presence of redundant genes generated by genome duplication and partly because a new set of genes generated is not fully compatible in the developmental process. Formation of an organism is accomplished by a complex interaction of different genes in the developmental process, and in this case all the genes in the genome must be compatible with one another. In other words, a genome has to have a specific combination of genes in order to produce a viable individual, just like a gene requiring a specific nucleotide sequence to produce a functional protein. This functional constraint of a genome is called the genomic constraint or genomic conservation. An arbitrarily produced gene combination in a polyploid genome is therefore expected to be deleterious. For this reason, the genes in a newly produced polyploid must be sorted out to have a new harmonious genome, and in this process some genes may be eliminated.

Of course, the genomic constraint required for producing a healthy organism does not seem to be very stringent as in the case of the requirement for the formation of a gene sequence. Experimental data suggest that slightly different genomic sequences produce equally viable individuals and therefore different genomic structures may coexist within a population. A good example is the copy number variation of olfactory receptor genes in human populations. As shown in Chapter 5, the difference in the number of olfactory receptor genes between two individuals can be as large as 60 in human populations, but this does not seem to affect the fitness of individuals seriously. It is now well known that copy number variation exists for the entire set of genes in the genome and this variation seems to be as important as protein sequence variation for generating phenotypic variation within species.

However, although the genomic constraints are quite loose within species, the genomes of different species are generally incompatible, and for this reason interspecific hybrids are generally sterile or inviable even if the parental species are closely related. If the two species are distantly related, even mating cannot occur indicating that the genomic incompatibility is very high.

Genomic compatibility appears to be less stringent in complex organisms than in simple organisms, because the former has a higher number of genes than the latter. Therefore, the number of gene copies for a gene family may change more extensively in the former than in the latter. This type of change is often observed in chemosensory and immune system genes and is related to adaption to specific lifestyle and environmental conditions. Therefore, copy number variation within species may be useful for species evolution, but the actual relationship of variation within and between species is not always very clear. However, it is clear that the primary driving force of this type of evolution is caused by gene duplication and deletion, which are a form of mutation.

Another important role of gene duplication is to form complex genetic systems that are composed of several different multigene families. Examples are the adaptive immune system (AIS) in vertebrates and the flowering system in plants. In the case of the AIS, the important component gene families are those of the immunoglobulin, MHC, T-cell receptors, etc. Each of these gene families apparently evolved independently to serve different purposes, and then later they started to interact collaboratively. In this case different gene families were recruited or co-opted to form an integrated functional genetic system, which is now called the AIS (see Chapter 5). A similar evolutionary mechanism seems to have operated in the evolution of the flowering system in plants. Evolution often occurs by deploying extra genes that are available when needed. This type of evolution is often called "evolution by tinkering" following the idea of Jacob (1977). It may also be called "evolution by co-option" of different genes or gene families.

Even in this case, evolutionary change is generally conservative. Certainly, at the foreign peptide-binding site of MHC genes or immunoglobulin genes the rate of amino acid substitution is enhanced by positive selection, but the general structure of MHC or immunoglobulin genes remains unchanged. Therefore, from the structural or functional point of view these genes are also conserved. Only when drastic structural changes occur are the conservative natures of gene or genetic systems broken up. The basic function of the AIS or the flowering system apparently has not changed for hundreds of millions of years. However, constraint-breaking evolution has occasionally occurred, as in the case of the loss of some light-chains of immunoglobulins or the development of unusually long-term maintenance (about 50 million years) of heavy-chain variable region gene polymorphism in rabbit immunoglobulin. It is interesting to note that many genetic systems, including meiosis and mitosis, are highly conserved and that constraint-breaking mutations or innovative mutations occur very rarely. This view is similar to constraint-breaking evolution that applies to protein or DNA sequence evolution.

Gene co-option occurs in several different ways in addition to the formation of new genetic systems mentioned above. One interesting case is the recruitment of one functional gene into another function. Some enzyme genes which are used primarily for catalytic function in various tissues have been recruited for the formation of eye crystallins in vertebrates. In this case the same gene is used for two different functions, and for that reason this type of gene co-option is called "gene-sharing" or "moonlighting." This co-option happens because many proteins are bifunctional or multifunctional (Piatigorsky 2007). In another case of gene co-option a trypsinogen protease has been structurally modified and is now used as an antifreeze protein in teleost fish of the Antarctic Ocean. Furthermore, it is known that a large number of chloroplast genes have been transferred to the host nucleus in some plants, and about half of them are now used for various biochemical functions of the host plants.

Another important factor in organismal evolution is changes in the pattern of gene expression. DNA sequences in the genome cannot do anything unless they are expressed properly in the process of development and form various phenotypic characters. Study of the molecular basis of morphogenesis was initiated only about three decades ago, and many important questions in this area remain unanswered. However, we now know the basic principles of developmental biology and how to investigate phenotypic evolution by using these principles. The importance of gene regulation in phenotypic evolution is clear if we consider the morphological differences between queen and worker honeybees. Both queen and worker bees are female, but the queen has a larger body size than the worker and produces abundant offspring. By contrast, worker bees are sterile and take care of the queen and raise the queen's offspring. Whether a female becomes a queen or a worker is determined by the amount of royal jelly given during embryogenesis. When the amount of royal jelly given is high, a queen is produced, but otherwise worker bees are born. A recent study has shown that royal jelly contains a protein, Royalactin, and this protein has an epigenetic effect and initiates the development of a queen. Therefore, the presence or absence of this protein determines whether an embryo develops into a queen or a worker bee. We now know that there are many examples in which the presence or absence of a particular protein generates substantial phenotypic differences. Phenotypic differentiation can also be generated by the differences in non-coding regions of DNA.

In general, the amount of a protein produced is controlled by other proteins including transcription factors, non-coding regions of DNA (e.g. *cis*-regulatory elements), and RNA molecules encoded by non-coding regions of DNA. The last group of regulatory elements includes microRNAs (miRNAs), small nuclear RNAs, retrotransposons, etc., and plays an important regulatory function, though the detail is still unclear. Formation of phenotypic characters is also controlled by various biochemical pathways, genetic regulatory networks, and other protein-protein interactions. Epigenetic control of development is very important, but currently it is not well understood. Furthermore, alternative splicing of introns also plays an important role in generating polymorphic proteins (isoforms) from the same locus. These polymorphic proteins (or presence and absence of a protein) are sometimes used for determining different developmental pathways, as in the case of sex determination in insects.

In the preceding chapters, we discussed various forms of evolution by mutation and selection. The conclusions we have reached may be summarized as follows. (1) Mutation is the source of all genetic variation upon which any form of evolution is dependent. Mutation is the change of genomic structure and includes nucleotide substitution, insertion/deletion, segmental gene duplication, genomic duplication, changes in gene regulatory systems, transpositions of genes, horizontal gene transfer, symbiosis, etc. (2) Natural selection is for saving advantageous mutations and eliminating harmful mutations. Selective advantage of a mutation is determined by the type of DNA change, and therefore natural selection is an evolutionary process initiated by mutation. It does not have any creative power unlike the statements made by some authors. However, selective advantage of a mutation is also dependent on the set of other genes and the environmental conditions, the latter varying from generation to generation. For this reason, it is very difficult to study the extent of natural selection in wild populations. (3) Evolution is a process of increase or decrease of organismal complexity and enhancement of phenotypic diversity among different species. It may or may not be associated with the increase of fitnesses (number of offspring per individual), and therefore evolution can occur by neutral genetic processes such as gene duplication and gene co-option as well as by natural selection. (4) A gene is not a random combination of nucleotides but a very specific arrangement of nucleotides that encodes a biochemically functional protein or RNA molecule. Because of this functional constraint, most mutations occurring in a gene are deleterious and eliminated by purifying selection. (5) For a gene to have a new function, constraint-breaking mutations caused by new combinations of harmonious genes and gene sequences are necessary. These mutations occur with a low frequency at functionally important sites. A gene cannot have any function without having interaction with other genes. Therefore, constraint-breaking mutation may be controlled by many gene loci. (6) A genome is an integrated and conserved set of genes that is capable of producing healthy organisms. The innovational change of phenotypic characters is generated when constraint-breaking mutations occur at the genomic level. There is a considerable degree of flexibility in genomic constraint so that diploid individuals with two different genomes can survive and reproduce without trouble within a species. This flexibility appears to generate a large amount of neutral variation in phenotypic characters. However, if two different populations are isolated for a long evolutionary time, interpopulational hybrids become inviable or sterile because of genomic incompatibility. This hybrid weakness occurs because the genomes of two different populations evolve independently and therefore the compatibility of genes between different populations gradually declines. No positive selection is necessary for the establishment of hybrid sterility. (7) Although any organism lives under ecological constraints, such constraints are not usually very strong. Therefore, most organisms can live in a range of ecological niches, which may be called the ecological survival range. For this reason, a species may flourish easily in a new territory to which it was transferred. (8) Evolution occurs primarily as a result of constraint-breaking mutations rather than a result of the struggle for existence. If a species moves to a new habitat (e.g. marine habitat to land), a radiational speciation may occur because of relaxation of purifying selection and some advantageous mutations for different new territories.

The above conclusions were derived primarily from recent molecular studies of evolution. Previously, a mutational process was a black box as mentioned earlier, and it was not clear whether two different alleles at a locus are derived by a single mutational change or several changes. In the case of phenotypic characters, the molecular basis of allelic differences is not always clear because they are often controlled by many loci. It is now possible to identify a major gene locus and study the allelic difference at the molecular level. For example, the PTC bitter taste polymorphism is controlled by two alleles in humans and chimpanzees, and previously the polymorphism was thought to be controlled by the same set of alleles. It was then conjectured that the polymorphism had been maintained by overdominant selection for a long time. Recent molecular studies have shown that the alleles in the two species are not the same and are derived by independent mutations in each species. This has made the overdominance hypothesis questionable.

The above example shows that clarification of the molecular basis of mutational events is important in the study of natural selection as well. As discussed in Chapters 3 and 8, the study of natural selection in wild populations is very difficult, and the conclusions derived have been based at least partially on speculative arguments. However, what we are really interested in is not natural selection but the genetic difference between two genotypes or between two different species. Why are humans and chimpanzees so different phenotypically? What is the molecular basis of the presence and absence of the rear fins in stickleback fish? Because we now have the proper molecular techniques, we should be able to answer these questions.

In this book I have emphasized the importance of mutation in evolution but have not excluded the role of natural selection. One might therefore argue that the theory presented here is essentially the same as Charles Darwin's theory of evolution because he also considered both factors. This is true, but Darwin overly emphasized the importance of selection apparently because he did not know how new variations are generated. For this reason, his theory is now widely known as the theory of natural selection. In fact, as in the case of neo-Darwinian theory, he assumed that natural populations contain all kinds of heritable variation, so that the only evolutionary force necessary is natural selection.

The neo-Darwinian view is still maintained by a majority of evolutionary biologists (e.g. Futuyma 2005; Ayala 2007; Bell et al. 2009). My view of evolution is different from neo-Darwinism. I emphasize the importance of knowing the molecular basis of mutation and the molecular mechanism of selective advantage of a new mutation. At the present time, many investigators are conducting statistical analyses of natural selection in the hope of identifying important genes that may distinguish between different species. In fact, thousands of such studies have been done to understand the morphological differences between humans and chimpanzees, and thousands of potentially relevant genes have been reported. Because we now have the complete genomic sequences for both humans and chimpanzees, it is not difficult to conduct such statistical studies for all protein loci under various assumptions. However, these studies have made little advance in our knowledge of the cause of phenotypic differences between the two species.

Neomutationism or the theory of mutation-driven evolution is different from the classical mutationism, because it covers not only genic mutations but also all kinds of genomic change including genome duplication. In neomutationism the molecular study of mutational change as well as the selective advantage of new mutations are emphasized. Therefore, the cause of mutation is no longer treated as a black box. For these reasons, neomutationism or the theory of mutation-driven evolution is applicable for much wider biological situations than classical mutationism and at the same time demands a more sophisticated molecular approach.

In this book I have emphasized the importance of using the developmental approach in the study of phenotypic evolution. This approach may be more time-consuming than statistical studies, but it will give the ultimate answer to our question. Of course, this view is not new and has been expressed by many developmental biologists. However, evolutionary biologists have been slow in adopting this approach partly because evolutionists have been accustomed to the idea of the importance of natural selection and partly because experimental studies of developmental processes are more difficult to do than statistical studies.

Historically, evolutionary biologists have exercised highly speculative arguments because the long-term evolutionary changes of populations cannot be observed in our lifetime. Speculative arguments are particularly common in the study of the evolution of behavioral characters partly because identification of the genes controlling these characters is difficult. However, recent progress in developmental biology has changed this situation. As discussed in the preceding chapter, even the gene for initiating the queen caste in honeybees has been identified. The developmental biology of the formation of the castes of queens, soldiers, and minor workers in an ant genus (*Pheidole*) of *Hymenoptera* has also been clarified. Therefore, the currently mysterious caste formation in hymenopteran insects may soon be clarified.

Vertebrate species have various forms of sex determination. Placental mammals and some teleost fishes have the XY (male)/XX (female) system, and birds have the ZW (female)/ZZ (male) system. In some reptile species males and females are determined by the temperature during the incubation period, whereas many fish species do not show any sex chromosomes, and sex is determined by a small number of genes. Furthermore, the XY/XX and the ZW/ZZ system co-exist as polymorphism in some frog species. Yet, the molecular basis of sex determination is more fundamental than the chromosomal basis and shows the evolution of sex more straightforwardly. Here we can see the power of molecular biology. The molecular basis of the evolution of animal eyes is much simpler than expected from the morphological variation of eyes in animals. Of course, the molecular study of phenotypic evolution has just begun. At the present time, we have little idea about the evolution of the mammalian brain or even less complicated characters such as parental care in some mammals. However, the evolution of these characters will be eventually clarified at the molecular level.

In the past there have been many mathematical studies of evolution to find some general rules that might apply for a large group of organisms. R. A. Fisher's fundamental theorem is a good example. However, the principles of biology are different from those of physics. In physics, Newton's laws of motion and gravity can explain the movement of all stars in the universe and all bodies on earth using mathematical terms. (A more accurate explanation is given by Einstein's relativity theory.) In biology, the fundamental processes are metabolism and reproduction, and these processes are explained by the principles of biochemistry and molecular biology. This does not mean that the mathematical approach is unnecessary. Actually, mathematical or statistical approaches are often helpful as long as the question and the model used are biologically meaningful. This is clear from the recent success of phylogenetic analysis of different species and different gene families (e.g. Swofford et al. 1996; Nei and Kumar 2000; Felsenstein 2004). However, we should be very cautious about mathematical work when there are no genetic data.

In the study of phenotypic evolution it is important to realize that there are two different evolutionary forces operating at the genomic level. One is the genome-conservation force that maintains the developmental integrity of genes within individuals and the reproductive unity of individuals within species or populations. This force occurs because for a healthy individual to be produced a beneficial interaction of a large number of genes is required and to maintain the reproductive unity of a population all individuals must be mutually compatible with respect to mating ability.

The other evolutionary force is the genomic diversification of different species. This occurs because many constraint-breaking mutations are species-specific and these mutations contribute to the diversification of different species. This would also promote the niche-filling evolution mentioned previously. Of course, if the environmental condition is the same for different species, these species may show some convergent evolution. For example, many meat-eating carnivores and blood-sucking bats are known to have lost sweet taste receptor genes (Jiang et al. 2012; Zhao et al. 2012).

In general, however, convergent evolution is rare, and the force of genomic diversification has been the main factor for generating extensive biodiversity in all kingdoms of organisms. Genomic diversification also often generates new ecological niches that are habitable for new species. This would increase the number of species exponentially as the genomic diversity expands, and this diversification of biodiversity is caused primarily by various forms of mutations occurring in the conservative genome. In other words, genomic conservation and constraint-breaking mutation is the ultimate source of all biological innovations and the enormous amount of biodiversity in this world. In this view of evolution there is no need of considering teleological elements.

APPENDIX

Mathematical Notes

A. Allele Frequency Changes Due to Natural Selection

Deterministic Models for Allele Frequency Changes

We consider a locus with two alleles, A_1 and A_2, in a large randomly mating diploid population and denote the frequencies of alleles A_1 and A_2 by x and y, respectively, where $y = 1 - x$. We also denote the relative fitnesses of genotypes A_1A_1, A_1A_2, and A_2A_2 by w_{11}, w_{12}, and w_{22}, respectively, as shown below.

Genotype	A_1A_1	A_1A_2	A_2A_2
Frequency	x^2	$2xy$	y^2
Fitness	w_{11}	w_{12}	w_{22}

In this case the frequency of allele A_1 in the next generation (x') is given by

$$x' = [x^2 w_{11} + (1/2)2xyw_{12}]/\bar{w}$$
$$= x(xw_{11} + yw_{12})/\bar{w}$$

where $\bar{w} = x^2 w_{11} + 2xyw_{12} + y^2 w_{22}$ is the mean fitness of the population. Therefore, the amount of change in allele frequency per generation becomes

$$\Delta x = x' - x$$

$$= x(1-x)[x(w_{11} - w_{12}) + (1-x)(w_{12} - w_{22})]/\bar{w} \qquad \text{(A1)}$$

Let us consider some special cases.

(1) Semidominant alleles ($w_{11} = 1$, $w_{12} = 1 - s$, $w_{22} = 1 - 2s$)

$$\Delta x = sx(1-x)/[1 - 2s(1-x)] \qquad \text{(A2)}$$

(2) Dominant advantageous allele ($w_{11} = w_{12} = 1$, $w_{22} = 1 - s$)

$$\Delta x = sx(1-x)^2/[1 - s(1-x)^2] \qquad \text{(A3)}$$

(3) Recessive advantageous allele ($w_{11} = 1$, $w_{12} = w_{22} = 1 - s$)

$$\Delta x = sx^2(1-x)/[1 - s(1-x^2)] \qquad \text{(A4)}$$

(4) Overdominant alleles ($w_{11} = 1 - s$, $w_{12} = 1$, $w_{22} = 1 - t$)

$$\Delta x = x(1-x)[t - (s+t)x]/(1 - sx^2 - t(1-x)^2) \qquad \text{(A5)}$$

Equilibrium Frequencies Due to Mutation and Selection

Equations (A1) to (A4) indicate that allele frequency x always increases and therefore the frequency eventually reaches 1, whatever the initial frequency is. In reality, however, allele A_2 may not be completely lost from the population because the A_1 A_2 mutation becomes significant when the frequency of A_2 becomes low. In this case allele A_2 may be maintained at a low frequency. The equilibrium frequency of allele A_2 caused by this mutation-selection balance can be obtained by considering the decrease of y by selection and the increase by mutation.

Let us consider a recessive deleterious mutation (A_2) with fitnesses $w_{11} = w_{12} = 1$ and $w_{22} = 1 - s$. In this case the frequency (y) of A_2 declines every generation by $\Delta y = -\Delta x = -sy^2(1-y)/(1-sy^2)$ from Equation (A3), which is approximately $-sy^2$ because y is small. However, the frequency of A_2 increases by mutation at a rate of $u(1-y) \approx u$ per generation. Therefore, the total change of y per generation (Δy) is $u - sy^2$. At equilibrium, this should be 0, so that the equilibrium frequency (y) of A_2 is given by

$$\hat{y} = \sqrt{u/s} \qquad \text{(A6)}$$

This formula has been used extensively to estimate the mutation rate for recessive genetic diseases in human populations.

In the case of a dominant deleterious mutation, selection occurs primarily in heterozygous conditions when y is small, and the frequency of mutant homozygotes is negligibly small. Therefore, the fitnesses and frequencies of A_1A_1, A_1A_2, and A_2A_2 may be written as follows.

Mutation-Driven Evolution. First Edition. Masatoshi Nei.

Genotype	A_1A_1	A_1A_2	A_2A_2
Fitness	1	$1 - h$	$1 - s$
Frequency	$1 - 2y$	$2y$	-

where $s \gg h \gg 0$. Therefore, the amount of change in y by selection is $-hy$ approximately because the frequency of A_2A_2 is virtually 0. At equilibrium this is balanced with the gain in y by mutation $u(1 - y) \approx u$, so that $u = hy$. The equilibrium frequency of y is then given by

$$\hat{y} = u/h \tag{A7}$$

Equation (A5) shows that an overdominant mutation (A_2) increases or decreases depending on the s and t values, but its frequency (y) eventually reaches an equilibrium value, which is given by

$$\hat{y} = t/(s+t) \tag{A8}$$

B. Allele Frequency Distributions under the Infinite-Site Model

The extent of nucleotide variation in the protein-coding region or the noncoding region of DNA can be studied by examining the number or the relative frequency of polymorphic sites with a given allele frequency (x) of the mutant allele in the population under the assumption of irreversible mutation (infinite-site model). Here we consider a DNA region consisting of n nucleotide sites and assume that mutation occurs from the ancestral nucleotide (say A) to a mutant allele with a frequency of μ per generation per nucleotide site. The total mutation rate per DNA region (v) is then $n\mu$. Therefore, if N is the effective population size, the total number of mutations that arise every generation in the population is $2Nn\mu = 2Nv$.

We now want to know the expected number of nucleotide sites where the mutant frequency is x. This distribution, $\Phi(x)$, is given by

$$\Phi(x) = \frac{2v}{V_{\delta x}G(x)} \int_0^1 G(z)dz \Big/ \int_0^1 G(z)dz \tag{B1}$$

where

$$G(x) = \exp[-\int (2M_{\delta x} / V_{\delta x})dx]$$

Here $M_{\delta x}$ and $V_{\delta x}$ are the mean and variance of the change in x per generation in diffusion approximations. This equation is due to Kimura (1964), but an elementary derivation is given by Nei (1975, pp. 119–121). When all mutations are neutral, we have

$$\Phi(x) = 4Nv/x \tag{B2}$$

(Wright 1938a). In general, it is difficult to have an analytical solution for $\Phi(x)$. In this case, however, it is possible to obtain $\Phi(x)$ by numerical integration. The results given in Fig. 2.5 are obtained in this way.

C. Temporal Fluctuation of Selection Coefficients

General Comments

For studying the effect of temporal fluctuation of a selection coefficient, two different mathematical approaches have been used. One is to use Avery's (1977) discrete time model and assume that the selection coefficient varies at random from generation to generation, and the other is to consider the competitive selection model by Mather (1969) and Nei (1971) to incorporate the fluctuation of selection coefficients under the assumption that the mean fitness $\bar{w}$ is equal to 1 every generation (Nei and Yokoyama 1976; Takahata and Kimura 1979). (The adult population size is assumed to be the same every generation because of the limited carrying capacity of the environment.)

In the first approach we consider three possible genotypes for a pair of alleles A_1 and A_2 and assume that the fitnesses for genotypes A_1A_1, A_1A_2, and A_2A_2 are given by $1 + s$, 1, and $1 - s$, respectively. We then assume that s fluctuates at random from generation to generation with mean $\bar{s}$ and variance V_s. In diffusion approximations to the discrete time processes it is customary to assume that N adults in the population produce a large number of offspring, selection acts on the offspring deterministically, and then a random sample of N individuals are drawn to form the adults for the next generation. Mathematically, this stochastic process of allele frequency change can be treated by the diffusion approximations if the mean and variance of allele frequency changes are derived (Crow and Kimura 1970; Nei 1975; Ewens 2004).

Let us denote the frequency of allele A_1 in an adult population of N individuals by x and assume that random mating occurs among the adult individuals. The allele frequency change of A_1 by natural selection is given by

$$\Delta x = sx(1-x)/[1-s(1-2x)] \tag{C1}$$

When s is small, this can be approximated by

$$\Delta x \approx sx(1-x) + s^2x(1-x)(1-2x) \tag{C2}$$

Therefore, the expectation of Δx over generations is given by

$$M_{\delta x} = E(\Delta x) = \bar{s}x(1-x) + V_s x(1-2x) \tag{C3}$$

where $\bar{s}$ and V_s are the mean and the variance of s respectively (Avery 1977; Gillespie 1991). By contrast, the variance of Δx is given by

$$V_\delta = E[(\Delta x)^2] = V_s x^2(1-x)^2 + x(1-x)/(2N) \qquad \text{(C4)}$$

approximately, where the term $x(1-x)/(2N)$ is the sampling error due to genetic drift (Avery 1977). Equations (C3) and (C4) can be used to study the allele frequency distribution, the probability of fixation of a mutant allele in the population, etc. (Crow and Kimura 1970).

When $\bar{s} = 0$, there is no directional selective force biologically. However, $M_{\delta x} = V_s x(1 - 2x)$ in (C3) is positive when $x < 0.5$ but negative when $x > 0.5$. Therefore, the random fluctuation of s is expected to generate a form of stabilizing selection similar to overdominant selection even if $\bar{s} = 0$. This occurs because heterozygotes have a constant fitness and the fitnesses of homozygotes have fluctuating selection coefficients. The above formulation has been taken to mean that the formula $M_{\delta x} = 0$ used by Wright (1948a) and Kimura (1954) is incorrect. However, this is due to a mathematical artifact because a slightly different modeling of selection gives different results. That is, if we denote the fitnesses of A_1A_1, A_1A_2, and A_2A_2 by $1 + 2s$, $1 + s$, and 1, respectively, then $M_{\delta x}$ becomes $-2V_s x^2(1 - x)$ when $\bar{s} = 0$. Therefore, allele A_1 behaves as though it were deleterious allele (Nei and Yokoyama 1976).

By contrast, if we denote the fitnesses of A_1A_1, A_1A_2, A_2A_2 by 1, $1 - s$, and $1 - 2s$, we have $M_{\delta x} = 2V_s x(1 - x)^2$. Therefore, allele A_1 now behaves as though it were an advantageous allele. These results are certainly unreasonable because the average fitnesses of three genotypes are equal to 1 when $\bar{s} = 0$.They are also inconsistent with Wright's (1948a) original idea that fluctuation of s is a cause of noise in allele frequency changes.

What is wrong with this modeling? One problem is that in this modeling $\bar{w}$ is dependent on the genotype frequencies and selection coefficient, and therefore the population size after selection varies with x and s in the deterministic treatment. In reality, however, the adult population size after selection remains nearly constant because population size is regulated due to the limited carrying capacity of the environment, and natural selection often occurs by competitive selection. Mathematical models based on this idea have been developed by Mather (1969) and Nei (1971). In this model, $\bar{w}$ is always 1, and the change in allele frequency per generation is given by

$$\Delta x = \bar{s}x(1-x) \qquad \text{(C5)}$$

Therefore, the mean and variance of Δx in diffusion approximations are given by

$$M_{\delta x} = \bar{s}x(1-x) \qquad \text{(C6)}$$

$$V_{\delta x} = V_s x^2(1-x)^2 + x(1-x)/(2N) \qquad \text{(C7)}$$

(Nei and Yokoyama 1976). We call this model the competitive selection model and the model leading to (C3) the stabilizing selection model.

D. Artificial Selection for Quantitative Characters

Quantitative characters usually show a normal distribution (see Fig. 2.10). The variance of a quantitative character is composed of genetic variance (V_G) and environmental variance (V_E), and the ratio $V_G/(V_G + V_E)$ is called heritability (h^2). Suppose that we measure a quantitative character, say the number of abdominal bristles in *Drosophila melanogaster*, for a large number of adult individuals and choose the upper 5 percent of males and females to be mated for the next generation. The difference between the mean bristle number of selected individuals and the mean number of all individuals is called the selection differential (S) (see Fig. 2.10). If the heritability of the character is high, the response to selection or genetic gain (ΔG) is expected to be high. We then have the relationship (see Falconer 1960).

$$\Delta G = h^2 S \qquad \text{(D1)}$$

Thus, if the mean bristle number of the parental population is 30, the mean of selected individuals is 40, and h^2 is 0.5, ΔG will be 5. In other words, the offspring generation is expected to have a mean bristle number of 35. By contrast, if there is no genetic component in the phenotypic variation, $h^2 = 0$, and ΔG becomes 0 as expected. If artificial selection is continued for consecutive generations, ΔG usually declines as the generations proceed because V_G gradually decreases. As mentioned in Chapter 2, however, long-term selection experiments have shown that selection is often effective even in later generations apparently because of new mutations.

E. Genetic Load

When a pair of alleles A_1 and A_2 are maintained by mutation-selection balance or by overdominant selection, less fit genotypes will survive or leave offspring with a lower probability than other genotypes. Therefore, a certain amount of genetic deaths are required to maintain the polymorphism. This amount of genetic deaths can be substantial depending on the type of selection, and for this reason the maintenance of polymorphism becomes difficult in species which do not have a high fertility excess, as in the case of Haldane's cost of natural selection. The amount of genetic deaths caused by selection is called the genetic load (Muller 1950). The genetic load L for a locus is defined as

$$L = (w_{\max} - \bar{w})/w_{\max} \qquad \text{(E1)}$$

where w_{max} is the fitness of the best genotype and $\bar{w}$ is the mean fitness of the population (Crow and Kimura 1970). Here we consider the genetic load for the polymorphism maintained by mutation-selection balance (mutation load) and by overdominant selection (segregation load).

Mutation Load

Let us consider the mutation load for the case of a recessive deleterious mutation with $w_{11} = w_{12} = 1$ and $w_{22} = 1 - s$. In this case $w_{max} = 1$, and $\bar{w} = 1 - s\hat{y}^2$. Therefore, the mutation load is

$$L = s\hat{y}^2 = u \tag{E2}$$

because $\hat{y}^2 = u/s$ (see Equation A6).

Similarly, the mean fitness for a dominant deleterious mutation is $1 - 2h\,\hat{y}$ and $\hat{y} = u/h$ from Equation (A7). Therefore, the mutation load is

$$L = 2h\hat{y} = 2u \tag{E3}$$

Deleterious mutations have various degrees of dominance (h). However, according to Simmons and Crow (1977), most mutations are slightly deleterious, h being equal to approximately 0.05. In this case it can be shown that the mutation load is approximately $2u$ for a locus. Therefore, when there are n loci at which deleterious mutations occur independently, the total genetic load is given by $L = \sum_{i=1}^{n} 2u_i$ where u_i is the mutation rate for the ith locus. The mean fitness of the population is then given by

$$\begin{aligned} \bar{w} &= \sum_{i=1}^{n}(1-2u_i) \\ &\approx e^{-\sum_i 2u_i} \\ &= e^{-L} \end{aligned} \tag{E4}$$

Therefore, if $u_i = 10^{-5}$ and $n = 30\,000$, $L = 0.6$ and $\bar{w} = 0.55$. This means that to maintain a constant population size the population must have an average fertility of at least $1/\bar{w} = 1.8$ per individual.

Segregation Load

Consider a pair of alleles, A_1 and A_2, and assume that the fitnesses of A_1A_1, A_1A_2, and A_2A_2 are $1 - s$, 1, and $1 - t$, respectively. In this case, the equilibrium frequencies of A_1 and A_2 are given by $\hat{x} = t/(s + t)$ and $\hat{y} = s(s + t)$, respectively (see Equation 2.7). Therefore, $w_{max} = w_1 = 1$, and the average fitness of the population is $\bar{w} = \hat{x}^2(1 - s) + 2\hat{x}\hat{y} + \hat{y}^2(1 - t) = 1 - s\hat{x}^2 - t\hat{y}^2$, which becomes $\bar{w} = 1 - st/(s + t)$. Therefore, the segregation load is given by

$$\begin{aligned} L &= (w_1 - \bar{w})/w_1 \\ &= st/(s+t) \end{aligned} \tag{E5}$$

Wright and Dobzhansky (1946) studied the frequency changes of inversion chromosomes Standard (ST) and Chiricahua (CH) in a laboratory (cage) population of *Drosophila pseudoobscura* and showed that the ST chromosome eventually reaches a frequency of about 70 percent. From the chromosome frequency changes over generations, they estimated the genotype frequencies as follows.

Genotype	ST/ST	ST/CH	CH/CH
Fitness	1 – 0.3	1	1 – 0.7

Therefore, the expected equilibrium frequency of the ST chromosome is 0.7/(0.3 + 0.7) = 0.7 from Equation (2.7). This polymorphism is expected to generate a genetic load of 0.21 (see Equation E5). In other words, at least 21 percent of individuals in the cage population are expected to die just to maintain the polymorphism.

What would be the genetic load when m different polymorphic alleles exist in the population? The mathematical formulation of this problem is somewhat complicated (Crow and Kimura 1970). Let us consider a simple case, where all heterozygotes A_iA_j have a fitness of 1 and homozygote A_iA_i has a fitness of $1 - s_i$. In this case it can be shown that the genetic load is expressed as

$$L = 1/\sum_{i=1}^{m}(1/s_i) = \frac{\tilde{s}}{m} \tag{E6}$$

where $\tilde{s}$ is the harmonic mean of s_i. When all s_i values are equal to s, L becomes s/m. This indicates that L generally decreases as m increases. In human populations HLA (MHC) polymorphism is believed to be maintained by overdominant selection, and the average s has been estimated to be 0.01 (Satta et al. 1994). In the HLA B locus, there seem to be about 25 alleles in many human populations (Roychoudhury and Nei 1988). Therefore, the genetic load required for the maintenance of this polymorphism is 0.01/25 = 0.0004 per locus. This genetic load is quite low compared with that for the inversion polymorphism mentioned above. The reason for this is that when m is large most individuals become heterozygous and the extent of selection is reduced.

F. Bayesian Method of Detecting Positively Selected Codons

The purpose of this method is to identify specific codon sites of nucleotide sequences at which positive selection is operating repeatedly. In this method the codon is considered as the unit of selection, and the ratio (w) of the rate of nonsynonymous substitution (r_N) to the rate of synonymous substitution (r_S) is estimated for each codon site. In the computation of w the following model of nucleotide substitution (codeml) is used. Let us consider a set of sequences of n homologous codons, and let π_j be the relative frequency of the jth codon. It is assumed that the instantaneous substitution rate (q_{ij}) from codon i to codon j ($i \neq j$) at a given codon site is given by the following equations.

$$q_{ij} = \begin{bmatrix} 0 & \text{if nucleotaltimge change occurs at two or} \\ & \text{more positions,} \\ \pi_j & \text{for synonymous transversion,} \\ k\pi_j & \text{for synonymous transition,} \\ w\pi_j & \text{for nonsynonymous transversion,} \\ wk\pi_j & \text{for nonsynonymous transition,} \end{bmatrix} \quad \text{(F1)}$$

where k is the transition/transversion rate ratio, and it may be written as α/β if the rates of transitional and transversional changes are α and β, respectively. This k is assumed to be the same for all codon sites and for the entire evolutionary time period.

In this substitution model, the rate of synonymous substitution (r_S) is assumed to be a neutral rate, and therefore $w < 1$, $w = 1$, and $w > 1$ represent purifying selection, neutral mutation, and positive selection, respectively. At each codon site, w is assumed to remain constant. However, it may vary among different codon sites, and our purpose is to find codon sites where the estimate ($\hat{w}$) of w is significantly higher than 1, and this is done by using the empirical Bayesian method (see Yang et al. 2005; Zhang et al. 2005). In this method one must consider the distributions of w for the neutral evolution model (null hypothesis) and for the selection model. For the null model (M_O), the uniform distribution of w for all or a portion of codon sites is usually used, and for the selection model (M_S) several different distributions such as the β distribution and the composite distributions of three or four different classes of w values are used. The selection model (M_S) that fits the data significantly better than M_O (by a likelihood ratio test) is chosen, and then the $\hat{w}$ value for each codon site is inferred by an empirical Bayesian method. If the $\hat{w}$ value at a codon site is significantly higher than 1, the site is assumed to be under positive selection.

Because this method often gives codon sites where $\hat{w}$ is significantly greater than 1, many biologists have preferred to use it. However, the credibility of this method has been questioned by a number of authors (e.g. Suzuki and Nei 2004; Yokoyama et al. 2008; Nei et al. 2010). First, the codeml model is not realistic for modeling protein evolution. In Table 4.3, I showed many examples in which protein function is changed by a few amino acid substitutions. In this case, if proper amino acids are placed in specific codon positions, an innovative function is often generated, and this set of amino acids undergoes no more changes until another new function is generated. In this case, the w value is expected to be small, though the few amino acid substitutions involved are important for protein evolution. A good example is the amino acids at positions 277 and 285 of the red and green opsins (visual proteins) in vertebrates. If the amino acids at positions 277 and 285 are tyrosine (Y) and threonine (T), respectively, the opsin will be red-color sensitive, and if they are phenylalanine (F) and alanine (A), the opsin will be green-color sensitive. This is virtually the same for all vertebrate species so far examined (Yokoyama 2008). Therefore, these two sites are expected to have a small w value (or 0), though they are evolutionarily very important.

Second, we never know the appropriate M_O and M_S models in real data analysis. Therefore, they are set up on an intuitive basis. For this reason, the results of the likelihood test are not reliable. In fact, even in computer simulations with all parameters specified, the Bayesian estimates of parameters are generally quite different from the true values. Thus, the estimate of w often becomes infinite and significant even if the true value is set to 1 (Nozawa et al. 2009a). It is also known that even if the codon at a given site is identical for all sequences used the w value can be significantly greater than 1 (Suzuki and Nei 2004).

Third, even in legitimate computer simulations, the distribution of Type I error (P value) can become U-shaped instead of the uniform distribution theoretically required (Nozawa et al. 2009a). Therefore, we never know the reliability of the test results when we analyze actual data with intuitively specified M_O and M_S models.

Because of these problems the only way to study adaptive evolution at the codon level would be to do experiments on reconstruction of ancient proteins (e.g. Jermann et al. 1995; Zhang and Rosenberg 2002; Yokoyama 2008). Experimental results so far obtained do not support Bayesian predictions (Fig. 4.10).

References

Abbot P, Abe J, Alcock J, Alizon S, Alpedrinha JA et al. 2011. Inclusive fitness theory and eusociality. Nature 471:E1–4; author reply E9–10.

Abzhanov A, Kuo WP, Hartmann C, Grant BR, Grant PR et al. 2006. The calmodulin pathway and evolution of elongated beak morphology in Darwin's finches. Nature 442:563–7.

Abzhanov A, Protas M, Grant BR, Grant PR, and Tabin CJ. 2004. Bmp4 and morphological variation of beaks in Darwin's finches. Science 305:1462–5.

Adams KL, and Wendel JF. 2005. Polyploidy and genome evolution in plants. Curr Opin Plant Biol 8:135–41.

Akey JM. 2009. Constructing genomic maps of positive selection in humans: where do we go from here? Genome Res 19:711–22.

Akey JM, Zhang G, Zhang K, Jin L, and Shriver MD. 2002. Interrogating a high-density SNP map for signatures of natural selection. Genome Res 12:1805–14.

Alekseyenko AA, Peng S, Larschan E, Gorchakov AA, Lee OK et al. 2008. A sequence motif within chromatin entry sites directs MSL establishment on the *Drosophila* X chromosome. Cell 134:599–609.

Allen E, Xie Z, Gustafson AM, Sung GH, Spatafora JW et al. 2004. Evolution of microRNA genes by inverted duplication of target gene sequences in *Arabidopsis thaliana*. Nat Genet 36:1282–90.

Allen GE. 1969. Hugo de Vries and the reception of the "mutation theory." J Hist Biol 2:55–87.

Allen GE. 1978. Thomas Hunt Morgan: the man and his science. Princeton University Press, Princeton.

Allison AC. 1954. Protection afforded by sickle-cell trait against subtertian malarial infection. Br Med J 1:290–4.

Ambros V, and Horvitz HR. 1984. Heterochronic mutants of the nematode *Caenorhabditis elegans*. Science 226:409–16.

Amores A, Suzuki T, Yan YL, Pomeroy J, Singer A et al. 2004. Developmental roles of pufferfish Hox clusters and genome evolution in ray-fin fish. Genome Res 14:1–10.

Anfinsen CB. 1959. Some approaches to the study of active centers. J Cell Comp Physiol 54:215–20.

Arnheim N. 1983. Concerted evolution of multigene families. In M. Nei, and R. K. Koehn, eds. Evolution of genes and proteins, pp. 39–61. Sinauer Assoc, Sunderland, MA.

Arthur W. 2011. Evolution: a developmental approach. Wiley-Blackwell, Oxford.

Avery PJ. 1977. The effect of random selection coefficients on populations of finite size—some particular models. Genet Res 29:97–112.

Avise JC, and Selander RK. 1972. Evolutionary genetics of cave-dwelling fishes of the genus *Astyanax*. Evolution 26:1–20.

Axtell MJ, and Bowman JL. 2008. Evolution of plant microRNAs and their targets. Trends Plant Sci 13:343–9.

Ayala FJ. 1986. On the virtues and pitfalls of the molecular evolutionary clock. J Hered 77:226–35.

Ayala FJ. 2007. Darwin's greatest discovery: design without designer. Proc Natl Acad Sci USA 104 Suppl 1:8567–73.

Ayala FJ, Powell JR, and Dobzhansky T. 1971. Polymorphisms in continental and island populations of *Drosophila willistoni*. Proc Natl Acad Sci USA 68:2480–3.

Badaeva ED, Dedkova OS, Gay G, Pukhalskyi VA, Zelenin AV et al. 2007. Chromosomal rearrangements in wheat: their types and distribution. Genome 50:907–26.

Bajaj M, Blundell TL, Horuk R, Pitts JE, Wood SP et al. 1986. Coypu insulin. Primary structure, conformation and biological properties of a hystricomorph rodent insulin. Biochem J 238:345–51.

Bakermans-Kranenburg MJ, and van IJzendoorn MH. 2006. Gene-environment interaction of the dopamine D4 receptor (DRD4) and observed maternal insensitivity predicting externalizing behavior in preschoolers. Dev Psychobiol 48:406–9.

Baker HG, and Stebbins GL (eds). 1965. The genetics of colonizing species. Academic Press, New York.

Barreiro LB, Laval G, Quach H, Patin E, and Quintana-Murci L. 2008. Natural selection has driven population differentiation in modern humans. Nat Genet 40:340–5.

Bartel DP. 2009. MicroRNAs: target recognition and regulatory functions. Cell 136:215–33.

Barton NH, and Charlesworth B. 1984. Genetic revolutions, founder effects, and speciation. Annu Rev Ecol Syst 15:133–64.

Bastow R, Mylne JS, Lister C, Lippman Z, Martienssen RA et al. 2004. Vernalization requires epigenetic silencing of FLC by histone methylation. Nature 427:164–7.

Bateson W. 1894. Materials for the study of variation. Macmillan, London.

Bateson W. 1902. Mendel's principles of heredity: a defence. Cambridge University Press, Cambridge.

Bateson W. 1909. Heredity and variation in modern lights. Cambridge University Press, Cambridge.

Baurle I, and Dean C. 2006. The timing of developmental transitions in plants. Cell 125:655–64.

Bayes JJ, and Malik HS. 2009. Altered heterochromatin binding by a hybrid sterility protein in *Drosophila* sibling species. Science 326:1538–41.

Beadle GW, and Tatum EL. 1941. Genetic control of biochemical reactions in *Neurospora*. Proc Natl Acad Sci USA 27:499–506.

Begun DJ, Holloway AK, Stevens K, Hillier LW, Poh YP et al. 2007. Population genomics: whole-genome analysis of polymorphism and divergence in *Drosophila simulans*. PLoS Biol 5:e310.

Bell G. 2010. Fluctuating selection: the perpetual renewal of adaptation in variable environments. Philos Trans R Soc Lond B Biol Sci 365:87–97.

Bell MA, Futuyma DJ, Eanes WF, and Levinton JS, eds. 2009. Evolution since Darwin: the first 150 years. Sinauer Associates, Sunderland, MA.

Bennett DC, and Lamoreux ML. 2003. The color loci of mice—a genetic century. Pigment Cell Res 16:333–44.

Benton MJ, Donoghue PCJ, and Asher RJ. 2009. Calibrating and constraining molecular clocks. In S. B. Hedges, and S. Kumar, eds. The timetree of life, pp. 35–86. Oxford University Press, New York.

Benzer S. 1967. Behavioral mutants of *Drosophila* isolated by countercurrent distribution. Proc Natl Acad Sci USA 58:1112–19.

Berezikov E, Thuemmler F, van Laake LW, Kondova I, Bontrop R et al. 2006. Diversity of microRNAs in human and chimpanzee brains. Nat Genet 38:1375–7.

Bergero R, and Charlesworth D. 2009. The evolution of restricted recombination in sex chromosomes. Trends Ecol Evol 24:94–102.

Beye M. 2004. The dice of fate: the csd gene and how its allelic composition regulates sexual development in the honey bee, *Apis mellifera*. Bioessays 26:1131–9.

Bikard D, Patel D, Le Mette C, Giorgi V, Camilleri C et al. 2009. Divergent evolution of duplicate genes leads to genetic incompatibilities within *A. thaliana*. Science 323:623–6.

Birky CW, Jr., and Skavaril RV. 1976. Maintenance of genetic homogeneity in systems with multiple genomes. Genet Res 27:249–65.

Bodmer WF, and Parsons PA. 1962. Linkage and recombination in evolution. Adv Genet 11:1–100.

Borges RM. 2008. The objection is sustained: A defence of the defense of beanbag genetics. Int J Epidemiol 37:451–4.

Borrello ME. 2005. The rise, fall and resurrection of group selection. Endeavour 29:43–7.

Boss PK, Bastow RM, Mylne JS, and Dean C. 2004. Multiple pathways in the decision to flower: enabling, promoting, and resetting. Plant Cell 16 Suppl:S18–31.

Bourke AFG. 2011. Principles of social evolution. Oxford University Press, New York.

Bowler PJ. 1983. Eclipse of Darwinism: Anti-Darwinian evolution theories in the decade around 1900. Johns Hopkins University, Baltimore.

Boycott AE, and Diver C. 1923. On the inheritance of sinistrality in *Limnaea peregra*. Proc R Soc Lond B Biol Sci 95:207–13.

Brakefield PM, Gates J, Keys D, Kesbeke F, Wijngaarden PJ et al. 1996. Development, plasticity and evolution of butterfly eyespot patterns. Nature 384:236–42.

Breitbart RE, Andreadis A, and Nadal-Ginard B. 1987. Alternative splicing: a ubiquitous mechanism for the generation of multiple protein isoforms from single genes. Annu Rev Biochem 56:467–95.

Brideau NJ, Flores HA, Wang J, Maheshwari S, Wang X et al. 2006. Two Dobzhansky-Muller genes interact to cause hybrid lethality in *Drosophila*. Science 314:1292–5.

Bridges CB. 1935. Salivary chromosome MAPS: with a key to the banding of the chromosomes of *Drosophila melanogaster*. J Hered 26:60–4.

Brouha B, Schustak J, Badge RM, Lutz-Prigge S, Farley AH et al. 2003. Hot L1s account for the bulk of retrotransposition in the human population. Proc Natl Acad Sci USA 100:5280–5.

Brown DD, and Sugimoto K. 1974. The structure and evolution of ribosomal and 5S DNAs in *Xenopus laevis* and *Xenopus mulleri*. Cold Spring Harb Symp Quant Biol 38:501–5.

Brown DD, Wensink PC, and Jordan E. 1972. A comparison of the ribosomal DNAs of *Xenopus laevis* and *Xenopus mulleri*: the evolution of tandem genes. J Mol Biol 63:57–73.

Brown JD, and O'Neill RJ. 2010. Chromosomes, conflict, and epigenetics: chromosomal speciation revisited. Annu Rev Genomics Hum Genet 11:291–316.

Brownell E, Krystal M, and Arnheim N. 1983. Structure and evolution of human and African ape rDNA pseudogenes. Mol Biol Evol 1:29–37.

Brues AM. 1969. Genetic load and its varieties. Science 164:1130–6.

Bryson V, and Vogel HJ. 1965. Evolving genes and proteins. Academic Press, New York.

Buck L, and Axel R. 1991. A novel multigene family may encode odorant receptors: a molecular basis for odor recognition. Cell 65:175–87.

Bull JJ. 1983. Evolution of sex determining mechanisms. Benjamin/Cummings, Menlo Park, CA.

Bulmer M. 2004. Did Jenkin's swamping argument invalidate Darwin's theory of natural selection? Brit Soc Histor Sci 37:281–97.

Bulmer MG, and Bull JJ. 1982. Models of polygenic sex determination and sex ratio control. Evolution 36:13–26.

Burglin TR. 1997. Analysis of TALE superclass homeobox genes (MEIS, PBC, KNOX, Iroquois, TGIF) reveals a novel domain conserved between plants and animals. Nucleic Acids Res 25:4173–80.

Burke GR, and Moran NA. 2011. Massive genomic decay in *Serratia symbiotica*, a recently evolved symbiont of aphids. Genome Biol Evol 3:195–208.

Caicedo AL, Williamson SH, Hernandez RD, Boyko A, Fledel-Alon A et al. 2007. Genome-wide patterns of nucleotide polymorphism in domesticated rice. PLoS Genet 3:1745–56.

Cain AJ, Cook LM, and Currey JD. 1990. Population size and morph frequency in a long term study of *Cepaea nemoralis*. P Roy Soc B-Biol Sci 240:231–50.

Cain AJ, and Sheppard PM. 1950. Selection in the polymorphic land snail *Cepaea nemoralis*. Heredity 4:275–94.

Cain AJ, and Sheppard PM. 1954. Natural selection in *Cepaea*. Genetics 39:89–116.

Carlson CS, Thomas DJ, Eberle MA, Swanson JE, Livingston RJ et al. 2005. Genomic regions exhibiting positive selection identified from dense genotype data. Genome Res 15:1553–65.

Carroll SB. 2005a. Endless forms most beautiful. Norton, New York.

Carroll SB. 2005b. Evolution at two levels: on genes and form. PLoS Biol 3:e245.

Carroll SB. 2008. Evo-devo and an expanding evolutionary synthesis: a genetic theory of morphological evolution. Cell 134:25–36.

Carroll SB, Grenier J, and Weatherbee SD. 2005. From DNA to diversity: molecular genetics and the evolution of animal design. Blackwell Publishing, Malden, MA.

Carson HL. 1971. Speciation and the founder principle. Stadler Symp 3:51–70.

Castle WE. 1903. Mendel's law of heredity. Science 18:396–406.

Caudy AA, and Pikaard CS. 2002. Xenopus ribosomal RNA gene intergenic spacer elements conferring transcriptional enhancement and nuclear dominance-like competition in oocytes. J Biol Chem 277:31577–84.

Cavalli-Sforza LL, and Bodmer WF. 1971. The genetics of human populations. W.H. Freeman, San Francisco.

Chakraborty R, Fuerst PA, and Nei M. 1980. Statistical studies on protein polymorphism in natural populations. III. Distribution of allele frequencies and the number of alleles per locus. Genetics 94:1039–63.

Chakraborty R, and Nei M. 1974. Dynamics of gene differentiation between incompletely isolated populations of unequal sizes. Theor Popul Biol 5:460–9.

Chakraborty R, and Nei M. 1977. Bottleneck effects on average heterozygosity and genetic distance with the stepwise mutation model. Evol 31:347–56.

Chandrasekaran S, Ament SA, Eddy JA, Rodriguez-Zas SL, Schatz BR et al. 2011. Behavior-specific changes in transcriptional modules lead to distinct and predictable neurogenomic states. Proc Natl Acad Sci USA 108:18020–5.

Charlesworth B. 1978. Model for evolution of Y chromosomes and dosage compensation. Proc Natl Acad Sci USA 75:5618–22.

Charlesworth B. 1991. The evolution of sex chromosomes. Science 251:1030–3.

Charlesworth B, and Charlesworth D. 2000. The degeneration of Y chromosomes. Philos Trans R Soc Lond B Biol Sci 355:1563–72.

Charlesworth B, and Charlesworth D. 2010. Elements of evolutionary genetics. Roberts & Company Publishers.

Chen J, Ding J, Ouyang Y, Du H, Yang J et al. 2008. A triallelic system of S5 is a major regulator of the reproductive barrier and compatibility of indica-japonica hybrids in rice. Proc Natl Acad Sci USA 105:11436–41.

Chen L, DeVries AL, and Cheng CH. 1997. Evolution of antifreeze glycoprotein gene from a trypsinogen gene in Antarctic notothenioid fish. Proc Natl Acad Sci USA 94:3811–16.

Chen X. 2005. MicroRNA biogenesis and function in plants. FEBS Lett 579:5923–31.

Cheng CHC, and Chen LB. 1999. Evolution of an antifreeze glycoprotein. Nature 401:443–4.

Chester M, Gallagher JP, Symonds VV, Cruz da Silva AV, Mavrodiev EV et al. 2012. Extensive chromosomal variation in a recently formed natural allopolyploid species, *Tragopogon miscellus* (Asteraceae). Proc Natl Acad Sci USA 109:1176–81.

Chou JY, Hung YS, Lin KH, Lee HY, and Leu JY. 2010. Multiple molecular mechanisms cause reproductive isolation between three yeast species. PLoS Biol 8:e1000432.

Christiansen FB, and Frydenberg O. 1973. Selection component analysis of natural polymorphisms using population samples including mother-offspring combinations. Theor Popul Biol 4:425–45.

Clark JB, Maddison WP, and Kidwell MG. 1994. Phylogenetic analysis supports horizontal transfer of P transposable elements. Mol Biol Evol 11:40-50.

Clark MA, Moran NA, and Baumann P. 1999. Sequence evolution in bacterial endosymbionts having extreme base compositions. Mol Biol Evol 16:1586–98.

Clarke B. 1971. Darwinian evolution of proteins. Science 168:1009–11.

Clarke CA, Clarke FMM, and Owen DF. 1991. Natural selection and the scarlet tiger moth, *Panaxia dominula*:

inconsistencies in the scoring of the heterozygote, *f. medionigra*. Proc Roy Soc B-Biol Sci 244:203–5.

Clarke CA, and Sheppard PM. 1966. A local survey of the distribution of the industrial melanic forms in the moth *Biston betularia* and estimates of the selective values of these in an industrial environment. Proc Roy Soc B-Biol Sci 165:424–39.

Clayton GA, and Robertson A. 1955. Mutation and quantitative variation. Amer Nat 89:151–8.

Cleland RE. 1923. Chromosome arrangements during meiosis in certain *Oenotherae*. Am Nat 57:562–6.

Cleland RE. 1972. *Oenothera*: cytogenetics and evolution. Academic Press, New York.

Committee on DNA forensic science: update-National Research Council. 1996. The evaluation of forensic DNA evidence. National Academy Press, Washington, DC.

Cook LM, and Jones DA. 1996. The *medionigra* gene in the moth *Panaxia dominula*: the case for selection. Philos Trans R Soc Lond B Biol Sci 351:1623–34.

Cordaux R, and Batzer MA. 2009. The impact of retrotransposons on human genome evolution. Nat Rev Genet 10:691–703.

Costa FF. 2007. Non-coding RNAs: lost in translation? Gene 386:1–10.

Cox EC, and Yanofsky C. 1967. Altered base ratios in the DNA of an *Escherichia coli* mutator strain. Proc Natl Acad Sci USA 58:1895–902.

Coyne JA, Barton NH, and Turelli M. 1997. Perspective: A critique of Sewall Wright's shifting balance theory of evolution. Evolution 51:643–71.

Coyne JA, Barton NH, and Turelli M. 2000. Is Wright's shifting balance process important in evolution? Evolution 54:306–17.

Coyne JA, and Orr HA. 2004. Speciation. Sinauer Associates, Sunderland, MA.

Crow JF. 1957. Genetics of insect resistance to chemicals. Ann Rev Ent 2:227–46.

Crow JF. 1968. The cost of evolution and genetic load. In K R Dronamraju, ed. Haldane and modern biology. John Hopkins Press, Baltimore.

Crow JF. 1970. Genetic loads and the cost of natural selection. In K I Kojima, ed. Mathematical topics in population genetics, pp. 128–77. Springer-Verlag, Berlin.

Crow JF. 1999. Hardy, Weinberg and language impediments. Genetics 152:821–5.

Crow JF. 2008. Commentary: Haldane and beanbag genetics. Int J Epidemiol 37:442–5.

Crow JF, and Kimura M. 1970. An introduction to population genetics theory. Harper & Row, New York.

Crow JF, and Morton NE. 1955. Measurement of gene frequency drift in small populations. Evolution 9:202–14.

Crow JF, and Temin RG. 1964. Evidence for the partial dominance of recessive lethal genes in natural populations of *Drosophila*. Am Nat 98:21–33.

Crozier RH, and Pamilo P. 1996. Evolution of social insect colonies. Sex allocation and kin selection. Oxford University Press, Oxford, UK.

Crumpacker DW, and Williams JS. 1973. Density, dispersion, and population structure in *Drosophila pseudoobscura*. Evol Monogr 43:499–538.

Daniels GR, and Deininger PL. 1985. Repeat sequence families derived from mammalian tRNA genes. Nature 317:819-822.

Darwin C. 1859. On the origin of species by means of natural selection, or the preservation of favoured races in the struggle for life. Murray, London.

Darwin C. 1868. The variation of plants and animals under domestication. Murray, London.

Darwin C. 1871. The descent of man. Murray, London.

Darwin C. 1872. The origin of species, 6th ed. Murray, London.

Das S, Nikolaidis N, and Nei M. 2009. Genomic organization and evolution of immunoglobulin kappa gene enhancers and kappa deleting element in mammals. Mol Immunol 46:3171–7.

Das S, Nozawa M, Klein J, and Nei M. 2008. Evolutionary dynamics of the immunoglobulin heavy chain variable region genes in vertebrates. Immunogenetics 60:47–55.

Davidson EH. 2006. The regulatory genome: Gene regulatory networks in development and evolution. Academic Press, London.

Davidson EH, and Erwin DH. 2006. Gene regulatory networks and the evolution of animal body plans. Science 311:796–800.

Davis BM. 1912. Genetical studies on *Oenothera* III. Further hybrids of *Oenothera biennis* and *O. grandiflora* that resemble *O. lamarckiana*. Am Nat 46:377–427.

Davis BM. 1943. An amphidiploid in the F_1 generation from the cross *Oenothera franciscana* x *Oenothera biennis*, and its progeny. Genetics 28:275–85.

Davis NM, Kurpios NA, Sun X, Gros J, Martin JF et al. 2008. The chirality of gut rotation derives from left-right asymmetric changes in the architecture of the dorsal mesentery. Dev Cell 15:134–45.

Davuluri RV, Suzuki Y, Sugano S, Plass C, and Huang TH. 2008. The functional consequences of alternative promoter use in mammalian genomes. Trends Genet 24:167–77.

Dawkins R. 1976. The selfish gene. Oxford University Press, New York.

Dawkins R. 1982. The extended phenotype: the gene as the unit of selection. Oxford University Press, San Francisco.

Dawkins R. 1987. The blind watchmaker. Norton, New York.

Dawkins R. 1997. Human chauvinism: a review of S. J. Gould's Full House. Evolution 51:1015–20.

Dayhoff MO. 1969. Atlas of protein sequence and structure, Volume 4. National Biomedical Research Foundation, Silver Springs, MD.

Dayhoff MO. 1972. Atlas of protein sequence and structure, Volume 5. National Biomedical Research Foundation, Silver Springs, MD.

De Bodt S, Maere S, and Van de Peer Y. 2005. Genome duplication and the origin of angiosperms. Trends Ecol Evol 20:591–7.

de Meaux J, Goebel U, Pop A, and Mitchell-Olds T. 2005. Allele-specific assay reveals functional variation in the *chalcone synthase* promoter of *Arabidopsis thaliana* that is compatible with neutral evolution. Plant Cell 17:676–90.

de Meaux J, Pop A, and Mitchell-Olds T. 2006. *Cis*-regulatory evolution of *chalcone-synthase* expression in the genus *Arabidopsis*. Genetics 174:2181–202.

de Vries H. 1901–1903. Die mutationstheorie. Vol. I and II. Von Veit, Leipzig.

de Vries H. 1909. The mutation theory: experiments and observations on the origin of species in the vegetable kingdom. Vol. I. The origin of species by mutation. English translation by Farmer, JB and Darbishire, AD. Open Court Publishing Company, Chicago.

de Vries H. 1910. The mutation theory: experiments and observations on the origin of species in the vegetable kingdom. Vol. II. The origin of varieties by mutation. English translation by Farmer, JB and Darbishire, AD. Open Court Publishing Company, Chicago.

de Vries H. 1912. Species and varieties: their origin by mutation. In D. T. MacDougal, ed. Genes, cells and organisms: great books in experimental biology. Open Court Publishing Company, Chicago.

De Winter W. 1997. The beanbag genetics controversy: towards a synthesis of opposing views of natural selection. Biol Philos 12:149–84.

Delneri D, Colson I, Grammenoudi S, Roberts IN, Louis EJ et al. 2003. Engineering evolution to study speciation in yeasts. Nature 422:68–72.

Desiderio UV, Zhu X, and Evans JP. 2010. ADAM2 interactions with mouse eggs and cell lines expressing alpha4/alpha9 (ITGA4/ITGA9) integrins: implications for integrin-based adhesion and fertilization. PLoS One 5:e13744.

Dickerson RE. 1971. The structures of cytochrome c and the rates of molecular evolution. J Mol Evol 1:26–45.

Dobzhansky T. 1937. Genetics and the origin of species. Columbia University Press, New York.

Dobzhansky T. 1951. Genetics and the origin of species, 2nd ed. Columbia University Press, New York.

Dobzhansky T. 1955. A review of some fundamental concepts and problems of population genetics. Cold Spring Harb Symp Quant Biol 20:1–15.

Dobzhansky T. 1970. Genetics of the evolutionary process. Columbia University Press, New York.

Dobzhansky T, and Spassky B. 1968. Genetics of natural populations. XL. Heterotic and deleterious effects of recessive lethals in populations of *Drosophila pseudoobscura*. Genetics 59:411–25.

Doherty PC, and Zinkernagel RM. 1975. Enhanced immunological surveillance in mice heterozygous at the H-2 gene complex. Nature 256:50–2.

Doolittle RF, and Blombaeck B. 1964. Amino-acid sequence investigations of fibrinopeptides from various mammals: Evolutionary implications. Nature 202:147–52.

Doolittle WF. 1999. Phylogenetic classification and the universal tree. Science 284:2124–28.

Dowdeswell WH, Fisher RA, and Ford EB. 1940. The quantitative study of populations in the Lepidoptera *I. Polyommatus icarus* Rott. Ann Eugenic 10:123–36.

Doyle JJ, Flagel LE, Paterson AH, Rapp RA, Soltis DE et al. 2008. Evolutionary genetics of genome merger and doubling in plants. Annu Rev Genet 42:443–61.

Dulac C, and Axel R. 1995. A novel family of genes encoding putative pheromone receptors in mammals. Cell 83:195–206.

Dunning Hotopp JC, Clark ME, Oliveira DCSG, Foster JM, Fischer P et al. 2007. Widespread lateral gene transfer from intracellular bacteria to multicellular eukaryotes. Science 317:1753–6.

East EM. 1910. Notes on an experiment concerning the nature of unit characters. Science 32:93–5.

Easteal S, Collet C, and Betty D. 1995. The mammalian molecular clock. Springer-Verlag, New York.

Edwards AC, Rollmann SM, Morgan TJ, and Mackay TF. 2006. Quantitative genomics of aggressive behavior in *Drosophila melanogaster*. PLoS Genet 2:e154.

Eickbush TH, and Eickbush DG. 2007. Finely orchestrated movements: evolution of the ribosomal RNA genes. Genetics 175:477–85.

Eirin-Lopez JM, Gonzalez-Tizon AM, Martinez A, and Mendez J. 2004. Birth-and-death evolution with strong purifying selection in the histone H1 multigene family and the origin of orphon H1 genes. Mol Biol Evol 21:1992–2003.

Eizirik E, Yuhki N, Johnson WE, Menotti-Raymond M, Hannah SS et al. 2003. Molecular genetics and evolution of melanism in the cat family. Curr Biol 13:448–53.

Emerson RA, and East EM. 1913. The inheritance of quantitative characters in maize. Bull Agric Exp Stn Nebr 2:1–120.

Emerson S. 1939. A preliminary survey of the *Oenothera organensis* population. Genetics 24:524–37.

ENCODE Project Consortium. 2007. Identification and analysis of functional elements in 1% of the human genome by the ENCODE pilot project. Nature 447:799–816.

ENCODE Project Consortium. 2012. An integrated encyclopedia of DNA elements in the human genome. Nature 489:57–74.

Endler JA. 1986. Natural selection in the wild. Princeton University Press, Princeton.

Esteves PJ, Lanning D, Ferrand N, Knight KL, Zhai SK et al. 2004. Allelic variation at the VHa locus in natural populations of rabbit (*Oryctolagus cuniculus*, L.). J Immunol 172:1044–53.

Evans JP, and Florman HM. 2002. The state of the union: the cell biology of fertilization. Nat Cell Biol 4 Suppl:s57–63.

Ewens WJ. 1967. A note on the mathematical theory of the evolution of dominance. Am Nat 101:35–40.

Ewens WJ. 1970. Remarks on the substitutional load. Theor Popul Biol 1:129–39.

Ewens WJ. 1972. The sampling theory of selectively neutral alleles. Theor Popul Biol 3:87–112.

Ewens WJ. 1989. An interpretation and proof of the fundamental theorem of natural selection. Theor Popul Biol 36:167–80.

Ewens WJ. 1993. Beanbag genetics and after. In P P Majumder, ed. Human population genetics, pp. 7–28. Plenum Press, New York.

Ewens WJ. 2000. A hundred years of population genetics theory. J Epidemiol Biostat 5:17–23.

Ewens WJ. 2004. Mathematical population genetics. Springer, New York.

Falconer DS. 1960. Introduction to quantitative genetics. Oliver and Boyd, Edinburgh.

Fares MA, Moya A, Escarmis C, Baranowski E, Domingo E et al. 2001. Evidence for positive selection in the capsid protein-coding region of the foot-and-mouth disease virus (FMDV) subjected to experimental passage regimens. Mol Biol Evol 18:10–21.

Fay JC, Wyckoff GJ, and Wu CI. 2002. Testing the neutral theory of molecular evolution with genomic data from *Drosophila*. Nature 415:1024–6.

Feder JN, Penny DM, Irrinki A, Lee VK, Lebron JA et al. 1998. The hemochromatosis gene product complexes with the transferrin receptor and lowers its affinity for ligand binding. Proc Natl Acad Sci USA 95:1472–7.

Feldman M, and Levy AA. 2012. Genome evolution due to allopolyploidization in wheat. Genetics 192:763–74.

Feldman MW. 1972. Selection for linkage modification. I. Random mating populations. Theor Popul Biol 3:324–46.

Felsenstein J. 1971. On the biological significance of the cost of gene substitution. Am Nat 105:1–11.

Felsenstein J. 1974. The evolutionary advantage of recombination. Genetics 78:737–56.

Felsenstein J. 2004. Inferring phylogenies. Sinauer Associates, Sunderland, MA.

Ferree PM, and Barbash DA. 2009. Species-specific heterochromatin prevents mitotic chromosome segregation to cause hybrid lethality in *Drosophila*. PLoS Biol 7:e1000234.

Figueroa F, Gunther E, and Klein J. 1988. MHC polymorphism pre-dating speciation. Nature 335:265–7.

Filipowicz W, Bhattacharyya SN, and Sonenberg N. 2008. Mechanisms of post-transcriptional regulation by microRNAs: are the answers in sight? Nat Rev Genet 9:102–14.

Filmore[J4] D. 2004. It's a GPCR world. Modern Drug Discovery 7:24–8.

Fisher RA. 1918. The correlation between relatives on the supposition of Mendelian inheritance. P Roy Soc Edinburgh 52:399–433.

Fisher RA. 1922. On the dominance ratio. P Roy Soc Edinburgh 42:321–41.

Fisher RA. 1928. The possible modifications of the wild type to recurrent mutations. Am Nat 62:115–26.

Fisher RA. 1930. The genetical theory of natural selection. Clarendon, Oxford.

Fisher, RA. 1931. The evolution of dominance. Biol Rev 6:345–68.

Fisher RA. 1935. The sheltering of lethals. Am Nat 69:446–55.

Fisher RA. 1941. Average excess and average effect of a gene substitution. Ann Eugenic 11:53–63.

Fisher RA. 1958. The genetical theory of natural selection, 2nd Ed. Dover Press, New York.

Fisher RA, and Ford EB. 1947. The spread of a gene in natural conditions in a colony of the moth *Panaxia dominula L.* Heredity 1:143–74.

Fisher RA, and Ford EB. 1950. The Sewall Wright effect. Heredity 4:117–9.

Fisher RA, Ford EB, and Huxley J. 1939. Taste-testing the Anthropoid Apes. Nature 144:750.

Fitch WM, Bush RM, Bender CA, and Cox NJ. 1997. Long term trends in the evolution of H(3) HA1 human influenza type A. Proc Natl Acad Sci USA 94:7712–18.

Flagel LE, and Wendel JF. 2009. Gene duplication and evolutionary novelty in plants. New Phytol 183:557–64.

Ford EB. 1964. Ecological genetics, Methuen, London.

Ford EB. 1975. Ecological genetics, 4th ed. Chapman and Hall, London.

Frank SA. 1991. Divergence of meiotic drive-suppression systems as an explanation for sex-biased hybrid sterility and inviability. Evolution 45:262–7.

Frank SA. 1998. Foundations of social evolution. Princeton University Press, Princeton, New Jersey.

Franklin I, and Lewontin RC. 1970. Is the gene the unit of selection? Genetics 65:707–34.

Freese E. 1962. On the evolution of base composition of DNA. J Theor Biol 3:82–101.

Freese E, and Yoshida A. 1965. The role of mutations in evolution. In V Bryson, and H J Vogel, eds. Evolving Genes and Proteins, pp. 341–55. Academic, New York.

Furuya EY, and Lowy FD. 2006. Antimicrobial-resistant bacteria in the community setting. Nat Rev Microbiol 4:36–45.

Futuyma DJ. 2005. Evolution. Sinauer, Sunderland, MA.

Galindo BE, Vacquier VD, and Swanson WJ. 2003. Positive selection in the egg receptor for abalone sperm lysin. Proc Natl Acad Sci USA 100:4639–43.

Gates RR. 1908. The chromosomes of *Oenothera*. Science 27:193–5.

Gayon J. 1998. Darwinism's struggle for survival: Heredity and the hypothesis of natural selection. Cambridge University Press, Cambridge.

Gehring WJ. 1998. Master control genes in development and evolution: The homebox story. Yale University Press, New Haven.

Gehring WJ. 2005. New perspectives on eye development and the evolution of eyes and photoreceptors. J Hered 96:171–84.

Gehring WJ. 2011. Chance and necessity in eye evolution. Genome Biol Evol 3:1053–66.

Gehring WJ, and Ikeo K. 1999. Pax 6: mastering eye morphogenesis and eye evolution. Trends Genet 15:371–7.

Gehring WJ, Kloter U, and Suga H. 2009. Evolution of the Hox gene complex from an evolutionary ground state. Curr Top Dev Biol 88:35–61.

Gempe T, and Beye M. 2010. Function and evolution of sex determination mechanisms, genes and pathways in insects. Bioessays 33:52–60.

Gerhart J, and Kirschner MW. 1997. Cells, embryos, and evolution. Blackwell Science, Malden, MA.

Gerstein MB, Bruce C, Rozowsky JS, Zheng D, Du J et al. 2007. What is a gene, post-ENCODE? History and updated definition. Genome Res 17:669–81.

Gibbs R A, Rogers J, Katze MG, Bumgarner R, Weinstock GM, et al. 2007. Evolutionary and biomedical insights from the rhesus macaque genome. Science 316:222–34.

Gilbert S. 2006. Developmental biology. Sinauer Assoc, Sunderland, MA.

Gilbert W. 1978. Why genes in pieces? Nature 271:501.

Gilbert W. 1986. Origin of life: the RNA world. Nature 319:618.

Gillespie JH. 1973. Natural selection with varying selection coefficients - a haploid model. Genet Res 21:115-120.

Gillespie JH. 1980. Protein polymorphism and the SAS-CFF model. Genetics 94:1089–90.

Gillespie JH. 1991. The causes of molecular evolution. Oxford University Press, Oxford.

Gimelbrant AA, Skaletsky H, and Chess A. 2004. Selective pressures on the olfactory receptor repertoire since the human-chimpanzee divergence. Proc Natl Acad Sci USA 101:9019–22.

Glusman G, Yanai I, Rubin I, and Lancet D. 2001. The complete human olfactory subgenome. Genome Res 11:685–702.

Go Y, and Niimura Y. 2008. Similar numbers but different repertoires of olfactory receptor genes in humans and chimpanzees. Mol Biol Evol 25:1897–907.

Gojobori T, and Nei M. 1984. Concerted evolution of the immunoglobulin VH gene family. Mol Biol Evol 1:195–212.

Goldman N, and Yang Z. 1994. A codon-based model of nucleotide substitution for protein-coding DNA sequences. Mol Biol Evol 11:725–36.

Gonzalez E, Kulkarni H, Bolivar H, Mangano A, Sanchez R et al. 2005. The influence of CCL3L1 gene-containing segmental duplications on HIV-1/AIDS susceptibility. Science 307:1434–40.

Gonzalez IL, and Sylvester JE. 2001. Human rDNA: evolutionary patterns within the genes and tandem arrays derived from multiple chromosomes. Genomics 73:255–63.

Gould SJ. 1980. The panda's thumb: more reflections in natural history. Norton, New York.

Gould SJ. 2002. The structure of evolutionary theory. Harvard University Press, Cambridge, MA.

Gould SJ, and Lewontin RC. 1979. The spandrels of San Marco and the Panglossian paradigm: a critique of the adaptationist programme. P Roy Soc B-Biol Sci 205:581–98.

Grande C, and Patel NH. 2009. Nodal signaling is involved in left-right asymmetry in snails. Nature 457:1007–11.

Grant BS, Owen DF, and Clarke CA. 1996. Parallel rise and fall of melanic peppered moths in America and Britain. J Hered 87:351–7.

Grant V. 1981. Plant speciation, 2nd ed. Columbia University Press, New York.

Graur D. 1985. Gene diversity in Hymenoptera. Evolution 39:190–9.

Graves JAM. 2008. Weird animal genomes and the evolution of vertebrate sex and sex chromosomes. Ann Rev Genet 42:565–86.

Grus WE, Shi P, Zhang YP, and Zhang J. 2005. Dramatic variation of the vomeronasal pheromone receptor gene repertoire among five orders of placental and marsupial mammals. Proc Natl Acad Sci USA 102:5767–72.

Grus WE, and Zhang J. 2009. Origin of the genetic components of the vomeronasal system in the common ancestor of all extant vertebrates. Mol Biol Evol 26:407–19.

Gu X, and Nei M. 1999. Locus specificity of polymorphic alleles and evolution by a birth-and-death process in mammalian MHC genes. Mol Biol Evol 16:147–56.

Haigh J. 1978. The accumulation of deleterious genes in a population—Muller's Ratchet. Theor Popul Biol 14:251–67.

Haldane JBS. 1922. Sex ratio and unisexual sterility in hybrid animals. Journal of Genetics 12:101–9.

Haldane JBS. 1924. The mathematical theory of natural and artificial selection. Part I. Trans Cambridge Philos Soc 23:19–41.

Haldane JBS. 1927. The mathematical theory of natural and artificial selection. Part V. P Camb Philol Soc 23:838–44.

Haldane JBS. 1932. The causes of evolution. Longmans and Green, London.

Haldane JBS. 1933. The part played by recurrent mutation in evolution. Am Nat 67:5–19.

Haldane JBS. 1955. Population genetics. Penguin New Biol 18:34–51.

Haldane JBS. 1957. The cost of natural selection. J Genet 55:511–24.

Haldane JBS. 1964. A defense of beanbag genetics. Perspect Biol Med 7:343–59.

Hall C, Brachat S, and Dietrich FS. 2005. Contribution of horizontal gene transfer to the evolution of *Saccharomyces cerevisiae*. Eukaryot Cell 4:1102–15.

Hamada H, Meno C, Watanabe D, and Saijoh Y. 2002. Establishment of vertebrate left-right asymmetry. Nat Rev Genet 3:103–13.

Hamers-Casterman C, Atarhouch T, Muyldermans S, Robinson G, Hamers C et al. 1993. Naturally occurring antibodies devoid of light chains. Nature 363:446–8.

Hamilton AJ, and Baulcombe DC. 1999. A species of small antisense RNA in posttranscriptional gene silencing in plants. Science 286:950–2.

Hamilton WD. 1964. The genetical evolution of social behavior, I and II. J Theor Biol 7:1–52.

Hanzawa Y, Money T, and Bradley D. 2005. A single amino acid converts a repressor to an activator of flowering. Proc Natl Acad Sci USA 102:7748–53.

Hao L, and Nei M. 2004. Genomic organization and evolutionary analysis of Ly49 genes encoding the rodent natural killer cell receptors: rapid evolution by repeated gene duplication. Immunogenetics 56:343–54.

Hao L, and Nei M. 2005. Rapid expansion of killer cell immunoglobulin-like receptor genes in primates and their coevolution with MHC Class I genes. Gene 347:149–59.

Hardy GH. 1908. Mendelian proportions in a mixed population. Science 28:49–50.

Harris H. 1966. Enzyme polymorphisms in man. P Roy Soc B-Biol Sci 164:298–310.

Hartl DL. 1969. Dysfunctional sperm production in *Drosophila melanogaster* males homozygous for the segregation distorter elements. Proc Natl Acad Sci USA 63:782–9.

Hartl DL, and Taubes CH. 1998. Towards a theory of evolutionary adaptation. Genetica 102–103:525–33.

Hawks J, Wang ET, Cochran GM, Harpending HC, and Moyzis RK. 2007. Recent acceleration of human adaptive evolution. Proc Natl Acad Sci USA 104:20753–8.

Hedge PJ, and Spratt BG. 1985. Amino acid substitutions that reduce the affinity of penicillin-binding protein 3 of *Escherichia coli* for cephalexin. Eur J Biochem 151:111–21.

Hedges SB, and Kumar S. 2009. The timetree of life. Oxford University Press, New York.

Hedrick PW. 2000. Genetics of poulations, 2nd ed. Jones and Barlett Publishers, Sudbury, MA.

Hedrick PW. 2002. Pathogen resistance and genetic variation at MHC loci. Evolution 56:1902–8.

Heimberg AM, Sempere LF, Moy VN, Donoghue PC, and Peterson KJ. 2008. MicroRNAs and the advent of vertebrate morphological complexity. Proc Natl Acad Sci USA 105:2946–50.

Henikoff S, and Malik HS. 2002. Centromeres: selfish drivers. Nature 417:227.

Hentschel CC, and Birnstiel ML. 1981. The organization and expression of histone gene families. Cell 25:301–13.

Hermisson J. 2009. Who believes in whole-genome scans for selection? Heredity 103:283–4.

Hill WG. 1982. Rates of change in quantitative traits from fixation of new mutations. Proc Natl Acad Sci USA 79:142–5.

Hill WG, and Robertson A. 1966. The effect of linkage on limits to artificial selection. Genet Res 8:269–94.

Hillis DM, and Green DM. 1990. Evolutionary changes of heterogametic sex in the phylogenetic history of amphibians. J Evol Biol 3:49-64.

Hirschberg J, and McIntosh L. 1983. Molecular basis of herbicide resistance in *Amaranthus hybridus*. Science 222:1346–9.

Hoegg S, and Meyer A. 2005. Hox clusters as models for vertebrate genome evolution. Trends Genet 21:421–4.

Hoekstra HE, and Coyne JA. 2007. The locus of evolution: evo devo and the genetics of adaptation. Evolution 61:995–1016.

Hoekstra HE, Hirschmann RJ, Bundey RA, Insel PA, and Crossland JP. 2006. A single amino acid mutation contributes to adaptive beach mouse color pattern. Science 313:101–4.

Hollox EJ, Huffmeier U, Zeeuwen PL, Palla R, Lascorz J et al. 2008. Psoriasis is associated with increased beta-defensin genomic copy number. Nat Genet 40:23–5.

Holt CA, and Childs G. 1984. A new family of tandem repetitive early histone genes in the sea urchin *Lytechinus pictus*: evidence for concerted evolution within tandem arrays. Nucleic Acids Res 12:6455–71.

Hood L, Campbell JH, and Elgin SC. 1975. The organization, expression, and evolution of antibody genes and other multigene families. Annu Rev Genet 9:305–53.

Hood L, Kronenberg M, and Hunkapiller T. 1985. T cell antigen receptors and the immunoglobulin supergene family. Cell 40:225–9.

Horuk R, Goodwin P, O'Connor K, Neville RW, Lazarus NR et al. 1979. Evolutionary change in the insulin receptors of hystricomorph rodents. Nature 279:439–40.

Hudson RR, Kreitman M, and Aguade M. 1987. A test of neutral molecular evolution based on nucleotide data. Genetics 116:153–9.

Hughes AL. 1999a. Phylogenies of developmentally important proteins do not support the hypothesis of two rounds of genome duplication early in vertebrate history. J Mol Evol 48:565–76.

Hughes AL. 1999b. Adaptive evolution of genes and genomes. Oxford University Press, New York.

Hughes AL. 2002. Evolution of the human killer cell inhibitory receptor family. Mol Phylogenet Evol 25:330–40.

Hughes AL. 2008. Near neutrality leading edge of the neutral theory of molecular evolution. Ann N Y Acad Sci 1133:162–79.

Hughes AL, and Friedman R. 2008. Codon-based tests of positive selection, branch lengths, and the evolution of mammalian immune system genes. Immunogenetics 60:495–506.

Hughes AL, and Nei M. 1988. Pattern of nucleotide substitution at major histocompatibility complex class I loci reveals overdominant selection. Nature 335:167–70.

Hughes AL, and Nei M. 1989. Nucleotide substitution at major histocompatibility complex class II loci: evidence for overdominant selection. Proc Natl Acad Sci USA 86:958–62.

Hughes AL, and Nei M. 1990. Evolutionary relationships of class II major-histocompatibility-complex genes in mammals. Mol Biol Evol 7:491–514.

Hughes AL, and Nei M. 1993. Evolutionary relationships of the classes of major histocompatibility complex genes. Immunogenetics 37:337–46.

Hughes AL, Packer B, Welch R, Bergen AW, Chanock SJ et al. 2003. Widespread purifying selection at polymorphic sites in human protein-coding loci. Proc Natl Acad Sci USA 100:15754–7.

Hughes AL, and Yeager M. 1998. Natural selection at major histocompatibility complex loci of vertebrates. Annu Rev Genet 32:415–35.

Hunt BG, Ometto L, Wurm Y, Shoemaker D, Yi SV et al. 2011. Relaxed selection is a precursor to the evolution of phenotypic plasticity. Proc Natl Acad Sci USA 108:15936–41.

Hurst LD, and Pomiankowski A. 1991. Causes of sex ratio bias may account for unisexual sterility in hybrids: a new explanation of Haldane's rule and related phenomena. Genetics 128:841–58.

Huxley JS. 1942. Evolution: the modern synthesis. Allen and Unwin, London.

Imaizumi Y, Nei M, and Furusho T. 1970. Variability and heritability of human fertility. Ann Hum Genet 33:251–9.

Ina Y, and Gojobori T. 1994. Statistical analysis of nucleotide sequences of the hemagglutinin gene of human influenza A viruses. Proc Natl Acad Sci USA 91:8388–92.

Ingram VM. 1961. Gene evolution and the haemoglobins. Nature 189:704–8.

Ingram VM. 1963. The hemoglobins in genetics and evolution. Columbia University Press, New York.

International HapMap Consortium. 2005. A haplotype map of the human genome. Nature 437:1299–320.

International HapMap 3 Consortium, Altshuler DM, Gibbs RA, Peltonen L, Altshuler DM et al. 2010. Integrating common and rare genetic variation in diverse human populations. Nature 467:52–8.

International HapMap Consortium, Frazer KA, Ballinger DG, Cox DR, Hinds DA et al. 2007. A second generation human haplotype map of over 3.1 million SNPs. Nature 449:851–61.

International Human Genome Sequencing Consortium. 2001. Initial sequencing and analysis of the human genome. Nature 409:860–921.

International Human Genome Sequencing Consortium. 2004. Initial sequencing and analysis of the human genome. Nature 409:860–921.

Itoh T, Martin W, and Nei M. 2002. Acceleration of genomic evolution caused by enhanced mutation rate in endocellular symbionts. Proc Natl Acad Sci USA 99:12944–8.

Jacob F. 1977. Evolution and tinkering. Science 196:1161–6.

Jacob F, and Monod J. 1961. Genetic regulatory mechanisms in the synthesis of proteins. J Mol Biol 3:318–56.

Jacobs EE, and Sanadi DR. 1960. The reversible removal of cytochrome c from mitochondria. J Biol Chem 235:531–4.

Jaenike J. 2001. Sex chromosome meiotic drive. Annu Rev Ecol Syst 32:25–49.

Javaux EJ, Knoll AH, and Walter MR. 2001. Morphological and ecological complexity in early eukaryotic ecosystems. Nature 412:66–9.

Jeffery CJ. 2003. Moonlighting proteins: old proteins learning new tricks. Trends Genet 19:415–17.

Jeffery WR. 2009. Regressive evolution in *Astyanax* cavefish. Annu Rev Genet 43:25–47.

Jeffreys AJ. 1979. DNA sequence variants in the G gamma-, A gamma-, delta- and beta-globin genes of man. Cell 18:1–10.

Jeffreys AJ. 2005. Genetic fingerprinting. Nat Med 11:1035–1039.

Jenkin F. 1867. (Review of) "The origin of species." N Brit Rev 46:277–318.

Jensen L. 1973. Random selective advantages of genes and their probabilities of fixation. Genet Res 21:215–19.

Jermann TM, Opitz JG, Stackhouse J, and Benner SA. 1995. Reconstructing the evolutionary history of the artiodactyl ribonuclease superfamily. Nature 374:57–9.

Jessen TH, Weber RE, Fermi G, Tame J, Braunitzer G. 1991. Adaptation of bird hemoglobins to high altitudes: Demonstration of molecular mechanism by protein engineering. Proc Natl Acad Sci USA 88:6519–22.

Jiang PH, Josue J, Li X, Glaser D, Li WH et al. 2012. Major taste loss in carnivorous mammals. Proc Natl Acad Sci USA 109:4956–61.

Jiao Y, Wang Y, Xue D, Wang J, Yan M et al. 2010. Regulation of *OsSPL14* by OsmiR156 defines ideal plant architecture in rice. Nat Genet 42:541–4.

Jiao Y, Wickett NJ, Ayyampalayam S, Chanderbali AS, Landherr L et al. 2011. Ancestral polyploidy in seed plants and angiosperms. Nature 473:97–100.

Johannsen W. 1909. Elemente der exakten Erblichkeitslehre. Gustav Fischer, Jena.

Jolles P, Schoentgen F, Jolles J, Dobson DE, Prager EM et al. 1984. Stomach lysozymes of ruminants. II. Amino acid sequence of cow lysozyme 2 and immunological comparisons with other lysozymes. J Biol Chem 259:11617–25.

Jones FC, Grabherr MG, Chan YF, Russell P, Mauceli E et al. 2012. The genomic basis of adaptive evolution in threespine sticklebacks. Nature 484:55-61.

Jones JS, Bryant SH, Lewontin RC, Moore JA, and Prout T. 1981. Gene flow and the geographical distribution of a molecular polymorphism in *Drosophila pseudoobscura*. Genetics 98:157–78.

Joyce GF. 2002. The antiquity of RNA-based evolution. Nature 418:214–21.

Joyce GF. 2004. Directed evolution of nucleic acid enzymes. Annu Rev Biochem 73:791–836.

Just W, Rau W, Vogel W, Akhverdian M, Fredga K et al. 1995. Absence of Sry in species of the vole *Ellobius*. Nat Genet 11:117–18.

Kajikawa M, and Okada N. 2002. LINEs mobilize SINEs in the eel through a shared 3′ sequence. Cell 111:433-444.

Kamakura M. 2011. Royalactin induces queen differentiation in honeybees. Nature 473:478–83.

Kamei N, and Glabe CG. 2003. The species-specific egg receptor for sea urchin sperm adhesion is EBR1, a novel ADAMTS protein. Genes Dev 17:2502–7.

Kasahara M, Hayashi M, Tanaka K, Inoko H, Sugaya K et al. 1996. Chromosomal localization of the proteasome Z subunit gene reveals an ancient chromosomal duplication involving the major histocompatibility complex. Proc Natl Acad Sci USA 93:9096–101.

Kasahara M, Suzuki T, and Pasquier LD. 2004. On the origins of the adaptive immune system: novel insights from invertebrates and cold-blooded vertebrates. Trends Immunol 25:105–11.

Kato Y, Kobayashi K, Watanabe H, and Iguchi T. 2011. Environmental sex determination in the branchiopod crustacean *Daphnia magna*: deep conservation of a Doublesex gene in the sex-determining pathway. PLoS Genet 7:e1001345.

Katz LA, Bornstein JG, Lasek-Nesselquist E, and Muse SV. 2004. Dramatic diversity of ciliate histone H4 genes revealed by comparisons of patterns of substitutions and paralog divergences among eukaryotes. Mol Biol Evol 21:555–62.

Kawaguti S, and Yamasu T. 1965. Electron microscopy on the symbiosis between an elysioid gastropod and chloroplasts from a green alga. J Biol Okayama Univ. II: 57–64.

Kawai A, Ishijima J, Nishida C, Kosaka A, Ota H et al. 2009. The ZW sex chromosomes of *Gekko hokouensis* (Gekkonidae, Squamata) represent highly conserved homology with those of avian species. Chromosoma 118:43–51.

Kedes LH. 1979. Histone genes and histone messengers. Annu Rev Biochem 48:837–70.

Keeling PJ, and Palmer JD. 2008. Horizontal gene transfer in eukaryotic evolution. Nat Rev Genet 9:605–18.

Keightley PD. 2012. Rates and fitness consequences of new mutations in humans. Genetics 190:295–304.

Keightley PD, Lercher MJ, and Eyre-Walker A. 2005. Evidence for widespread degradation of gene control regions in hominid genomes. PLoS Biol 3:e42.

Keller A, Zhuang H, Chi Q, Vosshall LB, and Matsunami H. 2007. Genetic variation in a human odorant receptor alters odour perception. Nature 449:468–72.

Kelley J, Walter L, and Trowsdale J. 2005. Comparative genomics of natural killer cell receptor gene clusters. PLoS Genet 1:129–39.

Kelley JL, Madeoy J, Calhoun JC, Swanson W, and Akey JM. 2006. Genomic signatures of positive selection in humans and the limits of outlier approaches. Genome Res 16:980–9.

Kellis M, Birren BW, and Lander ES. 2004. Proof and evolutionary analysis of ancient genome duplication in the yeast *Saccharomyces cerevisiae*. Nature 428:617–24.

Kettlewell HBD. 1955. Recognition of appropriate backgrounds by the pale and black phases of Lepidoptera. Nature 175:943–4.

Kettlewell HBD. 1973. The Evolution of Melanism: a study of recurring necessity; with special reference to industrial melanism in the Lepidoptera. Clarenden Press, Oxford.

Khaitovich P, Enard W, Lachmann M, and Paabo S. 2006. Evolution of primate gene expression. Nat Rev Genet 7:693–702.

Khakoo SI, Rajalingam R, Shum BP, Weidenbach K, Flodin L et al. 2000. Rapid evolution of NK cell receptor systems demonstrated by comparison of chimpanzees and humans. Immunity 12:687–98.

Kikkawa H. 1937. Spontaneous crossing over in the male of *Drosophila ananassae*. Zool Magaz (Tokyo) 49:159–60.

Kim DH, Doyle MR, Sung S, and Amasino RM. 2009. Vernalization: winter and the timing of flowering in plants. Annu Rev Cell Dev Biol 25:277–99.

Kim HN, and Yamazaki T. 2004. Nonconcerted evolution of histone 3 genes in a liverwort, *Conocephalum conicum*. Genes Genet Syst 79:331–44.

Kimura M. 1954. Process leading to quasi-fixation of genes in natural populations due to random fluctuation of selection intensities. Genetics 39:280–95.

Kimura M. 1956. A model of a genetic system which tends to closer linkage by natural selection. Evolution 10:278–87.

Kimura M. 1957. Some problems of stochastic processes in genetics. Ann Math Stat 28:882–901.

Kimura M. 1958. On the change of population fitness by natural selection. Heredity 12:145–67.

Kimura M. 1962. On the probability of fixation of mutant genes in a population. Genetics 47:713–19.

Kimura M. 1964. Diffusion models in population genetics. J. Appl. Prob. 1:177–232.

Kimura M. 1968a. Evolutionary rate at the molecular level. Nature 217:624–6.

Kimura M. 1968b. Genetic variability maintained in a finite population due to mutational production of neutral and nearly neutral isoalleles. Genet Res 11:247–69.

Kimura M. 1969. The rate of molecular evolution considered from the standpoint of population genetics. Proc Natl Acad Sci USA 63:1181–8.

Kimura M. 1977. Preponderance of synonymous changes as evidence for the neutral theory of molecular evolution. Nature 267:275–6.

Kimura M. 1983. The neutral theory of molecular evolution. Cambridge University Press, Cambridge.

Kimura M, and Crow JF. 1964. The number of alleles that can be maintained in a finite population. Genetics 49:725–38.

Kimura M, and Maruyama T. 1969. The substitutional load in a finite population. Heredity 24:101–14.

Kimura M, and Ohta T. 1971. Theoretical aspects of population genetics. Princeton Univ Press, Princeton, NJ.

King JL. 1967. Continuously distributed factors affecting fitness. Genetics 55:483–92.

King JL, and Jukes TH. 1969. Non-Darwinian evolution. Science 164:788–98.

King MC, and Wilson AC. 1975. Evolution at two levels in humans and chimpanzees. Science 188:107–16.

Kirschner MW, and Gerhart JC. 2005. The plausibility of life. Yale University Press, New Haven, CT.

Klein J, and Figueroa F. 1986. Evolution of the major histocompatibility complex. CRC Crit Rev Immunol 6:295–386.

Klein J, and Horejsi V. 1997. Immunology. Blackwell Science, Oxford.

Klein J, and Nikolaidis N. 2005. The descent of the antibody-based immune system by gradual evolution. Proc Natl Acad Sci USA 102:169–74.

Klein J, Sato A, and Nikolaidis N. 2007. MHC, TSP, and the origin of species: from immunogenetics to evolutionary genetics. Annu Rev Genet 41:281–304.

Klein J, and Takahata N. 2002. Where do we come from? The molecular evidence for human descent. Springer-Verlag, Berlin.

Kohne DE. 1970. Evolution of higher-organism DNA. Q Rev Biophys 3:327–75.

Kondrashov AS. 1995. Contamination of the genome by very slightly deleterious mutations: why have we not died 100 times over? J Theor Biol 175:583–94.

Koopman P, Munsterberg A, Capel B, Vivian N, and Lovell-Badge R. 1990. Expression of a candidate sex-determining gene during mouse testis differentiation. Nature 348:450–2.

Kosakovsky Pond SL, Frost SD, and Muse SV. 2005. HyPhy: hypothesis testing using phylogenies. Bioinformatics 21:676–9.

Kriener K, O'HUigin C, and Klein J. 2000a. Conversion or convergence? Introns of primate DRB genes tell the true story. In M. Kasahara, eds. Major histocompatibility complex; evolution, structure, and function. Spring-Verlag, Tokyo.

Kriener K, O'HUigin C, Tichy H, and Klein J. 2000b. Convergent evolution of major histocompatibility complex molecules in humans and New World monkeys. Immunogenetics 51:169–78.

Kubo K, Entani T, Takara A, Wang N, Fields AM et al. 2010. Collaborative non-self recognition system in S-RNase-based self-incompatibility. Science 330:796–9.

Kulski JK, Shiina T, Anzai T, Kohara S, and Inoko H. 2002. Comparative genomic analysis of the MHC: the evolution of class I duplication blocks, diversity and complexity from shark to man. Immunol Rev 190:95–122.

Kuroda R, Endo B, Abe M, and Shimizu M. 2009. Chiral blastomere arrangement dictates zygotic left-right asymmetry pathway in snails. Nature 462:790–4.

Kuroiwa A, Handa S, Nishiyama C, Chiba E, Yamada F et al. 2011. Additional copies of CBX2 in the genomes of males of mammals lacking SRY, the Amami spiny rat (*Tokudaia osimensis*) and the Tokunoshima spiny rat (*Tokudaia tokunoshimensis*). Chromosome Res 19:635–44.

Kuroiwa A, Ishiguchi Y, Yamada F, Shintaro A, and Matsuda Y. 2010. The process of a Y-loss event in an XO/XO mammal, the Ryukyu spiny rat. Chromosoma 119:519–26.

Kusano A, Staber C, Chan HY, and Ganetzky B. 2003. Closing the (Ran)GAP on segregation distortion in *Drosophila*. Bioessays 25:108–15.

Laird CD, McConaughy BL, and McCarthy BJ. 1969. Rate of fixation of nucleotide substitutions in evolution. Nature 224:149–54.

Lamarck JB. 1809. Philosophie Zoologique. Dentu, Paris.

Lamotte M. 1951. Research on the genetic structure of natural populations of *Cepaea nemoralis*. Annee Biol 55:39–49.

Lamotte M. 1959. Polymorphism of natural populations of *Cepaea nemoralis*. Cold Spring Harb Symp Quant Biol 24:65–86.

Lance VA. 2009. Is regulation of aromatase expression in reptiles the key to understanding temperature-dependent sex determination? J Exp Zool A Ecol Genet Physiol 311:314–22.

Lander ES, Linton LM, Birren B, Nusbaum C, Zody MC et al. 2001. Initial sequencing and analysis of the human genome. Nature 409:860–921.

Langley CH, and Fitch WM. 1974. An examination of the constancy of the rate of molecular evolution. J Mol Evol 3:161–77.

Lawlor DA, Ward FE, Ennis PD, Jackson AP, and Parham P. 1988. HLA-A and B polymorphisms predate the divergence of humans and chimpanzees. Nature 335:268–71.

Lederberg J, and Lederberg EM. 1952. Replica plating and indirect selection of bacterial mutants. J Bacteriol 63:399–406.

Lee AP, Koh EG, Tay A, Brenner S, and Venkatesh B. 2006. Highly conserved syntenic blocks at the vertebrate Hox loci and conserved regulatory elements within and outside Hox gene clusters. Proc Natl Acad Sci USA 103:6994–9.

Lee HY, Chou JY, Cheong L, Chang NH, Yang SY et al. 2008. Incompatibility of nuclear and mitochondrial genomes causes hybrid sterility between two yeast species. Cell 135:1065–73.

Lee RC, Feinbaum RL, and Ambros V. 1993. The C. elegans heterochronic gene *lin-4* encodes small RNAs with antisense complementarity to *lin-14*. Cell 75:843–54.

Lessard S. 1997. Fisher's fundamental theorem of natural selection revisited. Theor Popul Biol 52:119–36.

Levene H. 1953. Genetic equilibrium when more than one ecological niche is available. Amer Nat 87:131–3.

Lewis EB. 1951. Pseudoallelism and gene evolution. Cold Spring Harbor Symposium on Quantitative Biology 16:159–74.

Lewontin RC. 1974. The genetic basis of evolutionary change. Columbia University Press, New York.

Lewontin RC. 1978. Adaptation. Sci Am 239:212–22.

Lewontin RC, and Hubby JL. 1966. A molecular approach to the study of genic heterozygosity in natural populations. II. Amount of variation and degree of heterozygosity in natural populations of *Drosophila pseudoobscura*. Genetics 54:595–609.

Lewontin RC, and Kojima KI. 1960. The evolutionary dynamics of complex polymorphisms. Evolution 14:458–78.

Lewontin RC, and Krakauer J. 1973. Distribution of gene frequency as a test of the theory of the selective neutrality of polymorphisms. Genetics 74:175–95.

Lewontin RC, and Krakauer J. 1975. Testing the heterogeneity of F values. Genetics 80:397–8.

Lewontin RC, Rose S, and Kamin LJ. 1984. Not in our genes. Pantheon Books, New York.

Lewontin RC, and White MJD. 1960. Interaction between inversion polymorphisms of two chromosome pairs in the grasshopper, *Moraba scurra*. Evolution 14:116–29.

Li WH, Gojobori T, and Nei M. 1981. Pseudogenes as a paradigm of neutral evolution. Nature 292:237–9.

Li WH, and Nei M. 1977. Persistence of common alleles in two related populations or species. Genetics 86:901–14.

Li WH, Tanimura M, and Sharp PM. 1987. An evaluation of the molecular clock hypothesis using mammalian DNA sequences. J Mol Evol 25:330–42.

Liao BY, Weng MP, and Zhang J. 2010. Contrasting genetic paths to morphological and physiological evolution. Proc Natl Acad Sci USA 107:7353–8.

Liao D, Pavelitz T, and Weiner AM. 1998. Characterization of a novel class of interspersed LTR elements in primate genomes: structure, genomic distribution, and evolution. J Mol Evol 46:649–60.

Lin F, Xing K, Zhang J, and He X. 2012. Expression reduction in mammalian X chromosome evolution refutes Ohno's hypothesis of dosage compensation. Proc Natl Acad Sci USA. 109:11752–7.

Lin Z, Kong H, Nei M, and Ma H. 2006. Origins and evolution of the recA/RAD51 gene family: evidence for ancient gene duplication and endosymbiotic gene transfer. Proc Natl Acad Sci USA 103:10328–33.

Lin Z, Nei M, and Ma H. 2007. The origins and early evolution of DNA mismatch repair genes—multiple horizontal gene transfers and co-evolution. Nucleic Acids Res 35:7591–603.

Long Y, Zhao L, Niu B, Su J, Wu H et al. 2008. Hybrid male sterility in rice controlled by interaction between divergent alleles of two adjacent genes. Proc Natl Acad Sci USA 105:18871–6.

Lopez de Castro JA, Strominger JL, Strong DM, and Orr HT. 1982. Structure of crossreactive human histocompatibility antigens HLA-A28 and HLA-A2: possible implications for the generation of HLA polymorphism. Proc Natl Acad Sci USA 79:3813–17.

Lu J, Shen Y, Wu Q, Kumar S, He B et al. 2008. The birth and death of microRNA genes in *Drosophila*. Nat Genet 40:351–5.

Lucchesi JC, Kelly WG, and Panning B. 2005. Chromatin remodeling in dosage compensation. Annu Rev Genet 39:615–51.

Ludwig MZ, Patel NH, and Kreitman M. 1998. Functional analysis of *eve* stripe 2 enhancer evolution in *Drosophila*: rules governing conservation and change. Development 125:949–58.

Luria SE, and Delbruck M. 1943. Mutations of bacteria from virus sensitivity to virus resistance. Genetics 28:491–511.

Lutz AM. 1907. A preliminary note on the chromosomes of *Oenothera lamarckiana* and one of its mutants, *O. gigas*. Science 26:151–2.

Lynch M. 1987. The consequences of fluctuating selection for isozyme polymorphisms in *Daphnia*. Genetics 115:657-669.

Lynch M. 1996. Mutation accumulation in transfer RNAs: molecular evidence for Muller's ratchet in mitochondrial genomes. Mol Biol Evol 13:209–20.

Lynch M. 1997. Mutation accumulation in nuclear, organelle, and prokaryotic transfer RNA genes. Mol Biol Evol 14:914–25.

Lynch M. 2007. The origins of genome architecture. Sinauer, Sunderland, MA.

Lynch M, and Force AG. 2000. The origin of interspecific genomic incompatibility via gene duplication. Am Nat 156:590–605.

Lynch M, Sung W, Morris K, Coffey N, Landry CR et al. 2008. A genome-wide view of the spectrum of spontaneous mutations in yeast. Proc Natl Acad Sci USA 105:9272–7.

Lynch M, and Walsh B. 1998. Genetics and analysis of quantitative traits. Sinauer Associates, Sunderland, MA.

Lyttle TW. 1991. Segregation distorters. Annu Rev Genet 25:511–57.

Ma H. 2012. Endosymbiosis and photosynthetic animals. Molecular Evolution Forum: 9 May 2012. http://molecularevolutionforum.blogspot.co.uk/2012/05/endosymbiosis-and-photosynthetic.html.

Ma H, and dePamphilis C. 2000. The ABCs of floral evolution. Cell 101:5–8.

Macdowell EC. 1917. Bristle inheritance in *Drosophila*. II Selection. J Exp Zool 23:109–46.

Mackay TFC. 2010. Mutations and quantitative genetic variation: lessons from *Drosophila*. Philos T Roy Soc B 365:1229–39.

Mackay TFC, Lyman RF, and Lawrence F. 2005. Polygenic mutation in *Drosophila melanogaster*: mapping spontaneous mutations affecting sensory bristle number. Genetics 170:1723–35.

Magni GE. 1969. Spontaneous mutations. Proc Int Cong Genet 12 (3):247–59. Tokyo.

Maheshwari S, Wang J, and Barbash DA. 2008. Recurrent positive selection of the *Drosophila* hybrid incompatibility gene Hmr. Mol Biol Evol 25:2421–30.

Makalowski W. 2001. Are we polyploids? A brief history of one hypothesis. Genome Res 11:667-670.

Malecot G. 1948. Les mathematiques de l'heredite. Masson et Cie., Paris.

Malecot G. 1969. The mathematics of heredity. (Translated by D. M. Yermanos from "Les mathematiques de l'heredite"). Freeman, San Francisco.

Malnic B, Hirono J, Sato T, and Buck LB. 1999. Combinatorial receptor codes for odors. Cell 96:713–23.

Mandl B, Brandt WF, Superti-Furga G, Graninger PG, Birnstiel ML et al. 1997. The five cleavage-stage (CS) histones of the sea urchin are encoded by a maternally expressed family of replacement histone genes: functional equivalence of the CS H1 and frog H1M (B4) proteins. Mol Cell Biol 17:1189–200.

Mank JE. 2009. The W, X, Y and Z of sex-chromosome dosage compensation. Trends Genet 25:226–33.

Margoliash E. 1963. Primary structure and evolution of cytochrome C. Proc Natl Acad Sci USA 50:672–9.

Margoliash E, and Smith EL. 1965. Structural and functional aspects of cytochrome c in relation to evolution. In V Bryson, and H J Bogel, eds. Evolving genes and proteins. Academic Press, New York.

Martin W, Rujan T, Richly E, Hansen A, Cornelsen S et al. 2002. Evolutionary analysis of *Arabidopsis*, cyanobacterial, and chloroplast genomes reveals plastid phylogeny and thousands of cyanobacterial genes in the nucleus. Proc Natl Acad Sci USA 99:12246–51.

Maruyama T, and Crow JF. 1975. Heterozygous effects of x-ray induced mutations on viability of *Drosophila melanogaster*. Mutat Res 27:241–8.

Maruyama T, and Kimura M. 1980. Genetic variability and effective population size when local extinction and recolonization of subpopulations are frequent. Proc Natl Acad Sci USA 77:6710–14.

Maruyama T, and Nei M. 1981. Genetic variability maintained by mutation and overdominant selection in finite populations. Genetics 98:441–59.

Masly JP, Jones CD, Noor MA, Locke J, and Orr HA. 2006. Gene transposition as a cause of hybrid sterility in *Drosophila*. Science 313:1448–50.

Masternak K, Peyraud N, Krawczyk M, Barras E, and Reith W. 2003. Chromatin remodeling and extragenic transcription at the MHC class II locus control region. Nat Immunol 4:132–7.

Mather K. 1948. Biometrical genetics. Methuen, London.

Mather K. 1969. Selection through competition. Heredity 24:529–40.

Matsuda F, Ishii K, Bourvagnet P, Kuma K, Hayashida H et al. 1998. The complete nucleotide sequence of the human immunoglobulin heavy chain variable region locus. J Exp Med 188:2151–62.

Matsui A, Go Y, and Niimura Y. 2010. Degeneration of olfactory receptor gene repertories in primates: no direct link to full trichromatic vision. Mol Biol Evol 27:1192–200.

Matsuo Y, and Yamazaki T. 1989. tRNA derived insertion element in histone gene repeating unit of *Drosophila melanogaster*. Nucleic Acids Res 17:225–38.

Mattick JS. 2011. The central role of RNA in human development and cognition. FEBS Lett 585:1600–16.

Maxam AM, and Gilbert W. 1977. A new method for sequencing DNA. Proc Natl Acad Sci USA 74:560–4.

Maxson R, Cohn R, Kedes L, and Mohun T. 1983. Expression and organization of histone genes. Annu Rev Genet 17:239–77.

Mayer WE, Jonker M, Klein D, Ivanyi P, van Seventer G et al. 1988. Nucleotide sequences of chimpanzee MHC class I alleles: evidence for trans-species mode of evolution. EMBO J 7:2765–74.

Maynard Smith J. 1964. Group selection and kin selection. Nature 201:1145–7.

Maynard Smith J. 1968. "Haldane's dilemma" and the rate of evolution. Nature 219:1114–16.

Maynard Smith J. 1978. The evolution of sex. Cambridge University Press, New York.

Maynard Smith J. 1989. Evolutionary genetics. Oxford University Press, New York.

Mayr E. 1942. Systematics and the origin of species. Columbia University Press, New York.

Mayr E. 1954. Change of genetic environment and evolution. In J. Huxley, A. C. Hardy, and E. B. Ford, eds. Evolution as a process. Allen and Unwin, London.

Mayr E. 1959. Where are we? Cold Spring Harb Symp Quant Biol 24:1–24.

Mayr E. 1963. Animal species and evolution. Harvard University Press, Cambridge, MA.

Mayr E. 1965. Discussion. In V Bryson, and H J Vogel, eds. Evolving Genes and Proteins, pp. 293–4. Academic Press, New York.

Mayr E. 1970. Populations, species, and evolution. Harvard University Press, Cambridge, MA.

Mayr E. 1982. The growth of biological thought. Harvard University Press, Cambridge, MA.

Mayr E. 1997. The objects of selection. Proc Natl Acad Sci USA 94:2091–4.

McCarthy EM, Asmussen MA, and Anderson WW. 1995. A theoretical assessment of recombinational speciation. Heredity 74:502–9.

McConnell TJ, Talbot WS, McIndoe RA, and Wakeland EK. 1988. The origin of MHC class II gene polymorphism within the genus *Mus*. Nature 332:651–4.

McDonald JH, and Kreitman M. 1991. Adaptive protein evolution at the *Adh* locus in *Drosophila*. Nature 351:652–4.

McGowen MR, Clark C, and Gatesy J. 2008. The vestigial olfactory receptor subgenome of odontocete whales: phylogenetic congruence between gene-tree reconciliation and supermatrix methods. Syst Biol 57:574–90.

McKusick VA. 1986. Mendelian inheritance in man: catalogs of autosomal dominant, autosomal recessive, and X-linked phenotypes. John Hopkins University Press, Baltimore.

Mellor AL, Weiss EH, Ramachandran K, and Flavell RA. 1983. A potential donor gene for the bm1 gene conversion event in the C57BL mouse. Nature 306:792–5.

Mendel G. 1866. Versuche uber Pflanzenhybriden. Verh Naturforsch Ver Brunn 4:3–47.

Meyer BJ. 2005. X-Chromosome dosage compensation. WormBook 1:1–14.

Meyer BJ. 2010. Targeting X chromosomes for repression. Curr Opin Genet Dev 20:179–89.

Michaut L, Flister S, Neeb M, White KP, Certa U et al. 2003. Analysis of the eye developmental pathway in *Drosophila* using DNA microarrays. Proc Natl Acad Sci USA 100:4024–9.

Michelmore RW, and Meyers BC. 1998. Clusters of resistance genes in plants evolve by divergent selection and a birth-and-death process. Genome Res 8:1113–30.

Milkman RD. 1967. Heterosis as a major cause of heterozygosity in nature. Genetics 55:493–5.

Miura F, Tsukamoto K, Mehta RB, Naruse K, Magtoon W et al. 2010a. Transspecies dimorphic allelic lineages of the proteasome subunit beta-type 8 gene (PSMB8) in the teleost genus *Oryzias*. Proc Natl Acad Sci USA 107:21599–604.

Miura I, Ohtani H, Nakamura M, Ichikawa Y, and Saitoh K. 1998. The origin and differentiation of the heteromorphic sex chromosomes Z, W, X, and Y in the frog *Rana rugosa*, inferred from the sequences of a sex-linked gene, ADP/ ATP translocase. Mol Biol Evol 15:1612–19.

Miura I, Ohtani H, and Ogata M. 2012. Independent degeneration of W and Y sex chromosomes in frog *Rana rugosa*. Chromosome Res 20:47–55.

Miura K, Ikeda M, Matsubara A, Song XJ, Ito M et al. 2010b. *OsSPL14* promotes panicle branching and higher grain productivity in rice. Nat Genet 42:545–9.

Miura S, Zhang Z, and Nei M. 2013. Random fluctuation of selection coefficients and the extent of nucleotide variation in human populations. Proc Natl Acad Sci USA 110:10676–81.

Miyashita NT. 2001. DNA variation in the 5′ upstream region of the *Adh* locus of the wild plants *Arabidopsis thaliana* and *Arabis gemmifera*. Mol Biol Evol 18:164–71.

Miyata T, and Yasunaga T. 1981. Rapidly evolving mouse alpha-globin-related pseudo gene and its evolutionary history. Proc Natl Acad Sci USA 78:450–3.

Mizuta Y, Harushima Y, and Kurata N. 2010. Rice pollen hybrid incompatibility caused by reciprocal gene loss of duplicated genes. Proc Natl Acad Sci USA 107:20417–22.

Moran NA. 1996. Accelerated evolution and Muller's ratchet in endosymbiotic bacteria. Proc Natl Acad Sci USA 93:2873–8.

Moran NA, and Degnan PH. 2006. Functional genomics of Buchnera and the ecology of aphid hosts. Mol Ecol 15:1251–61.

Moran NA, and Jarvik T. 2010. Lateral transfer of genes from fungi underlies carotenoid production in aphids. Science 328:624–7.

Moran NA, Munson MA, Baumann P, and Ishikawa H. 1993. A molecular clock in endosymbiotic bacteria is calibrated using the insect hosts. P Roy Soc B-Biol Sci 253:167–71.

Morar M, and Wright GD. 2010. The genomic enzymology of antibiotic resistance. Annu Rev Genet 44:25–51.

Morgan TH. 1903. Evolution and adaptation. Macmillan, New York.

Morgan TH. 1916. A critique of the theory of evolution. Princeton University Press, Princeton.

Morgan TH. 1925. Evolution and genetics. Princeton University Press, Princeton.

Morgan TH. 1932. The scientific basis of evolution. W. W. Norton, New York.

Morgan TH, Sturtevant AH, Muller HJ, and Bridges CB. 1915. The mechanism of Mendelian heredity. Henry Holt and Company, New York.

Moriwaki D. 1940. Enhanced crossing over in the second chromosome of *Drosophila ananassae*. Japan J Genet 16:37–48.

Mortiz A. 2010. The origin of life. http://www.talkorigins.org/faqs/abioprob/originoflife.html.

Motulsky AG. 1964. Hereditary red cell traits and malaria. Am J Trop Med Hyg 13:Suppl147–58.

Mower JP, Touzet P, Gummow JS, Delph LF, and Palmer JD. 2007. Extensive variation in synonymous substitution rates in mitochondrial genes of seed plants. BMC Evol Biol 7:135.

Mukai T, and Burdick AB. 1959. Single gene heterosis associated with a second chromosome recessive lethal in *Drosophila melanogaster*. Genetics 44:211–32.

Mukai T, Chigusa SI, Mettler LE, and Crow JF. 1972. Mutation rate and dominance of genes affecting viability in *Drosophila melanogaster*. Genetics 72:335–5.

Muller HJ. 1914. A gene for the fourth chromosome of *Drosophila*. J Exp Zool 17:325–36.

Muller HJ. 1925. Why polyploidy is rarer in animals than in plants. Am Nat 59:346–53.

Muller HJ. 1929. The gene as the basis of life. Proc Internat'l Congr Plant Sciences 1:897–921.

Muller HJ. 1932. Some genetic aspects of sex. Am Nat 68:118–38.

Muller HJ. 1936. Bar duplication. Science 83:528–30.

Muller HJ. 1940. Bearing of the *Drosophila* work on systematics. Clarendon Press, Oxford.

Muller HJ. 1942. Isolating mechanisms, evolution and temperature. Biol Sympos 6:71–125.

Muller HJ. 1950. Our load of mutations. Am J Hum Genet 2:111–76.

Muller HJ. 1959. Advances in radiation mutagenesis through studies on *Drosophila*. In Progress in nuclear energy, pp. 146–60. Pergamon Press, New York.

Muller HJ. 1964. The relation of recombination to mutational advance. Mutat Res 106:2–9.

Muller HJ. 1967. The gene material as the initiator and the organizing basis of life. In A Brink, ed. Heritage from Mendel, pp. 419–48. The University of Wisconsin Press, Madison, WI.

Muller HJ, and Altenburg E. 1919. The rate of change of hereditary factors in *Drosophila*. Proc Soc Exp Biol Med 17:10–14.

Muse SV, and Gaut BS. 1994. A likelihood approach for comparing synonymous and nonsynonymous nucleotide substitution rates, with application to the chloroplast genome. Mol Biol Evol 11:715–24.

Myles S, Tang K, Somel M, Green RE, Kelso J et al. 2008. Identification and analysis of genomic regions with large between-population differentiation in humans. Ann Hum Genet 72:99–110.

Nachman MW. 2005. The genetic basis of adaptation: lessons from concealing coloration in pocket mice. Genetica 123:125–36.

Nachman MW, Hoekstra HE, and D'Agostino SL. 2003. The genetic basis of adaptive melanism in pocket mice. Proc Natl Acad Sci USA 100:5268–73.

Nagylaki T, and Petes TD. 1982. Intrachromosomal gene conversion and the maintenance of sequence homogeneity among repeated genes. Genetics 100:315–37.

Nam J, dePamphilis CW, Ma H, and Nei M. 2003. Antiquity and evolution of the MADS-box gene family controlling flower development in plants. Mol Biol Evol 20:1435–47.

Nam J, Dong P, Tarpine R, Istrail S, and Davidson EH. 2010. Functional *cis*-regulatory genomics for systems biology. Proc Natl Acad Sci USA 107:3930–5.

Nam J, Kim J, Lee S, An G, Ma H et al. 2004. Type I MADS-box genes have experienced faster birth-and-death evolution than type II MADS-box genes in angiosperms. Proc Natl Acad Sci USA 101:1910–15.

Nam J, and Nei M. 2005. Evolutionary change of the numbers of homeobox genes in bilateral animals. Mol Biol Evol 22:2386–94.

Nandi H. 1936. The chromosome morphology, secondary association and origin of cultivated rice. J Genet 33: 315–36.

Nathans J, Thomas D, and Hogness DS. 1986. Molecular genetics of human color vision: the genes encoding blue, green, and red pigments. Science 232:193–202.

Near TJ, Dornburg A, Kuhn KL, Eastman JT, Pennington JN et al. 2012. Ancient climate change, antifreeze, and the evolutionary diversification of Antarctic fishes. Proc Natl Acad Sci USA 109:3434–9.

Nei M. 1964. Effects of linkage and epistasis on the equilibrium frequencies of lethal genes. II. Numerical solutions. Jpn J Genet 39:7–25.

Nei M. 1967. Modification of linkage intensity by natural selection. Genetics 57:625–641.

Nei M. 1968. Evolutionary change of linkage intensity. Nature 218:1160–1.

Nei M. 1969a. Gene duplication and nucleotide substitution in evolution. Nature 221:40–2.

Nei M. 1969b. Linkage modifications and sex difference in recombination. Genetics 63:681–99.

Nei M. 1970. Accumulation of nonfunctional genes on sheltered chromosomes. Am Nat 104:311–22.

Nei M. 1971. Fertility excess necessary for gene substitution in regulated populations. Genetics 68:169–84.

Nei M. 1972. Genetic distance between populations. Am Nat 106:283–92.

Nei M. 1975. Molecular population genetics and evolution. American Elsevier Publishing Company, Inc., New York.

Nei M. 1976. Mathematical models of speciation and genetic distance. In S Karlin, and E Nevo, eds. Population Genetics and Ecology, pp. 723–65. Academic Press, New York.

Nei M. 1980a. Protein polymorphism and the SAS-CFF model. Genetics 94:1085–7.

Nei M. 1980b. Stochastic theory of population genetics and evolution., In C Barigozzi, ed. Vito Volterra Symposium of Mathematical Models in Biology, pp. 17–47. Springer-Verlag, Berlin.

Nei M. 1983. Genetic polymorphism and the role of mutation in evolution. In M Nei, and R K Koehn, eds. Evolution of genes and proteins, pp. 165–90. Sinauer Assoc., Sunderland.

Nei M. 1984. Genetic polymorphism and neomutationism. In G S Mani, ed. Evolutionary dynamics of genetic diversity, pp. 214–41. Springer-Verlag, Heidelberg.

Nei M. 1987. Molecular evolutionary genetics. Columbia University Press, New York.

Nei M. 2005. Selectionism and neutralism in molecular evolution. Mol Biol Evol 22:2318–42.

Nei M. 2007. The new mutation theory of phenotypic evolution. Proc Natl Acad Sci USA 104:12235–42.

Nei M. 2012. Soldier ants and caste evolution. Molecular Evolution Forum: 11 April 2012. http://molecularevolutionforum.blogspot.co.uk/2012/04/soldier-ants-and-caste-evolution.html.

Nei M, Fuerst PA, and Chakraborty R. 1976. Testing the neutral mutation hypothesis by distribution of single locus heterozygosity. Nature 262:491–3.

Nei M, and Graur D. 1984. Extent of protein polymorphism and the neutral mutation theory. Evol Biol 17:73–118.

Nei M, Gu X, and Sitnikova T. 1997. Evolution by the birth-and-death process in multigene families of the vertebrate immune system. Proc Natl Acad Sci USA 94:7799–806.

Nei M, and Hughes AL. 1992. Balanced polymorphism and evolution by the birth-and-death process in the MHC loci. In K Tsuji, M Aizawa, and T Sasazuki, eds. Proceedings of the 11th Histocompatibility Workshop and Conference, pp. 27–38. Oxford University Press, Oxford.

Nei M, Kojima KI, and Schaffer HE. 1967. Frequency changes of new inversions in populations under mutation-selection equilibria. Genetics 57:741–50.

Nei M, and Kumar S. 2000. Molecular evolution and phylogenetics. Oxford University Press, Oxford.

Nei M, and Maruyama T. 1975. Lewontin-Krakauer test for neutral genes. Genetics 80:395.

Nei M, Maruyama T, and Chakraborty R. 1975. The bottleneck effect and genetic variability in populations. Evolution 29:1–10.

Nei M, Maruyama T, and Wu CI. 1983. Models of evolution of reproductive isolation. Genetics 103:557–79.

Nei M, Niimura Y, and Nozawa M. 2008. The evolution of animal chemosensory receptor gene repertoires: roles of chance and necessity. Nat Rev Genet 9:951–63.

Nei M, and Nozawa M. 2011. Roles of mutation and selection in speciation: from Hugo de Vries to the modern genomic era. Genome Biol Evol 3:812–29.

Nei M, Rogozin IB, and Piontkivska H. 2000. Purifying selection and birth-and-death evolution in the ubiquitin gene family. Proc Natl Acad Sci USA 97:10866–71.

Nei M, and Rooney AP. 2005. Concerted and birth-and-death evolution of multigene families. Annu Rev Genet 39:121–52.

Nei M, and Roychoudhury AK. 1972. Gene differences between Caucasian, Negro, and Japanese populations. Science 177:434–6.

Nei M, and Roychoudhury AK. 1973. Probability of fixation of nonfunctional genes at duplicate loci. Am Nat 107:362–72.

Nei M, Suzuki Y, and Nozawa M. 2010. The neutral theory of molecular evolution in the genomic era. Annu Rev Genom Hum G 11:265–89.

Nei M, and Yokoyama S. 1976. Effects of random fluctuation of selection intensity on genetic variability in a finite population. Jpn J Genet 51:355–69.

Nei M, and Zhang J. 1998. Molecular origin of species. Science 282:1428–9.

Nenoi M, Mita K, Ichimura S, and Kawano A. 1998. Higher frequency of concerted evolutionary events in rodents than in man at the polyubiquitin gene VNTR locus. Genetics 148:867–76.

Nguyen DK, and Disteche CM. 2006. Dosage compensation of the active X chromosome in mammals. Nat Genet 38:47–53.

Nielsen R, Hellmann I, Hubisz M, Bustamante C, and Clark AG. 2007. Recent and ongoing selection in the human genome. Nat Rev Genet 8:857–68.

Niimura Y, and Nei M. 2005. Evolutionary dynamics of olfactory receptor genes in fishes and tetrapods. Proc Natl Acad Sci USA 102:6039–44.

Nikaido M, Rooney AP, and Okada N. 1999. Phylogenetic relationships among cetartiodactyls based on insertions of short and long interpersed elements: hippopotamuses are the closest extant relatives of whales. Proc Natl Acad Sci USA 96:10261–6.

Nikolaidis N, Makalowska I, Chalkia D, Makalowski W, Klein J et al. 2005. Origin and evolution of the chicken leukocyte receptor complex. Proc Natl Acad Sci USA 102:4057–62.

Nilsson-Ehle H. 1909. Kreuzungsuntersuchungen an Hafer und Weizen. Lunds Universitets Arsskrift. 5:1–122.

Novick A, and Szilard L. 1950. Experiments with the chemostat on spontaneous mutations of bacteria. Proc Natl Acad Sci USA 36:708–19.

Nowak MA, Tarnita CE, and Wilson EO. 2010. The evolution of eusociality. Nature 466:1057–62.

Nowak MA, Tarnita CE, and Wilson EO. 2011. Nowak et al. reply. Nature 471:E9–10.

Nozawa M, Kawahara Y, and Nei M. 2007. Genomic drift and copy number variation of sensory receptor genes in humans. Proc Natl Acad Sci USA 104:20421–6.

Nozawa M, Miura S, and Nei M. 2010. Origins and evolution of microRNA genes in *Drosophila* species. Genome Biol Evol 2:180–9.

Nozawa M, Suzuki Y, and Nei M. 2009a. Response to Yang et al.: Problems with Bayesian methods of detecting positive selection at the DNA sequence level. Proc Natl Acad Sci USA 106 e96.

Nozawa M, Suzuki Y, and Nei M. 2009b. Reliabilities of identifying positive selection by the branch-site and the site-prediction methods. Proc Natl Acad Sci USA 106 (16):6700–5.

O'Hara RB. 2005. Comparing the effects of genetic drift and fluctuating selection on genotype frequency changes in the scarlet tiger moth. P Roy Soc B-Biol Sci 272:211–17.

Ochiai K, Yamanaka T, Kuimura K, and Sawada O. 1959. Inheritance of drug resistance (and its transfer) between *Shigella* strains and between *Shigella* and *E.coli* strains (in Japanese). Hihon Iji Shimpor 1861:34.

Oetting WS, Garrett SS, Brott M, and King RA. 2005. P gene mutations associated with oculocutaneous albinism type II (OCA2). Hum Mutat 25:323.

Ohno S. 1967. Sex chromosomes and sex-linked genes. Springer-Verlag, New York.

Ohno S. 1970. Evolution by gene duplication. Springer, Berlin.

Ohno S. 1972a. So much "junk" DNA in our genome. Brookhaven Symp Biol 23:366–70.

Ohno S. 1972b. An argument for the genetic simplicity of man and other mammals. J Human Evol 1:651–62.

Ohno S. 1998. The notion of the Cambrian pananimalia genome and a genomic difference that separated vertebrates from invertebrates. Prog Mol Subcell Biol 21:97–117.

Ohta T. 1971. Associative overdominance caused by linked detrimental mutations. Genet Res 18:277–86.

Ohta T. 1972. Fixation probability of a mutant influenced by random fluctuation of selection intensity. Genet Res 19:33–8.

Ohta T. 1973. Slightly deleterious mutant substitutions in evolution. Nature 246:96–8.

Ohta T. 1974. Mutational pressure as the main cause of molecular evolution and polymorphism. Nature 252:351–4.

Ohta T. 1982. Allelic and nonallelic homology of a supergene family. Proc Natl Acad Sci USA 79:3251–4.

Ohta T. 1983. On the evolution of multigene families. Theor Popul Biol 23:216–40.

Ohta T. 1992. The nearly neutral theory of molecular evolution. Annu Rev Ecol Syst 23:263–86.

Ohta T. 2002. Near-neutrality in evolution of genes and gene regulation. Proc Natl Acad Sci USA 99:16134–7.

Oka HI. 1953. The mechanisms of sterility in the intervarietal hybrids. Phylogenetic differentiation of cultivated rice. VI. (In Japanese with English summary.) Japan J Breed 2:217–24.

Oka HI. 1957. Genic analysis for the sterility of hybrids between distantly related varieties of cultivated rice. J Genet 55:397–409.

Oka HI. 1974. Analysis of genes controlling f(1) sterility in rice by the use of isogenic lines. Genetics 77:521–34.

Oliver PL, Goodstadt L, Bayes JJ, Birtle Z, Roach KC et al. 2009a. Accelerated evolution of the Prdm9 speciation gene across diverse metazoan taxa. PLoS Genet 5:e1000753.

Oliver SN, Finnegan EJ, Dennis ES, Peacock WJ, and Trevaskis B. 2009b. Vernalization-induced flowering in cereals is associated with changes in histone methylation at the VERNALIZATION1 gene. Proc Natl Acad Sci USA 106:8386–91.

Oliveri P, Tu Q, and Davidson EH. 2008. Global regulatory logic for specification of an embryonic cell lineage. Proc Natl Acad Sci USA 105:5955–62.

Opazo JC, Palma RE, Melo F, and Lessa EP. 2005. Adaptive evolution of the insulin gene in caviomorph rodents. Mol Biol Evol 22:1290–8.

Orgel LE. 2004. Prebiotic chemistry and the origin of the RNA world. Crit Rev Biochem Mol Biol 39:99–123.

Orr HA. 1995. The population genetics of speciation: the evolution of hybrid incompatibilities. Genetics 139:1805–13.

Orr HA. 1996. Dobzhansky, Bateson, and the genetics of speciation. Genetics 144:1331–5.

Orr HA, and Kim Y. 1998. An adaptive hypothesis for the evolution of the Y chromosome. Genetics 150:1693–8.

Ossowski S, Schneeberger K, Lucas-Lledo JI, Warthmann N, Clark RM et al. 2010. The rate and molecular spectrum of spontaneous mutations in *Arabidopsis thaliana*. Science 327:92–4.

Ota T, and Nei M. 1994. Divergent evolution and evolution by the birth-and-death process in the immunoglobulin VH gene family. Mol Biol Evol 11:469–82.

Pal C, and Hurst LD. 2003. Evidence for co-evolution of gene order and recombination rate. Nat Genet 33:392–5.

Palmer JD, and Logsdon JM, Jr. 1991. The recent origins of introns. Curr Opin Genet Dev 1:470–7.

Pamilo P, Nei M, and Li WH. 1987. Accumulation of mutations in sexual and asexual populations. Genet Res 49:135–46.

Park Y, and Kuroda MI. 2001. Epigenetic aspects of X-chromosome dosage compensation. Science 293:1083–5.

Pavelitz T, Rusche L, Matera AG, Scharf JM, and Weiner AM. 1995. Concerted evolution of the tandem array encoding primate U2 snRNA occurs in situ, without changing the cytological context of the RNU2 locus. EMBO J 14:169–77.

Payer B, and Lee JT. 2008. X chromosome dosage compensation: how mammals keep the balance. Annu Rev Genet 42:733–72.

Payne F. 1918. The effect of artificial selection on bristle number in *Drosophila ampelophila* and its interpretation. Proc Natl Acad Sci USA 4:55–8.

Perutz MF. 1983. Species adaptation in a protein molecule. Mol Biol Evol 1:1–28.

Perutz MF, Bauer C, Gros G, Leclercq F, Vandecasserie C et al. 1981. Allosteric regulation of crocodilian haemoglobin. Nature 291:682–4.

Peter IS, and Davidson EH. 2011. Evolution of gene regulatory networks controlling body plan development. Cell 144:970–85.

Petschow D, Wurdinger I, Baumann R, Duhm J, Braunitzer G et al. 1977. Causes of high blood O_2 affinity of animals living at high altitude. J Appl Physiol 42:139–43.

Pettigrew JD. 1999. Electroreception in monotremes. J Exp Biol 202:1447–54.

Phadnis N, and Orr HA. 2009. A single gene causes both male sterility and segregation distortion in *Drosophila* hybrids. Science 323:376–9.

Piatigorsky J. 2007. Gene sharing and evolution: the diversity of protein functions. Harvard University Press, Cambridge, MA.

Piatigorsky J, O'Brien WE, Norman BL, Kalumuck K, Wistow GJ et al. 1988. Gene sharing by delta-crystallin and argininosuccinate lyase. Proc Natl Acad Sci USA 85:3479–83.

Piccinini M, Kleinschmidt T, Jurgens KD, and Braunitzer G. 1990. Primary structure and oxygen-binding properties of the hemoglobin from guanaco (*Lama guanacoe*, Tylopoda). Biol Chem Hoppe Seyler 371:641–8.

Piontkivska H, Rooney AP, and Nei M. 2002. Purifying selection and birth-and-death evolution in the histone H4 gene family. Mol Biol Evol 19:689–97.

Podlaha O, and Zhang J. 2010. Pseudogenes and their evolution. In Encyclopedia of Life Sciences (eLS). John Wiley & Sons, Ltd., Chichester.

Ponicsan SL, Kugel JF, and Goodrich JA. 2010. Genomic gems: SINE RNAs regulate mRNA production. Curr Opin Genet Dev 20:149–55.

Pouteau S, Carre I, Gaudin V, Ferret V, Lefebvre D et al. 2008. Diversification of photoperiodic response patterns in a collection of early-flowering mutants of *Arabidopsis*. Plant Physiol 148:1465–73.

Presgraves DC. 2008. Sex chromosomes and speciation in *Drosophila*. Trends Genet 24:336–43.

Presgraves DC, Balagopalan L, Abmayr SM, and Orr HA. 2003. Adaptive evolution drives divergence of a hybrid inviability gene between two species of *Drosophila*. Nature 423:715–19.

Presgraves DC, and Stephan W. 2007. Pervasive adaptive evolution among interactors of the *Drosophila* hybrid inviability gene, *Nup96*. Mol Biol Evol 24:306–14.

Price GR. 1972. Fisher's "fundamental theorem" made clear. Ann Hum Genet 36:129–40.

Protas M, Conrad M, Gross JB, Tabin C, and Borowsky R. 2007. Regressive evolution in the Mexican cave tetra, *Astyanax mexicanus*. Curr Biol 17:452–4.

Provine WB. 1971. The origins of theoretical population genetics. University Chicago Press, Chicago.

Provine WB. 1980. Genetics. In E Mayr and W B Provine, eds. The evolutionary synthesis, pp. 51–8. Harvard University Press, Cambridge, MA.

Provine WB. 1986. Sewall Wright and evolutionary biology. University Chicago Press, Chicago.

Provine WB. 2004. Ernst Mayr: Genetics and speciation. Genetics 167:1041–6.

Purugganan MD. 2000. The molecular population genetics of regulatory genes. Mol Ecol 9:1451–61.

Rajakumar R, San Mauro D, Dijkstra MB, Huang MH, Wheeler DE et al. 2012. Ancestral developmental potential facilitates parallel evolution in ants. Science 335:79–82.

Rajalingam R, Parham P, and Abi-Rached L. 2004. Domain shuffling has been the main mechanism forming new hominoid killer cell Ig-like receptors. J Immunol 172:356–69.

Ramsey M, and Crews D. 2009. Steroid signaling and temperature-dependent sex determination—Reviewing the evidence for early action of estrogen during ovarian determination in turtles. Semin Cell Dev Biol 20:283–92.

Raymond CS, Shamu CE, Shen MM, Seifert KJ, Hirsch B et al. 1998. Evidence for evolutionary conservation of sex-determining genes. Nature 391:691–5.

Redon R, Ishikawa S, Fitch KR, Feuk L, Perry GH et al. 2006. Global variation in copy number in the human genome. Nature 444:444–54.

Rees JL. 2003. Genetics of hair and skin color. Annu Rev Genet 37:67–90.

Reid JB, and Ross JJ. 2011. Mendel's genes: toward a full molecular characterization. Genetics 189:3–10.

Renner O. 1917. Versuche uber die gametische Konstitution der Oenotheren. Zeitchr ind Abst—u Vererb 18:121–294.

Rensing SA, Lang D, Zimmer AD, Terry A, Salamov A et al. 2008. The *Physcomitrella* genome reveals evolutionary insights into the conquest of land by plants. Science 319:64–9.

Rice WR. 1984. Sex-chromosomes and the evolution of sexual dimorphism. Evolution 38:735–2.

Rice WR. 1996. Evolution of the Y sex chromosome in animals. BioScience 46:331–43.

Ridley M. 2003. Evolution, 3rd ed. Wiley-Blackwell.

Rieseberg LH. 1997. Hybrid origins of plant species. Annual Review of Ecology and Systematics 28:359–89.

Rieseberg LH. 2001. Chromosomal rearrangements and speciation. Trends Ecol Evol 16:351–8.

Rieseberg LH, and Willis JH. 2007. Plant speciation. Science 317:910–14.

Robertson A. 1967. The nature of quantitative genetics. In R A Brink, eds. Heritage from Mendel, pp. 265–80. University of Wisconsin Press, Madison, WI.

Robertson A. 1975a. Gene frequency distributions as a test of selective neutrality. Genetics 81:775–85.

Robertson A. 1975b. Letters to the editors: Remarks on the Lewontin-Krakauer test. Genetics 80:396.

Robinett CC, O'Connor A, and Dunaway M. 1997. The repeat organizer, a specialized insulator element within the intergenic spacer of the Xenopus rRNA genes. Mol Cell Biol 17:2866–75.

Robinson GE, Grozinger CM, and Whitfield CW. 2005. Sociogenomics: social life in molecular terms. Nat Rev Genet 6:257–70.

Roelofs WL, and Rooney AP. 2003. Molecular genetics and evolution of pheromone biosynthesis in Lepidoptera. Proc Natl Acad Sci USA 100:14599.

Rokas A. 2008. The origins of multicellularity and the early history of the genetic toolkit for animal development. Annu Rev Genet 42:235–51.

Ronshaugen M, Biemar F, Piel J, Levine M, and Lai EC. 2005. The *Drosophila* microRNA iab-4 causes a dominant homeotic transformation of halteres to wings. Genes Dev 19:2947–52.

Rooney AP. 2009. Evolution of moth sex pheromone desaturases. Ann N Y Acad Sci 1170:506–10.

Rooney AP, Piontkivska H, and Nei M. 2002. Molecular evolution of the nontandemly repeated genes of the histone 3 multigene family. Mol Biol Evol 19:68–75.

Rouse GW, Goffredi SK, and Vrijenhoek RC. 2004. *Osedax*: bone-eating marine worms with dwarf males. Science 305:668–71.

Roychoudhury AK, and Nei M. 1988. Human polymorphic genes. Oxford University Press, New York.

Rumpho ME, Pelletreau KN, Moustafa A, and Bhattacharya D. 2011. The making of a photosynthetic animal. J Exp Biol 214:303–11.

Rutter M. 2006. Genes and behavior: nature-nurture interplay explained. Blackwell Publishing.

Sabeti PC, Reich DE, Higgins JM, Levine HZ, Richter DJ et al. 2002. Detecting recent positive selection in the human genome from haplotype structure. Nature 419:832–7.

Sabeti PC, Schaffner SF, Fry B, Lohmueller J, Varilly P et al. 2006. Positive natural selection in the human lineage. Science 312:1614–20.

Sabeti PC, Varilly P, Fry B, Lohmueller J, Hostetter E et al. 2007. Genome-wide detection and characterization of positive selection in human populations. Nature 449:913–18.

Sakai K. 1935. Chromosome study of *Oryza sativa L.* I. The secondary association of the meiotic chromosomes. Japan J Genet 44:149–56.

Sakamoto K, and Okada N. 1985. Rodent type 2 Alu family, rat identifier sequence, rabbit C family, and bovine or goat 73-bp repeat may have evolved from tRNA genes. J Mol Evol 22:134–140.

Salvini-Plawen L, and Mayr E. 1961. On the evolution of photoreceptors and eyes. Evol Biol 10:207–63.

Sandler L, Hiraizumi Y, and Sandler I. 1959. Meiotic drive in natural populations of *Drosophila melanogaster*. I. The cytogenetic basis of segregation-distortion. Genetics 44:233–50.

Sandmann T, Girardot C, Brehme M, Tongprasit W, Stolc V et al. 2007. A core transcriptional network for early mesoderm development in *Drosophila melanogaster*. Genes Dev 21:436–49.

Sanger F, Nicklen S, and Coulson AR. 1977. DNA sequencing with chain-terminating inhibitors. Proc Natl Acad Sci USA 74:5463–7.

Santiago E, Albornoz J, Dominguez A, Toro MA, and Lopez-Fanjul C. 1992. The distribution of spontaneous mutations on quantitative traits and fitness in *Drosophila melanogaster*. Genetics 132:771–81.

Sarre SD, Ezaz T, and Georges A. 2011. Transitions between sex-determining systems in reptiles and amphibians. Annu Rev Genom Hum G 12:391–406.

Sasaki T, Nishihara H, Hirakawa M, Fujimura K, Tanaka M et al. 2008. Possible involvement of SINEs in mammalian-specific brain formation. Proc Natl Acad Sci USA 105:4220–4225.

Sato A, Tichy H, O'HUigin C, Grant PR, Grant BR et al. 2001. On the origin of Darwin's finches. Mol Biol Evol 18:299–311.

Satta Y, O'HUigin C, Takahata N, and Klein J. 1994. Intensity of natural selection at the major histocompatibility complex loci. Proc Natl Acad Sci USA 91:7184–8.

Sawamura K, Yamamoto MT, and Watanabe TK. 1993. Hybrid lethal systems in the *Drosophila melanogaster* species complex. II. The Zygotic hybrid rescue (*Zhr*) gene of *D. melanogaster*. Genetics 133:307–13.

Sawyer LA, Hennessy JM, Peixoto AA, Rosato E, Parkinson H et al. 1997. Natural variation in a *Drosophila* clock gene and temperature compensation. Science 278:2117–20.

Sawyer SA, Parsch J, Zhang Z, and Hartl DL. 2007. Prevalence of positive selection among nearly neutral amino acid replacements in *Drosophila*. Proc Natl Acad Sci USA 104:6504–10.

Scannell DR, Byrne KP, Gordon JL, Wong S, and Wolfe KH. 2006. Multiple rounds of speciation associated with reciprocal gene loss in polyploid yeasts. Nature 440:341–5.

Scannell DR, Frank AC, Conant GC, Byrne KP, Woolfit M et al. 2007. Independent sorting-out of thousands of duplicated gene pairs in two yeast species descended from a whole-genome duplication. Proc Natl Acad Sci USA 104:8397–402.

Schartl M. 2004. A comparative view on sex determination in medaka. Mech Dev 121:639–45.

Sebat J, Lakshmi B, Troge J, Alexander J, Young J et al. 2004. Large-scale copy number polymorphism in the human genome. Science 305:525–8.

Sella G, Petrov DA, Przeworski M, and Andolfatto P. 2009. Pervasive natural selection in the *Drosophila* genome? PLoS Genet 5:e1000495.

Shapiro JA, Huang W, Zhang C, Hubisz MJ, Lu J et al. 2007. Adaptive genic evolution in the *Drosophila* genomes. Proc Natl Acad Sci USA 104:2271–76.

Shapiro MD, Marks ME, Peichel CL, Blackman BK, Nereng KS et al. 2004. Genetic and developmental basis of evolutionary pelvic reduction in threespine sticklebacks. Nature 428:717–23.

Sharp PM, and Li WH. 1987. The codon Adaptation Index—a measure of directional synonymous codon usage bias, and its potential applications. Nucleic Acids Res 15:1281–95.

Shaw A, Fortes PA, Stout CD, and Vacquier VD. 1995. Crystal structure and subunit dynamics of the abalone sperm lysin dimer: egg envelopes dissociate dimers, the monomer is the active species. J Cell Biol 130:1117–25.

Shaw CR. 1965. Electrophoretic variation in enzymes. Science 149:936–43.

Shepherd GM. 2004. The human sense of smell: are we better than we think? PLoS Biol 2:E146.

Sheppard PM, and Cook LM. 1962. The manifold effects of the *medionigra* gene of the moth *Panaxia dominula* and the maintenance of a polymorphism. Heredity 17:415–26.

Shoemaker CM, and Crews D. 2009. Analyzing the coordinated gene network underlying temperature-dependent sex determination in reptiles. Semin Cell Dev Biol 20:293–303.

Shoemaker-Daly CM, Jackson K, Yatsu R, Matsumoto Y, and Crews D. 2010. Genetic network underlying temperature-dependent sex determination is endogenously regulated by temperature in isolated cultured *Trachemys scripta* gonads. Dev Dyn 239:1061–75.

Simmons MJ, and Crow JF. 1977. Mutations affecting fitness in *Drosophila* populations. Annu Rev Genet 11:49–78.

Simpson GG. 1944. Tempo and mode in evolution. Columbia University Press, New York.

Simpson GG. 1949. The meaning of evolution. Yale University Press, New Haven.

Simpson GG. 1953. The major features of evolution. Yale University Press, New Haven.

Simpson GG. 1964. Organisms and molecules in evolution. Science 146:1535–8.

Simpson GG, Gendall AR, and Dean C. 1999. When to switch to flowering. Annu Rev Cell Dev Biol 15:519–50.

Sims GE, and Kim SH. 2011. Whole-genome phylogeny of *Escherichia coli/Shigella* group by feature frequency profiles (FFPs). Proc Natl Acad Sci USA 108:8329–34.

Sinclair AH, Berta P, Palmer MS, Hawkins JR, Griffiths BL et al. 1990. A gene from the human sex-determining region encodes a protein with homology to a conserved DNA-binding motif. Nature 346:240–4.

Singer MF. 1982. SINEs and LINEs: highly repeated short and long interspersed sequences in mammalian genomes. Cell 28:433-434.

Skibinski DOF, and Ward RD. 1981. Relationships between allozyme heterozygosity and rates of divergence. Genet Res 38:71–92.

Slightom JL, Blechl AE, and Smithies O. 1980. Human fetal G gamma- and A gamma-globin genes: complete nucleotide sequences suggest that DNA can be exchanged between these duplicated genes. Cell 21:627–38.

Sloan DB, MacQueen AH, Alverson AJ, Palmer JD, and Taylor DR. 2010. Extensive loss of RNA editing sites in rapidly evolving *Silene* mitochondrial genomes: selection vs retroprocessing as the driving force. Genetics 185:1369–80.

Smith CA, Roeszler KN, Ohnesorg T, Cummins DM, Farlie PG et al. 2009. The avian Z-linked gene DMRT1 is required for male sex determination in the chicken. Nature 461:267–71.

Smith GP. 1976. Evolution of repeated DNA sequences by unequal crossover. Science 191:528–35.

Smith NG, and Eyre-Walker A. 2002. Adaptive protein evolution in *Drosophila*. Nature 415:1022–4.

Sokolowski MB. 1998. Genes for normal behavioral variation: recent clues from flies and worms. Neuron 21:463–6.

Stark A, Lin MF, Kheradpour P, Pedersen JS, Parts L et al. 2007. Discovery of functional elements in 12 *Drosophila* genomes using evolutionary signatures. Nature 450:219–32.

Stearns S, and Hoekstra R. 2005. Evolution: an introduction, 2nd ed. Oxford University Press, Oxford.

Stebbins GL. 1950. Variation and evolution in plants. Columbia University Press, New York.

Stebbins GL. 1966. Processes of organic evolution. Prentice-Hall, Englewood Cliffs, New Jersey.

Stefani G, and Slack FJ. 2008. Small non-coding RNAs in animal development. Nat Rev Mol Cell Biol 9:219–30.

Stephens SG. 1951. Possible significance of duplication in evolution. Adv Genet 4:247–65.

Stern C. 1962. William Weinberg. Genetics 47:1–5.

Stoltzfus A. 2006. Mutationism and the dual causation of evolutionary change. Evol Dev 8:304–17.

Storset AK, Slettedal IO, Williams JL, Law A, and Dissen E. 2003. Natural killer cell receptors in cattle: a bovine killer cell immunoglobulin-like receptor multigene family contains members with divergent signaling motifs. Eur J Immunol 33:980–90.

Strickberger MW. 1996. Evolution, 2nd ed. Jones & Bartlett, Sudbury, MA.

Sturtevant AH. 1923. Inheritance of direction of coiling in *Limnaea*. Science 58:269–70.

Sturtevant AH. 1925. The effects of unequal crossing over at the Bar locus in *Drosophila*. Genetics 10:117–47.

Sturtevant AH, and Dobzhansky T. 1936. Inversions in the third chromosome of wild races of *Drosophila pseudoobscura*, and their use in the study of the history of the species. Proc Natl Acad Sci USA 22:448–50.

Su C, and Nei M. 1999. Fifty-million-year-old polymorphism at an immunoglobulin variable region gene locus in the rabbit evolutionary lineage. Proc Natl Acad Sci USA 96:9710–15.

Su C, and Nei M. 2001. Evolutionary dynamics of the T-cell receptor VB gene family as inferred from the human and mouse genomic sequences. Mol Biol Evol 18:503–13.

Sueoka N. 1962. On the genetic basis of variation and heterogeneity of DNA base composition. Proc Natl Acad Sci USA 48:582–92.

Sun S, Ting CT, and Wu CI. 2004. The normal function of a speciation gene, *Odysseus*, and its hybrid sterility effect. Science 305:81–3.

Sunyaev S, Ramensky V, Koch I, Lathe W, III, Kondrashov AS et al. 2001. Prediction of deleterious human alleles. Hum Mol Genet 10:591–7.

Sutou S, Mitsui Y, and Tsuchiya K. 2001. Sex determination without the Y chromosome in two Japanese rodents *Tokudaia osimensis osimensis* and *Tokudaia osimensis spp.* Mamm Genome 12:17–21.

Suzuki Y, and Gojobori T. 1999. A method for detecting positive selection at single amino acid sites. Mol Biol Evol 16:1315–28.

Suzuki Y, and Nei M. 2002. Simulation study of the reliability and robustness of the statistical methods for detecting positive selection at single amino acid sites. Mol Biol Evol 19:1865–9.

Suzuki Y, and Nei M. 2004. False-positive selection identified by ML-based methods: examples from the *Sig1* gene of the diatom *Thalassiosira weissflogii* and the tax gene of a human T-cell lymphotropic virus. Mol Biol Evol 21:914–21.

Sved JA. 1968. Possible rates of gene substitution in evolution. Am Nat 102:283–93.

Sved JA, Reed TE, and Bodmer WF. 1967. The number of balanced polymorphisms that can be maintained in a natural population. Genetics 55:469–81.

Swofford DL, Olsen GJ, Waddell PJ, and Hillis DM. 1996. Phylogenetic inferences. In D. M. Hillis, C. Mortiz, and B. K. Mable, eds. Molecular systematics, 2nd ed, pp. 407–514. Sinauer Associates, Sunderland, MA.

Sykes R. 2010. The 2009 Garrod lecture: the evolution of antimicrobial resistance: a Darwinian perspective. J Antimicrob Chemoth 65:1842–52.

Syvanen M. 1985. Cross-species gene-transfer—implications for a new theory of evolution. J Theor Biol 112:333–43.

Taft RJ, Pheasant M, and Mattick JS. 2007. The relationship between non-protein-coding DNA and eukaryotic complexity. Bioessays 29:288–99.

Tajima F. 1989. Statistical method for testing the neutral mutation hypothesis by DNA polymorphism. Genetics 123:585–95.

Takahata N. 1982. Sexual recombination under the joint effects of mutation, selection, and random sampling drift. Theor Popul Biol 22:258–77.

Takahata N, and Kimura M. 1979. Genetic variability maintained in a finite population under mutation and autocorrelated random fluctuation of selection intensity. Proc Natl Acad Sci USA 76:5813–17.

Takahata N, and Nei M. 1990. Allelic genealogy under overdominant and frequency-dependent selection and polymorphism of major histocompatibility complex loci. Genetics 124:967–78.

Takezaki N, and Nei M. 2009. Genomic drift and evolution of microsatellite DNAs in human populations. Mol Biol Evol 26:1835–1840.

Takezaki N, Nei M, and Tamura K. 2010. POPTREE2: Software for constructing population trees from allele frequency data and computing other population statistics with Windows interface. Mol Biol Evol 27:747–752.

Tanabe Y, Hasebe M, Sekimoto H, Nishiyama T, Kitani M et al. 2005. Characterization of MADS-box genes in charophycean green algae and its implication for the evolution of MADS-box genes. Proc Natl Acad Sci USA 102:2436–41.

Tanaka T, and Nei M. 1989. Positive Darwinian selection observed at the variable-region genes of immunoglobulins. Mol Biol Evol 6:447–59.

Tang S, and Presgraves DC. 2009. Evolution of the *Drosophila* nuclear pore complex results in multiple hybrid incompatibilities. Science 323:779–82.

Tanzer A, Amemiya CT, Kim CB, and Stadler PF. 2005. Evolution of microRNAs located within Hox gene clusters. J Exp Zoolog B Mol Dev Evol 304:75–85.

Tanzer A, and Stadler PF. 2004. Molecular evolution of a microRNA cluster. J Mol Biol 339:327–35.

Tao Y, Hartl DL, and Laurie CC. 2001. Sex-ratio segregation distortion associated with reproductive isolation in *Drosophila*. Proc Natl Acad Sci USA 98:13183–8.

Tauber CA, and Tauber MJ. 1977. Sympatric speciation based on allelic changes at three loci: evidence from natural populations in two habitats. Science 197:1298–9.

Taylor JS, and Raes J. 2004. Duplication and divergence: the evolution of new genes and old ideas. Annu Rev Genet 38:615–43.

Templeton AR. 1980. The theory of speciation via the founder principle. Genetics 94:1011–38.

Templeton AR. 2008. The reality and importance of founder speciation in evolution. Bioessays 30:470–9.

Theissen G. 2001. Development of floral organ identity: stories from the MADS house. Curr Opin Plant Biol 4:75–85.

Tindall BJ, Grimont PA, Garrity GM, and Euzeby JP. 2005. Nomenclature and taxonomy of the genus *Salmonella*. Int J Syst Evol Microbiol 55:521–24.

Ting CT, Tsaur SC, Wu ML, and Wu CI. 1998. A rapidly evolving homeobox at the site of a hybrid sterility gene. Science 282:1501–4.

Tonegawa S. 1983. Somatic generation of antibody diversity. Nature 302:575–81.

Torrents D, Suyama M, Zdobnov E, and Bork P. 2003. A genome-wide survey of human pseudogenes. Genome Res 13:2559–67.

Trevaskis B, Hemming MN, Dennis ES, and Peacock WJ. 2007. The molecular basis of vernalization-induced flowering in cereals. Trends Plant Sci 12:352–7.

Trowsdale J, Barten R, Haude A, Stewart CA, Beck S et al. 2001. The genomic context of natural killer receptor extended gene families. Immunol Rev 181:20–38.

True JR, and Carroll SB. 2002. Gene co-option in physiological and morphological evolution. Annu Rev Cell Dev Bi 18:53–80.

Tsantes C, and Steiper ME. 2009. Age at first reproduction explains rate variation in the strepsirrhine molecular clock. Proc Natl Acad Sci USA 106:18165–70.

Tsong AE, Tuch BB, Li H, and Johnson AD. 2006. Evolution of alternative transcriptional circuits with identical logic. Nature 443:415–20.

Uddin M, Goodman M, Erez O, Romero R, Liu G et al. 2008. Distinct genomic signatures of adaptation in pre- and postnatal environments during human evolution. Proc Natl Acad Sci USA 105:3215–20.

Ullu E, and Tschudi C. 1984. Alu sequences are processed 7SL RNA genes. Nature 312:171–172.

van den Berg TK, Yoder JA, and Litman GW. 2004. On the origins of adaptive immunity: innate immune receptors join the tale. Trends Immunol 25:11–16.

Velasco R, Zharkikh A, Affourtit J, Dhingra A, Cestaro A et al. 2010. The genome of the domesticated apple (*Malus x domestica* Borkh.). Nat Genet 42:833–9.

Verhulst EC, Beukeboom LW, and van de Zande L. 2010a. Maternal control of haplodiploid sex determination in the wasp *Nasonia*. Science 328:620–3.

Verhulst EC, van de Zande L, and Beukeboom LW. 2010b. Insect sex determination: it all evolves around transformer. Curr Opin Genet Dev 20:376–83.

Vogel C, and Chothia C. 2006. Protein family expansions and biological complexity. PLoS Comput Biol 2:e48.

Voight BF, Kudaravalli S, Wen X, and Pritchard JK. 2006. A map of recent positive selection in the human genome. PLoS Biol 4:e72.

Wade MJ, and Goodnight CJ. 1991. Wright's shifting balance theory: an experimental study. Science 253:1015–18.

Wagner A. 2008. Neutralism and selectionism: a network-based reconciliation. Nat Rev Genet 9:965–74.

Wallace AG, Detweiler D, and Schaeffer SW. 2011. Evolutionary history of the third chromosome gene arrangements of *Drosophila pseudoobscura* inferred from inversion breakpoints. Mol Biol Evol 28:2219–29.

Wallace AR. 1889. Darwinism. An exposition of the theory of natural selection. Macmillian, London.

Wallace B. 1966. Natural and radiation-induced chromosomal polymorphism in *Drosophila*. Mutat Res 3:194–200.

Wang ET, Kodama G, Baldi P, and Moyzis RK. 2006. Global landscape of recent inferred Darwinian selection for *Homo sapiens*. Proc Natl Acad Sci USA 103:135–40.

Wang Z, and Zhang J. 2011. Impact of gene expression noise on organismal fitness and the efficacy of natural selection. Proc Natl Acad Sci USA 108:E67–76.

Watanabe T. 1963. Episome-mediated transfer of drug resistance in Enterobacteriaceae. VI. High-frequency resistance transfer system in *Escherichia coli*. J Bacteriol 85:788–94.

Waterston RH, Lindblad-Toh K, Birney E, Rogers J, Abril JF, et al. 2002. Initial sequencing and comparative analysis of the mouse genome. Nature 420:520–62.

Watson JD, and Crick FH. 1953a. Genetical implications of the structure of deoxyribonucleic acid. Nature 171: 964–7.

Watson JD, and Crick FH. 1953b. The structure of DNA. Cold Spring Harb Symp Quant Biol 18:123–31.

Weigel D, and Meyerowitz EM. 1994. The ABCs of floral homeotic genes. Cell 78:203–9.

Weinberg W. 1908. Uber den Nachweis der Vererbung beim Menschen. Jahresh Verein f Vaterl Naturk Wuerttemb 64:368–82.

Weinberg W. 1910. Weitere Beitrage zur Theorie der Verebung. (Translated into English by K. Meyer). Arch Rassen Ges Biol 7:35–49.

Weinberg W. 1963. On the demonstration of heredity in man. (Translated from "Uber den Nachweis der Vererbung beim Menschen"). In S H Boyer, ed. Papers on human genetics, pp. 4–15. Prentice-Hall, Englewood Cliffs, NJ.

Weinberg W. 1984. Further contributions to the theory of inheritance. (Translated by K. Meyer from "Weitere Beitrage zur Theorie der Verebung"). In W. G. Hill, eds. Quantitative genetics. Part I: Explanation and analysis of continous variation, pp. 42–57. Van Nostrand Reinhold, New York.

Weiner AM. 2000. Do all SINEs lead to LINEs? Nat Genet 24:332–333.

Weiss EH, Mellor A, Golden L, Fahrner K, Simpson E et al. 1983. The structure of a mutant H-2 gene suggests that the generation of polymorphism in H-2 genes may occur by gene conversion-like events. Nature 301:671–4.

Wen YZ, Zheng LL, Liao JY, Wang MH, Wei Y et al. 2011. Pseudogene-derived small interference RNAs regulate gene expression in African *Trypanosoma brucei*. Proc Natl Acad Sci USA 108:8345–50.

Wendel JF. 2000. Genome evolution in polyploids. Plant Mol Biol 42:225–49.

Wernegreen JJ, and Moran NA. 1999. Evidence for genetic drift in endosymbionts (*Buchnera*): analyses of protein-coding genes. Mol Biol Evol 16:83–97.

Werth CR, and Windham MD. 1991. A model for divergent, allopatric speciation of polyploid pteridophytes resulting from silencing of duplicate-gene expression. Am Nat 137:515–26.

West-Eberhard MJ. 2003. Developmental plasticity and evolution. Oxford University Press, New York.

White MJD. 1969. Chromosomal rearrangements and speciation in animals. Annu Rev Genet 3:75.

Wightman B, Ha I, and Ruvkun G. 1993. Posttranscriptional regulation of the heterochronic gene *lin-14* by *lin-4* mediates temporal pattern formation in *C. elegans*. Cell 75:855–62.

Wilkens SA. 2002. The evolution of developmental pathways. Sinauer Assoc, Sunderland, MA.

Wilkins AS. 1995. Moving up the hierarchy: a hypothesis on the evolution of a genetic sex determination pathway. Bioessays 17:71–7.

Williams GC. 1966. Adaptation and natural selection: a critique of some current evolutionary thought. Princeton University Press, Princeton, N.J.

Wills C. 1981. Genetic variability. Oxford University Press, Oxford.

Wilson AC, Carlson SS, and White TJ. 1977. Biochemical evolution. Annu Rev Biochem 46:573–639.

Wilson DS, and Szostak JW. 1999. In vitro selection of functional nucleic acids. Annu Rev Biochem 68:611–47.

Wilson DS, and Wilson EO. 2007. Rethinking the theoretical foundation of sociobiology. Q Rev Biol 82:327–48.

Wilson EO. 1975. Sociobiology: the new synthesis. The Belknap Press of Harvard University Press, Cambridge, MA.

Wilson EO. 1978. On human nature. Harvard University Press, Cambridge, MA.

Wilson EO. 2008. One giant leap: how insects achieved altruism and colonial life. BioScience 58:17–25.

Winge O. 1927. The location of eighteen genes in *Lebistes reticulatus*. J Genet 18:1–43.

Woese CR. 1983. The primary lines of descent and the universal ancestor In D S Bendell, ed. Evolution: From molecules to men, pp. 209–33. Cambridge University Press, Cambridge.

Wolfe KH, Li WH, and Sharp PM. 1987. Rates of nucleotide substitution vary greatly among plant mitochondrial, chloroplast, and nuclear DNAs. Proc Natl Acad Sci USA 84:9054–8.

Wolfe KH, and Shields DC. 1997. Molecular evidence for an ancient duplication of the entire yeast genome. Nature 387:708–13.

Wood TE, Takebayashi N, Barker MS, Mayrose I, Greenspoon PB et al. 2009. The frequency of polyploid speciation in vascular plants. Proc Natl Acad Sci USA 106:13875–9.

Wooding S, Bufe B, Grassi C, Howard MT, Stone AC et al. 2006. Independent evolution of bitter-taste sensitivity in humans and chimpanzees. Nature 440:930–4.

Wray GA. 2007. The evolutionary significance of *cis*-regulatory mutations. Nature 8:206–16.

Wright S. 1916. An intensive study of the inheritance of color and other coat characters in guinea pigs with special reference to graded variation. Carnegie Inst Washington, Publ 241:59–160.

Wright S. 1921. Systems of Mating. I. The biometric relations between parent and offspring. Genetics 6:111–23.

Wright S. 1927. The effects in combination of the major color-factors of the guinea pig. Genetics 12:530–69.

Wright S. 1929a. Fisher's theory of dominance. Am Nat 63:274–9.

Wright S. 1929b. The evolution of dominance: Comment on Dr. Fisher's reply. Am Nat 63:556–61.

Wright S. 1931. Evolution in Mendelian populations. Genetics 16:97–159.

Wright S. 1932. The roles of mutation, inbreeding, crossbreeding, and selection in evolution. Proc 6th Int Cong Genet 1:356–66.

Wright S. 1937. The distribution of gene frequencies in populations. Proc Natl Acad Sci USA 23:307–20.

Wright S. 1938a. Size of population and breeding structure in relation to evolution. Science 87:430–1.

Wright S. 1938b. The distribution of gene frequencies under irreversible mutation. Proc Natl Acad Sci USA 24:253–9.

Wright S. 1939. The distribution of self-sterility alleles in populations. Genetics 24:538–52.

Wright S. 1941. On the probability of fixation of reciprocal translocations. Am Nat 75:513–22.

Wright S. 1948a. Genetics of populations. Encyclopedia Britannica 10:111–12.

Wright S. 1948b. On the roles of directed and random changes in gene frequency in the genetics of populations. Evolution 2:279–94.

Wright S. 1951. Fisher and Ford on "The Sewall Wright effect". Am Sci 39:452–8.

Wright S. 1960. "Genetics and twentieth century Darwinism"—A review and discussion. Am J Hum Genet 12:365–72.

Wright S. 1968. Evolution and the genetics of populations. University of Chicago Press, Chicago.

Wright S. 1969. Evolution and the genetics of populations. Vol. II: The theory of gene frequencies. University of Chicago Press, Chicago.

Wright S. 1977. Evolution and the genetics of populations. Vol. III: Experimental results and evolutionary deductions. University of Chicago Press, Chicago.

Wright S, and Dobzhansky T. 1946. Genetics of natural populations. Xii. Experimental reproduction of some of the changes caused by natural selection in certain populations of *Drosophila pseudoobscura*. Genetics 31: 125–56.

Wu CI, and Hammer M. 1991. Molecular evolution of ultraselfish genes of meiotic drive systems. In Selander

PK, Clark AG, and Whittam TS, eds. Evolution at the molecular level. Sinauer Associates, Sunderland, MA.

Wu CI, and Li WH. 1985. Evidence for higher rates of nucleotide substitution in rodents than in man. Proc Natl Acad Sci USA 82:1741–5.

Wu CI, Lyttle TW, Wu ML, and Lin GF. 1988. Association between a satellite DNA sequence and the responder of segregation distorter in *D. melanogaster*. Cell 54:179–89.

Wu CI, and Ting CT. 2004. Genes and speciation. Nat Rev Genet 5:114–22.

Wyman AR, and White R. 1980. A highly polymorphic locus in human DNA. Proc Natl Acad Sci USA 77:6754–6758.

Xiao S, Emerson B, Ratanasut K, Patrick E, O'Neill C et al. 2004. Origin and maintenance of a broad-spectrum disease resistance locus in *Arabidopsis*. Mol Biol Evol 21:1661–72.

Xu G, Guo C, Shan H, and Kong H. 2012. Divergence of duplicate genes in exon-intron structure. Proc Natl Acad Sci USA 109:1187–92.

Xu G, Ma H, Nei M, and Kong H. 2009. Evolution of F-box genes in plants: different modes of sequence divergence and their relationships with functional diversification. Proc Natl Acad Sci USA 106:835–40.

Yamagata Y, Yamamoto E, Aya K, Win KT, Doi K et al. 2010. Mitochondrial gene in the nuclear genome induces reproductive barrier in rice. Proc Natl Acad Sci USA 107:1494–9.

Yamamoto F, and Hakomori S. 1990. Sugar-nucleotide donor specificity of histo-blood group A and B transferases is based on amino acid substitutions. J Biol Chem 265:19257–62.

Yamamoto Y, Stock DW, and Jeffery WR. 2004. Hedgehog signaling controls eye degeneration in blind cavefish. Nature 431:844–7.

Yamazaki T. 1971. Measurement of fitness at the esterase-5 locus in *Drosophila pseudoobscura*. Genetics 67: 579–603.

Yamazaki T, and Maruyama T. 1972. Evidence for the neutral hypothesis of protein polymorphism. Science 178:56–8.

Yan L, Fu D, Li C, Blechl A, Tranquilli G et al. 2006. The wheat and barley vernalization gene *VRN3* is an orthologue of *FT*. Proc Natl Acad Sci USA 103:19581–6.

Yang Z. 2007. PAML 4: phylogenetic analysis by maximum likelihood. Mol Biol Evol 24:1586–91.

Yang Z, and dos Reis M. 2011. Statistical properties of the branch-site test of positive selection. Mol Biol Evol 28:1217–28.

Yang Z, Wong WS, and Nielsen R. 2005. Bayes empirical Bayes inference of amino acid sites under positive selection. Mol Biol Evol 22:1107–18.

Yokoyama R, and Yokoyama S. 1990. Convergent evolution of the red- and green-like visual pigment genes in fish, *Astyanax fasciatus*, and human. Proc Natl Acad Sci USA 87:9315–18.

Yokoyama S. 2008. Evolution of dim-light and color vision pigments. Annu Rev Genom Hum G 9:259–82.

Yokoyama S, and Nei M. 1979. Population dynamics of sex-determining alleles in honeybees and self-incompatibility alleles in plants. Genetics 91:609–26.

Yokoyama S, and Radlwimmer FB. 2001. The molecular genetics and evolution of red and green color vision in vertebrates. Genetics 158:1697–710.

Yokoyama S, Tada T, Zhang H, and Britt L. 2008. Elucidation of phenotypic adaptations: molecular analyses of dim-light vision proteins in vertebrates. Proc Natl Acad Sci USA 105:13480–5.

Yoshida S, Maruyama S, Nozaki H, and Shirasu K. 2010. Horizontal gene transfer by the parasitic plant *Striga hermonthica*. Science 328:1128.

Yoshimoto S, Okada E, Umemoto H, Tamura K, Uno Y et al. 2008. A W-linked DM-domain gene, *DM-W*, participates in primary ovary development in *Xenopus laevis*. Proc Natl Acad Sci USA 105:2469–74.

Young JM, Friedman C, Williams EM, Ross JA, Tonnes-Priddy L et al. 2002. Different evolutionary processes shaped the mouse and human olfactory receptor gene families. Hum Mol Genet 11:535–46.

Yu Q, Colot HV, Kyriacou CP, Hall JC, and Rosbash M. 1987. Behaviour modification by in vitro mutagenesis of a variable region within the period gene of *Drosophila*. Nature 326:765–9.

Yuan YX, Wu J, Sun RF, Zhang XW, Xu DH et al. 2009. A naturally occurring splicing site mutation in the *Brassica rapa FLC1* gene is associated with variation in flowering time. J Exp Bot 60:1299–308.

Zhang J. 2000. Protein-length distributions for the three domains of life. Trends Genet 16:107–9.

Zhang J. 2006. Parallel adaptive origins of digestive RNases in Asian and African leaf monkeys. Nat Genet 38:819–23.

Zhang J, Dyer KD, and Rosenberg HF. 2000. Evolution of the rodent eosinophil-associated RNase gene family by rapid gene sorting and positive selection. Proc Natl Acad Sci USA 97:4701–6.

Zhang J, and Nei M. 1996. Evolution of Antennapedia-class homeobox genes. Genetics 142:295–303.

Zhang J, Nielsen R, and Yang Z. 2005. Evaluation of an improved branch-site likelihood method for detecting positive selection at the molecular level. Mol Biol Evol 22:2472–9.

Zhang J, and Rosenberg HF. 2002. Complementary advantageous substitutions in the evolution of an antiviral

RNase of higher primates. Proc Natl Acad Sci USA 99:5486–91.

Zhang J, Rosenberg HF, and Nei M. 1998. Positive Darwinian selection after gene duplication in primate ribonuclease genes. Proc Natl Acad Sci USA 95:3708–13.

Zhang X, and Firestein S. 2002. The olfactory receptor gene superfamily of the mouse. Nat Neurosci 5:124–33.

Zhao HB, Xu D, Zhang SY, and Zhang JZ. 2012. Genomic and genetic evidence for the loss of umami taste in bats. Genome Biol Evol 4:73–9.

Zhuang H, Chien MS, and Matsunami H. 2010. Dynamic functional evolution of an odorant receptor for sex-steroid-derived odors in primates. Proc Natl Acad Sci USA 106:21247–51.

Zimmer EA, Martin SL, Beverley SM, Kan YW, and Wilson AC. 1980. Rapid duplication and loss of genes coding for the alpha chains of hemoglobin. Proc Natl Acad Sci USA 77:2158–62.

Zimmerman EC. 1960. Possible evidence of rapid evolution in Hawaiian moths. Evol 14:137–8.

Zuckerkandl E, and Pauling L. 1962. Molecular disease, evolution, and genetic heterogeneity. In M. Kasha, and B. Pullman, eds. Horizons in Biochemistry, pp. 189–225. Academic Press, New York.

Zuckerkandl E, and Pauling L. 1965. Evolutionary divergence and convergence in proteins. In V Bryson, and H J Vogel, eds, pp. 97–166. Evolving genes and proteins Academic Press, New York.

Author Index

Mutation-Driven Evolution. First Edition. Masatoshi Nei.

Subject Index

Mutation-Driven Evolution. First Edition. Masatoshi Nei.

The manufacturer's authorised representative in the EU for product safety is Oxford University Press España S.A. of El Parque Empresarial San Fernando de Henares, Avenida de Castilla, 2 - 28830 Madrid (www.oup.es/en or product.safety@oup.com). OUP España S.A. also acts as importer into Spain of products made by the manufacturer.

Printed and bound by CPI Group (UK) Ltd, Croydon, CR0 4YY

06/07/2026

02157639-0001